AF581884

New Training Strategies and Evaluation Methods for Improving Health and Physical Performance

New Training Strategies and Evaluation Methods for Improving Health and Physical Performance

Editors

Catarina Nunes Matias
Stefania Toselli
Cristina Monteiro
Francesco Campa

MDPI • Basel • Beijing • Wuhan • Barcelona • Belgrade • Manchester • Tokyo • Cluj • Tianjin

Editors
Catarina Nunes Matias
Bettery Life Lab
Universidade Lusófona
Lisbon
Portugal

Stefania Toselli
DIBINEM
University of Bologna
BOLOGNA
Italy

Cristina Monteiro
Laboratory of Physiology and Biochemistry of Exercise,
Faculdade de Motricidade Humana
Universidade de Lisboa
Lisbon
Portugal

Francesco Campa
Department of Biomedical Sciences
University of Padova
Padova
Italy

Editorial Office
MDPI
St. Alban-Anlage 66
4052 Basel, Switzerland

This is a reprint of articles from the Special Issue published online in the open access journal *International Journal of Environmental Research and Public Health* (ISSN 1660-4601) (available at: www.mdpi.com/journal/ijerph/special_issues/New_Training_Strategies_Evaluation_Methods_Improving_Health_Physical_Performance).

For citation purposes, cite each article independently as indicated on the article page online and as indicated below:

LastName, A.A.; LastName, B.B.; LastName, C.C. Article Title. *Journal Name* **Year**, *Volume Number*, Page Range.

ISBN 978-3-0365-4252-2 (Hbk)
ISBN 978-3-0365-4251-5 (PDF)

Cover image courtesy of Francesco Campa

Contents

About the Editors

Catarina Nunes Matias

Catarina N. Matias has a Masters degree in biochemistry and a Ph.D. (Summa cum laude) in Human kinetics in the Specialty of Physical Activity and Health at the Faculty of Human Kinetics at the University of Lisbon. Catarina has extensive experience in teaching, in the first cycle of studies at the University level within the curricular units of Biochemistry, Nutrition and Exercise, and Exercise Physiology, as well as in the advanced modules of the University masters and doctorates, either by the Faculty of Human Kinetics of the University of Lisbon or by the Faculty of Physical Education and Sport of the Lusófona University. This activity resulted in the supervision of seven masters and one doctorate. Parallel to the academics, Catarina developed a scientific approach which resulted in collaborations and coordination of 15 research projects, 5 scientific peer awards and distinctions, 3 dissertations, 1 international book chapter, 2 research reports, 76 scientific articles published in international peered review journals with impact factor, 69 scientific abstracts published in international journals with impact factor, and in the book of abstracts of national and international scientific congresses of specialty, and 62 oral or panel communications at international scientific congresses. Catarina is currently a researcher at CIDEFES, a research center at the Faculty of Physical Education and Sport at the University of Lusófona, where she is also an Assistant Professor. In addition, Catarina ensures the Coordination of Development & Research Laboratories, as well as Biochemistry, Body Composition and Metabolism, and Performance Laboratories at Bettery, S.A . The areas of expertise includes sports biochemistry, sports nutrition, metabolic regulation in exercise and immune response to exercise and training, with multiple scientific outputs and masters'thesis supervised within this fields.

Stefania Toselli

Stefania Toselli is associate professor of the Department of Biomedical and Neuromotor Sciences, Bologna, Italy. The main field of her research is represented by human biology, considering both adults and subjects during growth. In particular, the considered aspects regard auxology, secular trend, weight status, body composition (both considering methodological aspects and variability in different populations), somatotype, and body image, with specific reference to environmental and physical activity influence.

Cristina Monteiro

Cristina Monteiro has a degree in Biochemistry and a Ph.D. in Human Movement Sciences and is an Assistant Professor at the Faculty of Human Kinetics of the University of Lisbon teaching mainly exercise biochemistry and nutrition. Her scientific activity has been integrated in the Interdisciplinary Center for the Study of Human Performance (CIPER) of the Faculty of Human Kinetics. She was a member of the research team of three funded projects, two by the FCT, one by the Portuguese-French Scientific and Technical Cooperation Program. Additionally to this, her participation in several interdisciplinary investigations was central to the development and integration of her areas of expertise in the sports sciences and health fields, making the connection between them.

The areas of expertise includes sports biochemistry, sports nutrition, metabolic regulation in exercise and immune response to exercise and training, redox balance in exercise and disease and magnesium metabolism, with multiple scientific outputs and masters'thesis supervised within this fields. Recently she has supervised two Ph.D. thesis, one exploring the immune response to acute exercise, including the recovery period throughout 24 h, and training; and one other that sought to compare the effect of several leucine metabolites, namely two commercially available forms of beta-hydroxy-beta-methylbutyrate and leucic acid, on resistance training-induced changes induced on performance, body composition, and biochemical markers of muscle damage, inflammation and anabolic and catabolic hormones. These works have produced several research outputs in high level journals and show her previous experience in applying follow up studies with athletes in the field and randomized control trails to test the effects of nutritional supplements.

Francesco Campa

Francesco Campa is a researcher at the Department of Biomedical Sciences at the University of Padova (Italy), received his Ph.D. in pharmacology and toxicology, human development, and movement sciences from the University of Bologna. He has published one book on sports anthropometry, several book chapters, and over 70 peer-reviewed journal and conference papers (https://orcid.org/0000-0002-3028-7802), with an H-index of 16. His main interests are sports anthropometry, bioimpedance vector analysis, and body composition optimization applied to sports performance.

Preface to "New Training Strategies and Evaluation Methods for Improving Health and Physical Performance"

Physical activity is among the most effective methods for improving health, body composition, and physical function, and its practice is suitable for every population. Its benefits are known for sedentary individuals who, by initiating sport, improve their physical condition by reducing risk factors. Active training is also encouraged for the general population who need to maintain an optimal level of fitness, as well as for athletes who want to achieve high performance during competitive periods. Even young people benefit from sports practice, growing into healthy young adults with important implications for their psychological and social development. In the last few years, the scope of research in sports has become very wide and detailed, laying the foundations for the development of innovative training methods and new evaluation approaches aimed at improving health, body composition, and performance. Contemporary researchers have contributed to the field of body composition research in the development of new measurement methods and training strategies. The aforementioned aspects have laid the foundations for the development of innovative techniques and new evaluation approaches aimed at improving and assessing body composition and sports performance. In these contexts, the bioelectrical impedance analysis was proposed as a valid method to quantify body composition elements (e.g., fat and fat-free mass, body fluids, muscle mass) and is based on predictive equations or the qualitative interpretation of the raw data. On the other hand, innovative training strategies aimed at improving body composition and performance have been presented. The aim of this Special Issue was to propose, on the basis of the evidence that the current literature provides, new training techniques and specific evaluation methods for the different populations practicing physical activity.

Catarina Nunes Matias, Stefania Toselli, Cristina Monteiro, and Francesco Campa

Editors

International Journal of
Environmental Research and Public Health

MDPI

Review

Exercise Dose Equalization in High-Intensity Interval Training: A Scoping Review

Tom Normand-Gravier [1,2], Florian Britto [1,3], Thierry Launay [1,3], Andrew Renfree [4], Jean-François Toussaint [1,2,5] and François-Denis Desgorces [1,2,*]

1 Université Paris Cité, 75015 Paris, France; tom.normand-gravier@etu.u-paris.fr (T.N.-G.); florian.britto@u-paris.fr (F.B.); thierry.launay@u-paris.fr (T.L.); jean-francois.toussaint@aphp.fr (J.-F.T.)
2 URP 7329-IRMES (Institute for Research in Medicine and Epidemiology of Sport), INSEP, 75012 Paris, France
3 Institute Cochin, U1016 INSERM, 75014 Paris, France
4 School of Sport & Exercise Science, University of Worcester, Worcester WR2 6AJ, UK; a.renfree@worc.ac.uk
5 CIMS, Hôtel-Dieu, Assistance Publique–Hôpitaux de Paris, 75004 Paris, France
* Correspondence: francois.desgorces@u-paris.fr

Abstract: Based on comparisons to moderate continuous exercise (MICT), high-intensity interval training (HIIT) is becoming a worldwide trend in physical exercise. This raises methodological questions related to equalization of exercise dose when comparing protocols. The present scoping review aims to identify in the literature the evidence for protocol equalization and the soundness of methods used for it. PubMed and Scopus databases were searched for original investigations comparing the effects of HIIT to MICT. A total of 2041 articles were identified, and 169 were included. Of these, 98 articles equalized protocols by utilizing energy-based methods or exercise volume (58 and 31 articles, respectively). No clear consensus for protocol equalization appears to have evolved over recent years. Prominent equalization methods consider the exercise dose (i.e., energy expenditure/production or total volume) in absolute values without considering the nonlinear nature of its relationship with duration. Exercises resulting from these methods induced maximal exertion in HIIT but low exertion in MICT. A key question is, therefore, whether exercise doses are best considered in absolute terms or relative to individual exercise maximums. If protocol equalization is accepted as an essential methodological prerequisite, it is hypothesized that comparison of program effects would be more accurate if exercise was quantified relative to intensity-related maximums.

Keywords: training programs; physical activity; effort; patients; athletes

Citation: Normand-Gravier, T.; Britto, F.; Launay, T.; Renfree, A.; Toussaint, J.-F.; Desgorces, F.-D. Exercise Dose Equalization in High-Intensity Interval Training: A Scoping Review. *Int. J. Environ. Res. Public Health* **2022**, *19*, 4980. https://doi.org/10.3390/ijerph19094980

Academic Editor: Catarina Nunes Matias

Received: 21 March 2022
Accepted: 17 April 2022
Published: 20 April 2022

1. Introduction

Exercise is both described and prescribed on the basis of two main variables: intensity (i.e., level of muscular activity) and volume (e.g., duration, distance or number of repetitions of an interval or set, and of the entire session) [1,2]. Notably for the interval exercise modality, these major variables also depend on possible recovery pauses within the exercise bout, inducing a third exercise variable, called by some authors "exercise density" (i.e., work/recovery ratio but also intensity level of the recovery) [1,3,4]. For quantifying and designating the overall exercise performed, authors can use generic terms accounting for all exercise variables, such as exercise dose in exercise-induced health studies [1] or training load for athlete monitoring purposes [5–7]. Defining effort as what is required to achieve a task in line with individual maximal capacities [8], exercise dose and training load might refer to the quantity of exercise-induced effort [5,6].

The control and calibration of training protocols should be a prerequisite in exercise and sport science studies, and insufficient consideration of this may result in confusion regarding exercise program effects [9,10]. Manipulation of training variables (volume, intensity and density) might ensure that the effort level generated by two protocols being compared is similar, or in other terms, that their exercise dose is equalized. However, methodologically these comparisons are not easy to conduct. Targeting large populations, recommendations for physical activity frequently use absolute values for intensity or, sometimes, exercise durations characterized by large intensity ranges (e.g., light, moderate, vigorous) that could complicate the quantification of an individualized and unique dose value [1]. Viana et al. suggested that conclusions about high-intensity interval training (HIIT) remain difficult to draw because of insufficient control of the numerous exercise variables [11]. Recently, the lack of protocol equalization in HIIT and moderate-intensity continuous training (MICT) has been suggested to represent a possible methodological bias limiting studies' conclusions [12]. Comments on this paper suggest that consensus was not reached in the methods used for equalizing protocols nor, more surprisingly, in the necessity for equalizing them [13]. Limits and issues raised by equalization methods based on energy expenditure, although largely recommended, have only been recently reported [14,15]. Similar debates on adequate terms to use and on quantification methods are currently in progress regarding the training-load concept [2,7,16]. Therefore, we suggest that equalization of training doses should be a methodological prerequisite before comparing the effects of different training protocols and is therefore a major challenge facing exercise physiologists and sport scientists.

HIIT may be defined as repeated short-to-long exercise bouts performed at an intensity between 80% and 120% of maximum aerobic power (oxygen consumption or equivalent heart rate) [11]. Recently, the use of HIIT has been proposed as a method for improving quality of life of older people and for rehabilitation of patients suffering from several pathologies, such as cardiovascular diseases. As HIIT has become a real worldwide trend for exercise practice and sport sciences, this has increased the need for accurate equalization of training protocol doses in order to compare their efficiency [11,17,18]. Furthermore, we propose that HIIT studies display most of the characteristics necessary to understand the issues of exercise dose quantification and protocol equalization: (i) high number of studies published; (ii) changes in exercise variables; (iii) methods for equalization already developed and discussed.

The present scoping review aims to identify in the literature the evidence for protocol equalization and the soundness of methods used for it [19].

2. Materials and Methods

The latest methodological guidance for scoping reviews was followed, leading to completing the checklist of the Preferred Reporting Items for Systematic Reviews for scoping reviews (Supplementary File S1) [20–22].

2.1. Search Strategy

We analyzed published studies on electronic databases until 30 November 2020 without restriction set on the publication year. PubMed and SCOPUS databases were explored using a keyword search strategy for 'High-intensity interval training' with a first filter step used for including studies that were: written in English; randomized controlled trials, clinical trials or from journal articles; based on human subjects. A second step was based on abstract screening to select studies comparing HIIT to another type of training program and to retain only chronic training programs. When the information was missing in the abstract, the authors searched for it in the whole article. Because variables measured to control exercise do not correspond between sprint interval training and HIIT, the last step consisted of retaining studies focusing on HIIT (80–120% of VO_2max or equivalent) and excluding sprint interval training (intensity higher than 120%) [11]. All duplicate studies and protocols were excluded; if the same experimental protocol was used for several articles, only the first published was retained. Finally, studies were sorted according to publication year and type of subjects observed: (i) patients or older people; (ii) untrained; (iii) trained. All search results were extracted and imported into a reference manager (Zotero, version 5.0.96.3). No included studies were authored by any of the review authors, thereby limiting possible conflicts of interest.

2.2. Assessment of Reporting Quality

The reporting quality of studies was assessed using items specific to the research field. Most of them originated from a modified version of the Downs and Black checklist, resulting in eight assessment criteria (Supplementary File S2) [23]. Studies reporting quality were scored on a scale from '0' (unable to determine, or no) to '1' (yes) for each item. Scores were allocated on the basis of good (6–8), moderate (3–5) and poor (0–2) methodological reporting quality.

2.3. Terms Used and Methods Applied for Protocols Equalization

Articles' methods sections were analyzed, and the proportion of studies that equalized doses of training protocols and methods used for equalizing were recorded. If any information necessary for protocol equalization was not included in a study's methods section, they were considered as not equalized. In line with this search, articles were analyzed to determine terms used to describe how exercise-induced effort was quantified (e.g., exercise dose or exercise volume) and the equalization process (e.g., equated protocols or matched training).

To assess the soundness of the methods used for protocol equalization, the exercise details were extracted from the articles, specifically exercise volume (duration, distance or number of repetitions for session and for each interval or set), intensity (varied metrics in absolute or relative values), recovery (duration and intensity if necessary) and exercise type (running, walking, cycling or resistance training).

2.4. Statistical Analysis

The present study is largely descriptive, and quantifies proportions (%) of studies that equalized training protocols and identified methods used for equalization. Differences in reporting-quality methodology between studies that equalized protocols and those that did not, and between subject populations were assessed by using one-way analysis of variance for total score and Pearson's Chi-2 test for each criterion assessed. The evolution of dose equalization over the years was observed by linear regression analysis of percentage of studies equalizing doses. Statistical analysis was performed with R software (version 3.6.2), and statistical significance was set at $p < 0.05$.

3. Results

The identification process described in Figure 1 resulted in 169 studies being included in the review. The complete list of articles retained is presented in the Supplementary File S3.

We aimed first to document the equalization of exercise protocols in studies comparing HIIT and other exercise types. We also aimed to highlight if protocol equalization was associated with a better-quality study design and/or if it was specific to recent studies.

The assessment of methodological reporting quality of these articles was moderate, but with poor quality for calculations of statistical power, and moderate for group homogeneity and for groups matched by physical condition (Table 1). Matching by "subjects' physical condition" was the only criteria that led to a significant difference between types of subjects observed by studies ($p < 0.001$)

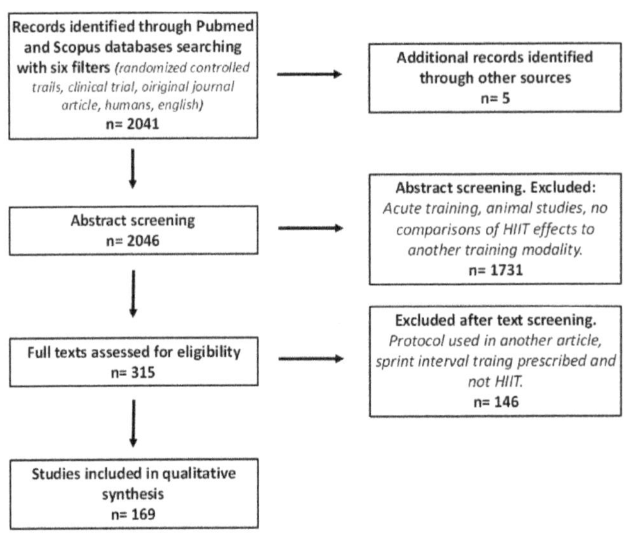

Figure 1. Phases of study selection during data collection.

Table 1. Reporting quality expressed through positive assessment of studies according to protocol equalization processes (middle of table) and population observed (bottom of table). Total score expressed as mean and standard deviations.

	Total Score	Recruitment in Same Population (%)	Subjects Ramdomization (%)	Physical Condition Matching (%)	Training Direct Supervision (%)	Exercise Control (%)	Adherence to Training (%)	Subjects Follow-Up (%)	Statistical Power (%)
Total (n = 169)	5.1 ± 1.5	47.6	70.8	53.0	73.8	88.7	61.3	84.5	29.8
Equalized protocols (n = 98)	5.2 ± 1.5	46.4	71.1	51.5	75.3	91.7	61.9	89.7	28.9
Non-equalized protocols (n = 71)	5.0 ± 1.6	49.3	70.4	54.9	71.8	84.5	60.5	77.4	31.0
Older people and patients (n = 99)	5.1 ± 1.7	50.5	67.0	44.4	76.7	85.8	62.6	85.8	29.3
Untrained (n = 41)	5.3 ± 1.3	52.5	85.0	47.5	75.0	92.5	57.5	87.5	32.5
Trained (n = 29)	5.0 ± 1.2	34.5	62.1	89.6 *	62.1	93.1	62.1	75.9	24.1

* significant differences with other groups of subjects ($p < 0.05$).

The most-frequently occurring terms used for the process of protocol equalization (total n = 98) were as follows: matched protocols (n = 44); equalized (or equated, equal, equivalent, n = 10); isocaloric (or isoenergetic, n = 8); The most-frequently used terms to designate what had been equalized were: total work (or external, mechanical, n = 26); workload (or training load, n = 29); exercise volume (or total volume, n = 13); exercise dose (or effort, n = 4). Protocol equalization did not evolve clearly over time, but there was a trend for a reduction in the proportion of equalizing studies ($R^2 = 0.21$, $p = 0.06$; Figure 2) and an increase in the absolute number of studies that equalized protocols ($R^2 = 0.59$, $p = 0.01$). No differences were observed in studies' reporting quality between those that equalized protocol doses and those that did not ($p = 0.1$).

The distribution of studies based on equalized and non-equalized protocols, and associated methods for quantifying exercise doses are shown in Figure 3. Studies observing patients and older people equalized protocols at 58.7%, compared to 62.5% in untrained subjects and 51.7% in trained, without significant differences between groups ($p = 0.09$). Training protocols differed between studies; however, typical HIIT exercises were identified among all studies (Figure 3).

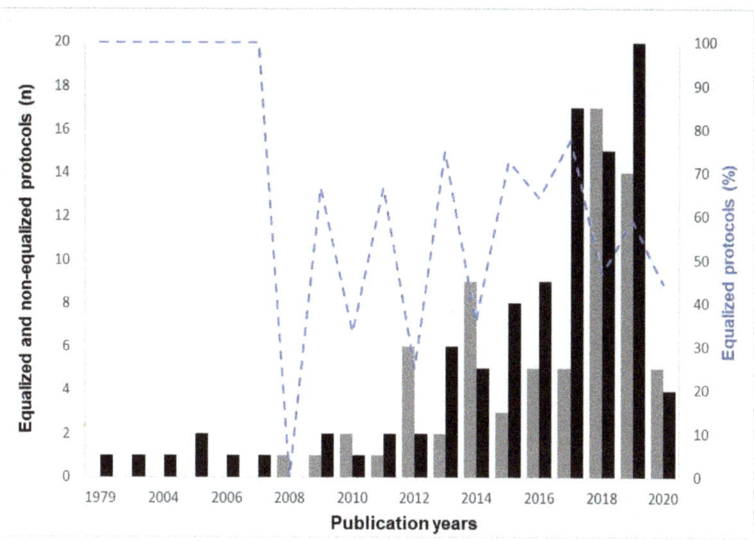

Figure 2. Percentage and absolute number of studies using equalized protocols (dashed blue line and black bars, respectively) and number of studies without using equalization of protocols (grey bars) from 1979 to November 2020.

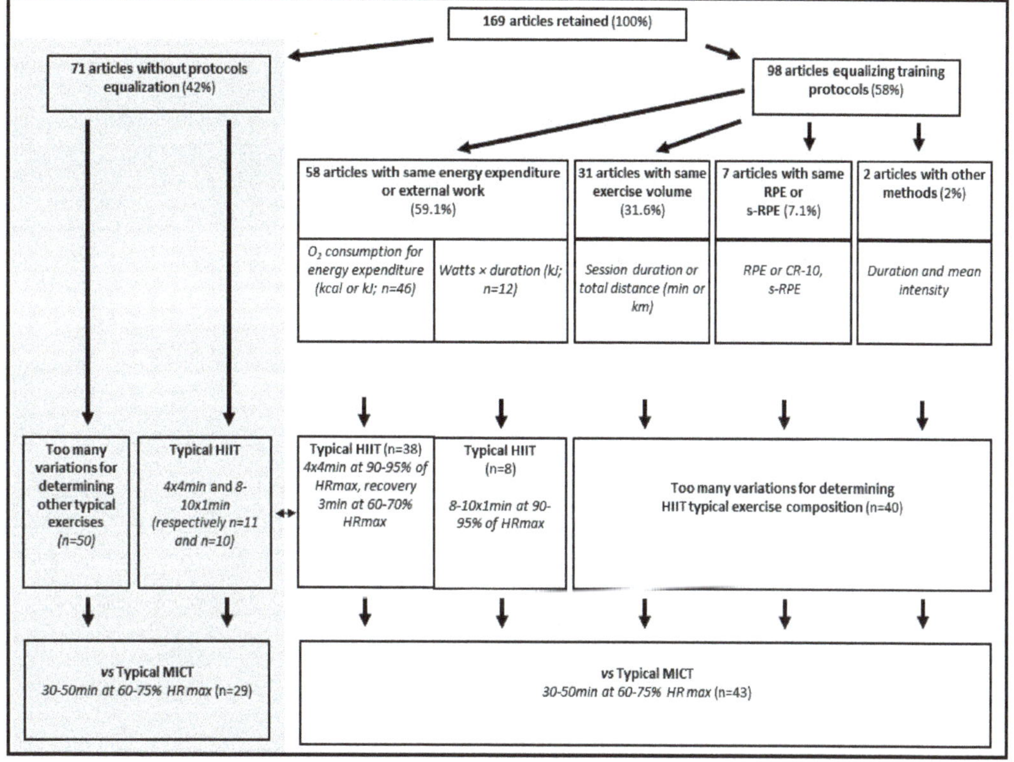

Figure 3. From article-selection process to equalization methods and exercise sessions.

4. Discussion

We aimed to determine whether researchers, when comparing HIIT to other types of programs, had utilized equalized protocols. Although most studies equalized protocols, a substantial number of studies did not. For designating what was equalized, authors mainly focused on actual measures performed (e.g., total work, energy expenditure or exercise volume) rather than using a more generic term (e.g., exercise dose, training load or effort). Energy-based methods were prominently used for equalizing protocols, whereas methods based on exercise volume and perceived exertion appeared markedly less frequently.

Among the 169 studies included in this review paper, most equalized their protocols (58%), whilst 42% did not. Consensus for protocol equalization is not apparent, and the protocol equalization rate has not evolved significantly since the first paper published in 1979. In addition, data did not show differences according to populations observed. This is in line with the assessment of reporting quality, which did not differentiate studies according to protocol equalization or populations observed. Satisfactorily, "exercise control" and "direct supervision" criteria of reporting quality achieved the highest assessments. Among studies that did not equalize protocols, twenty-one compared HIIT with typical MICT programs (Figure 3) that had been designated by previous studies to be equal based on energy expenditure or production [24]. Therefore, although protocol equalization was not reported in the methods section of these studies, it had possibly been achieved anyway, thereby increasing the proportion of protocols actually equalized.

Vollard and Metcalf [13] argued that the key advantage of HIIT is time efficiency. MICT requires more prolonged exercise duration than HIIT, and it could be presumed as self-evident that MICT is not as effective if exercise duration is short. However, for a given exercise duration, because of higher intensity, HIIT induces a greater exercise dose than MICT. If the aim is to demonstrate the positive effects of HIIT despite a short exercise duration, such demonstration could be achieved without requiring comparison with another training program. Conversely, when comparing programs' effects on performance improvement or biological parameters, if the higher exercise intensity of HIIT is not counterbalanced by a lower exercise volume, responses may have originated from the higher intensity, but also simply from a greater exercise dose. This methodological point was accounted for by 98 studies that attempted to equalize protocols.

In some studies, training protocols were partly equalized by prescribing similar total exercise durations. Such a method is in line with population-based studies that quantify physical activity through time spent in light/moderate/strenuous intensity ranges without aiming to compare the particular effects of these intensity levels [1]. Using session durations to equalize protocols corresponded to physical activity recommendations for health and wellbeing (e.g., three sessions of 30–45 min per week for HIIT and moderate intensities) [1]. During HIIT, high-intensity activity itself could not account for the entire 30–45 min of the session: 10–20 min of high-intensity exercise was paired with low-intensity exercise for the remaining 10–20 min. Therefore, protocols equalized by similar durations compared MICT to mixed MICT and HIIT, but studies did not describe the rationale underpinning the selection of exercise durations for different intensities. Equalization by total volume does not consider the slope of the relationship between intensity and duration, and even less the nonlinearity of this relationship. Consequently, the absolute value of exercise duration was equalized, but not the combination of the exercise variables. If expressed relative to respective maximums, durations prescribed by HIIT programs were markedly higher than for MICT. In these studies, responses to training might be due to changes in intensity or to changes in exercise dose. Furthermore, by proposing similar exercise durations, these protocols cancelled the time gains expected from HIIT [13].

The primary methods used for protocol equalization were energy-based. Most studies measured exercise-induced energy expenditure through oxygen consumption, while some others measured external work based on power output and exercise duration [12,24,25]. Energy expenditure methods typically incorporated both exercise and recovery periods, while methods based on external work only considered exercise bouts. That is quite surprising as the typical HIIT exercise utilized in studies based on external work (i.e., 8–10 × 1 min at 90–95% HRmax, 1–2 min recovery) were characterized by short–moderate recovery pauses, allowing maintenance of a high level of physiological stress [26]. Furthermore, exercise-induced excess post-oxygen consumption is largely influenced by exercise intensity and may be prolonged for many hours [27]. Although some authors suggest that exercise-induced energy expenditure should also account for exercise-induced excess post-oxygen consumption, this point may require more careful attention in HIIT studies that focus on the effects of changes in both intensity and interval volumes [1,12].

Energy-based methods for quantifying exercise consider the human ability for energy expenditure or external work to be similar whatever the exercise intensity. For several decades, models of the intensity–volume relationship have described a hyperbolic pattern, with maximal exercise volumes dramatically decreasing with increases in intensity [28–30]. By extension, maximal energy expenditure/external work follows the same pattern [31]. Thanks to recovery pauses, for a given intensity level, interval exercise allows accumulation of more exercise than continuous exercise and, consequently, greater energy expenditure [26]. The typical 4 × 4 min session is likely to be performed at a higher intensity level than a 16 min exercise performed in continuous modality [30]. Seiler et al. reported that the maximal tolerable intensity for 4 × 4 min was 94 ± 2% of maximal heart rate when interspersed with 2 min passive recovery [32]; in HIIT studies, an active recovery (3 min at 70% HRmax) was added to this maximal effort. Conversely, because of the nonlinear relationship between exercise intensity and energy expenditure, typical MICT exercise appears to be far from the exercise dose performed during typical HIIT. In fact, in typical MICT, 30–45 min is prescribed at 65–75% HRmax, an intensity that can be maintained for several hours before exhaustion. It may be assumed that the typical HIIT exercise resulting from the energy-based equalizing method reached a maximum of energy expenditure and was exhausting, while MICT represented relatively easy training. This assumption is supported by significantly higher ratings of perceived exertion (RPE) following HIIT sessions [25,33,34], and some authors argued that energy-based methods for equalization underestimate the work that athletes are able to perform at lower intensities [32,35]. Such differences in session-induced exertion should be considered as a possible methodological bias that is likely to become more pronounced with increases in intensity differences between programs. HIIT-induced dose could represent the maximum tolerable (or excessive) training stimulus, whereas MICT dose could be low or insufficient. Finally, despite the popularity of equalization methods based on energy expenditure, its soundness and relevancy are still questioned [13,15].

Finally, six studies used RPE to equalize protocols, and only one used the session-RPE-based method for training-load quantification (i.e., duration × RPE of the session). It seems that studies equalizing protocols by using RPE were composed of varied exercise modalities (e.g., running, resistance exercise or skating) [36–38]. RPE is not only influenced by exercise intensity [39], as exercise duration [40,41], interval volume [42], exercise modality [43] and recovery periods [44] have also been reported to significantly influence RPE. Finally, RPE appears to be influenced by all exercise variables and, consequently, might represent a subjective assessment of the exercise dose. Previous studies have shown that it provides similar session assessments to exercise volume expressed relative to maximum for the considered intensity level [4,40]. Conversely, training load based on RPE might account twice for the exercise volume (i.e., in duration and RPE itself), inflating the calculated load for prolonged sessions [4,5]. In line with studies that have used RPE for protocol equalization, some authors have suggested that RPE alone is therefore preferable for exercise quantification, thereby avoiding the overexpression of volume [4,45].

We acknowledge that the present study may have overlooked some published papers, as it was only conducted on two literature databases and only considered original experimental investigations. Based on the numerous studies utilizing equalization of protocols and researcher support for equalization, it seems that, although the need for equalization is not debated per se, the soundness of methods for equalizing is [1,5,7,12]. In addition, generic terms that designate the quantity of exercise-induced effort (i.e., exercise dose, internal training load) and associated quantification methods (e.g., RPE) may be considered to account for individual maximal capacities in the exercise considered [5–8]. In essence, this is not the case for energy expenditure/production or exercise volume. Finally, the main methodological issue is whether to quantify the exercise—whatever the method—in absolute values or relative to individual maximums for the considered exercise. Lack of consideration of the slope and nonlinearity of the energy–duration or of the intensity–duration relationship is questionable. As proposed recently for training-load quantification and by one study among the 169 retained [4,46], we hypothesize that exercise quantified relative to maximum energy expenditure/external work, or exercise volume, for specified intensity levels will allow more precise program comparisons. This may also be the case when dose is assessed via perceived exertion.

Although scoping reviews can be the first step before systematic review or meta-analysis on the topic, and even if only equalized protocols were retained, results of studies comparing HIIT vs. MICT should be interpreted carefully because of the uncertain accuracy of equalization methods mainly used [19].

5. Conclusions

In HIIT studies, no clear consensus for protocol equalization appears to exist, and there has been no evolution in practices over time. If the scientific community supports this methodological prerequisite, it may assist with the assessment of methodology reporting quality.

Equalization based on exercise duration does not consider all the variables composing exercise-induced effort. Primary equalization methods consider energy expenditure/external work in raw values without considering the slope and the nonlinear nature of its relationship with duration. Exercises resulting from these quantification methods induced maximal exertion in HIIT exercises but low exertion in MICT. Evidently, the main issue is whether to consider exercise dose in absolute values or relative to individual exercise maximums. It is hypothesized that comparison of program effects would be more accurate if the exercise (e.g., exercise volume, energy) was expressed relative to intensity-related maximums (e.g., perceived exertion, exercise volume relative to maximum).

Supplementary Materials: The following supporting information can be downloaded at: https://www.mdpi.com/article/10.3390/ijerph19094980/s1, Supplementary File S1: PRISMA checklist for scoping reviews. Supplementary File S2: Checklist for the assessment of the methodological quality for HIIT studies, adapted from Downs and Black (1998). Supplementary File S3: Complete list of articles retained according to dose equalization methods.

Author Contributions: T.N.-G. and F.-D.D. conceived the research question and designed the study protocol. T.N.-G. conducted the initial database search and screening and synthesized and analyzed relevant data. T.N.-G. and F.-D.D. conducted the full-text screening and drafted the paper. F.-D.D., T.L. and F.B. conducted the study quality assessment and data extraction. T.L., F.B., A.R. and J.-F.T. contributed throughout the review, starting from conceptualizing to editing subsequent drafts of the manuscript. All authors have read and agreed to the published version of the manuscript.

Funding: This research received no external funding.

Institutional Review Board Statement: Not applicable.

Informed Consent Statement: Not applicable.

Data Availability Statement: Not applicable.

Conflicts of Interest: The authors declare no conflict of interest.

References

1. Wasfy, M.M.; Baggish, A.L. Exercise Dose in Clinical Practice. *Circulation* **2016**, *133*, 2297–2313. [CrossRef] [PubMed]
2. Staunton, C.A.; Abt, G.; Weaving, D.; Wundersitz, D.W.T. Misuse of the Term "load" in Sport and Exercise Science. *J. Sci. Med. Sport* **2021**, S1440-2440(21)00212-7. [CrossRef] [PubMed]

3. Taha, T.; Thomas, S.G. Systems Modelling of the Relationship between Training and Performance. *Sports Med.* **2003**, *33*, 1061–1073. [CrossRef] [PubMed]
4. Desgorces, F.-D.; Hourcade, J.-C.; Dubois, R.; Toussaint, J.-F.; Noirez, P. Training Load Quantification of High Intensity Exercises: Discrepancies between Original and Alternative Methods. *PLoS ONE* **2020**, *15*, e0237027. [CrossRef] [PubMed]
5. Foster, C.; Rodriguez-Marroyo, J.A.; de Koning, J.J. Monitoring Training Loads: The Past, the Present, and the Future. *Int. J. Sports Physiol. Perform.* **2017**, *12*, S22–S28. [CrossRef]
6. Lambert, M.I.; Borresen, J. Measuring Training Load in Sports. *Int. J. Sports Physiol. Perform.* **2010**, *5*, 406–411. [CrossRef]
7. Gronwald, T.; Törpel, A.; Herold, F.; Budde, H. Perspective of Dose and Response for Individualized Physical Exercise and Training Prescription. *J. Funct. Morphol. Kinesiol.* **2020**, *5*, E48. [CrossRef]
8. Steele, J. What Is (Perception of) Effort? Objective and Subjective Effort during Attempted Task Performance. *PsyArXiv* **2020**. [CrossRef]
9. Burd, N.A.; Mitchell, C.J.; Churchward-Venne, T.A.; Phillips, S.M. Bigger Weights May Not Beget Bigger Muscles: Evidence from Acute Muscle Protein Synthetic Responses after Resistance Exercise. *Appl. Physiol. Nutr. Metab. Physiol. Appl. Nutr. Metab.* **2012**, *37*, 551–554. [CrossRef]
10. Montero, D.; Lundby, C. Refuting the Myth of Non-Response to Exercise Training: "non-Responders" Do Respond to Higher Dose of Training. *J. Physiol.* **2017**, *595*, 3377–3387. [CrossRef]
11. Viana, R.B.; de Lira, C.A.B.; Naves, J.P.A.; Coswig, V.S.; Del Vecchio, F.B.; Ramirez-Campillo, R.; Vieira, C.A.; Gentil, P. Can We Draw General Conclusions from Interval Training Studies? *Sports Med.* **2018**, *48*, 2001–2009. [CrossRef]
12. Andreato, L.V. High-Intensity Interval Training: Methodological Considerations for Interpreting Results and Conducting Research. *Trends Endocrinol. Metab. TEM* **2020**, *31*, 812–817. [CrossRef]
13. Vollaard, N.B.J.; Metcalfe, R.S. Those Apples Don't Taste Like Oranges! Why "Equalising" HIIT and MICT Protocols Does Not Make Sense. *Trends Endocrinol. Metab. TEM* **2021**, *32*, 131–132. [CrossRef]
14. Hansen, D.; Abreu, A.; Ambrosetti, M.; Cornelissen, V.; Gevaert, A.; Kemps, H.; Laukkanen, J.A.; Pedretti, R.; Simonenko, M.; Wilhelm, M.; et al. Exercise Intensity Assessment and Prescription in Cardiovascular Rehabilitation and beyond: Why and How: A Position Statement from the Secondary Prevention and Rehabilitation Section of the European Association of Preventive Cardiology. *Eur. J. Prev. Cardiol.* **2022**, *29*, 230–245. [CrossRef]
15. Yan, Z.-W.; Li, W.-G. Battle between High-Intensity Interval Training and Moderate-Intensity Continuous Training in Cardiac Rehabilitation Practice. *Eur. J. Prev. Cardiol.* **2022**, *96*, 1533–1544. [CrossRef]
16. Passfield, L.; Murias, J.M.; Sacchetti, M.; Nicolò, A. Validity of the Training-Load Concept. *Int. J. Sports Physiol. Perform.* **2022**, *17*, 507–514. [CrossRef]
17. MacInnis, M.J.; Gibala, M.J. Physiological Adaptations to Interval Training and the Role of Exercise Intensity. *J. Physiol.* **2017**, *595*, 2915–2930. [CrossRef]
18. Atakan, M.M.; Li, Y.; Koşar, Ş.N.; Turnagöl, H.H.; Yan, X. Evidence-Based Effects of High-Intensity Interval Training on Exercise Capacity and Health: A Review with Historical Perspective. *Int. J. Environ. Res. Public. Health* **2021**, *18*, 7201. [CrossRef]
19. Munn, Z.; Peters, M.D.J.; Stern, C.; Tufanaru, C.; McArthur, A.; Aromataris, E. Systematic Review or Scoping Review? Guidance for Authors When Choosing between a Systematic or Scoping Review Approach. *BMC Med. Res. Methodol.* **2018**, *18*, 143. [CrossRef]
20. Peters, M.D.J.; Godfrey, C.M.; Khalil, H.; McInerney, P.; Parker, D.; Soares, C.B. Guidance for Conducting Systematic Scoping Reviews. *Int. J. Evid. Based Healthc.* **2015**, *13*, 141–146. [CrossRef]
21. Peters, M.D.J.; Marnie, C.; Colquhoun, H.; Garritty, C.M.; Hempel, S.; Horsley, T.; Langlois, E.V.; Lillie, E.; O'Brien, K.K.; Tunçalp, Ö.; et al. Scoping Reviews: Reinforcing and Advancing the Methodology and Application. *Syst. Rev.* **2021**, *10*, 263. [CrossRef] [PubMed]
22. Tricco, A.C.; Lillie, E.; Zarin, W.; O'Brien, K.K.; Colquhoun, H.; Levac, D.; Moher, D.; Peters, M.D.J.; Horsley, T.; Weeks, L.; et al. PRISMA Extension for Scoping Reviews (PRISMAScR): Checklist and Explanation. *Ann. Intern. Med.* **2018**, *169*, 467–473. [CrossRef] [PubMed]
23. Downs, S.H.; Black, N. The Feasibility of Creating a Checklist for the Assessment of the Methodological Quality Both of Randomised and Non-Randomised Studies of Health Care Interventions. *J. Epidemiol. Community Health* **1998**, *52*, 377–384. [CrossRef] [PubMed]
24. Helgerud, J.; Høydal, K.; Wang, E.; Karlsen, T.; Berg, P.; Bjerkaas, M.; Simonsen, T.; Helgesen, C.; Hjorth, N.; Bach, R.; et al. Aerobic High-Intensity Intervals Improve VO2max More than Moderate Training. *Med. Sci. Sports Exerc.* **2007**, *39*, 665–671. [CrossRef]
25. Lanzi, S.; Codecasa, F.; Cornacchia, M.; Maestrini, S.; Capodaglio, P.; Brunani, A.; Fanari, P.; Salvadori, A.; Malatesta, D. Short-Term HIIT and Fat Max Training Increase Aerobic and Metabolic Fitness in Men with Class II and III Obesity. *Obesity* **2015**, *23*, 1987–1994. [CrossRef]
26. Billat, L.V. Interval Training for Performance: A Scientific and Empirical Practice. Special Recommendations for Middle- and Long-Distance Running. Part I: Aerobic Interval Training. *Sports Med.* **2001**, *31*, 13–31. [CrossRef]
27. LaForgia, J.; Withers, R.T.; Gore, C.J. Effects of Exercise Intensity and Duration on the Excess Post-Exercise Oxygen Consumption. *J. Sports Sci.* **2006**, *24*, 1247–1264. [CrossRef]

28. Desgorces, F.D.; Berthelot, G.; Dietrich, G.; Testa, M.S.A. Local Muscular Endurance and Prediction of 1 Repetition Maximum for Bench in 4 Athletic Populations. *J. Strength Cond. Res.* **2010**, *24*, 394–400. [CrossRef]
29. Monod, H.; Scherrer, J. Capacity for static work in a synergistic muscular group in man. *Comptes Rendus Seances Soc. Biol. Fil.* **1957**, *151*, 1358–1362.
30. Péronnet, F.; Thibault, G. Mathematical Analysis of Running Performance and World Running Records. *J. Appl. Physiol.* **1989**, *67*, 453–465. [CrossRef] [PubMed]
31. Billat, L.V.; Koralsztein, J.P.; Morton, R.H. Time in Human Endurance Models. From Empirical Models to Physiological Models. *Sports Med.* **1999**, *27*, 359–379. [CrossRef] [PubMed]
32. Seiler, S.; Jøranson, K.; Olesen, B.V.; Hetlelid, K.J. Adaptations to Aerobic Interval Training: Interactive Effects of Exercise Intensity and Total Work Duration. *Scand. J. Med. Sci. Sports* **2013**, *23*, 74–83. [CrossRef] [PubMed]
33. Klonizakis, M.; Moss, J.; Gilbert, S.; Broom, D.; Foster, J.; Tew, G.A. Low-Volume High-Intensity Interval Training Rapidly Improves Cardiopulmonary Function in Postmenopausal Women. *Menopause* **2014**, *21*, 1099–1105. [CrossRef] [PubMed]
34. Paolucci, E.M.; Loukov, D.; Bowdish, D.M.E.; Heisz, J.J. Exercise Reduces Depression and Inflammation but Intensity Matters. *Biol. Psychol.* **2018**, *133*, 79–84. [CrossRef]
35. Sandbakk, Ø.; Sandbakk, S.B.; Ettema, G.; Welde, B. Effects of Intensity and Duration in Aerobic High-Intensity Interval Training in Highly Trained Junior Cross-Country Skiers. *J. Strength Cond. Res.* **2013**, *27*, 1974–1980. [CrossRef]
36. Soh, S.-H.; Joo, M.C.; Yun, N.R.; Kim, M.-S. Randomized Controlled Trial of the Lateral Push-Off Skater Exercise for High-Intensity Interval Training vs. Conventional Treadmill Training. *Arch. Phys. Med. Rehabil.* **2020**, *101*, 187–195. [CrossRef]
37. Álvarez, C.; Ramírez-Campillo, R.; Ramírez-Vélez, R.; Izquierdo, M. Effects and Prevalence of Nonresponders after 12 Weeks of High-Intensity Interval or Resistance Training in Women with Insulin Resistance: A Randomized Trial. *J. Appl. Physiol.* **2017**, *122*, 985–996. [CrossRef]
38. Wens, I.; Dalgas, U.; Vandenabeele, F.; Grevendonk, L.; Verboven, K.; Hansen, D.; Eijnde, B.O. High Intensity Exercise in Multiple Sclerosis: Effects on Muscle Contractile Characteristics and Exercise Capacity, a Randomised Controlled Trial. *PloS ONE* **2015**, *10*, e0133697. [CrossRef]
39. Eston, R. Use of Ratings of Perceived Exertion in Sports. *Int. J. Sports Physiol. Perform.* **2012**, *7*, 175–182. [CrossRef]
40. Joseph, T.; Johnson, B.; Battista, R.A.; Wright, G.; Dodge, C.; Porcari, J.P.; de Koning, J.J.; Foster, C. Perception of Fatigue during Simulated Competition. *Med. Sci. Sports Exerc.* **2008**, *40*, 381–386. [CrossRef]
41. Noakes, T.D. Linear Relationship between the Perception of Effort and the Duration of Constant Load Exercise That Remains. *J. Appl. Physiol.* **2004**, *96*, 1571–1573. [CrossRef] [PubMed]
42. Barroso, R.; Salgueiro, D.F.; do Carmo, E.C.; Nakamura, F.Y. The Effects of Training Volume and Repetition Distance on Session Rating of Perceived Exertion and Internal Load in Swimmers. *Int. J. Sports Physiol. Perform.* **2015**, *10*, 848–852. [CrossRef] [PubMed]
43. Hourcade, J.-C.; Noirez, P.; Sidney, M.; Toussaint, J.-F.; Desgorces, F.D. Effects of Intensity Distribution Changes on Performance and on Training Loads Quantification. *Biol. Sport* **2018**, *35*, 67–74. [CrossRef]
44. Seiler, S.; Hetlelid, K.J. The Impact of Rest Duration on Work Intensity and RPE during Interval Training. *Med. Sci. Sports Exerc.* **2005**, *37*, 1601–1607. [CrossRef]
45. Agostinho, M.F.; Philippe, A.G.; Marcolino, G.S.; Pereira, E.R.; Busso, T.; Candau, R.B.; Franchini, E. Perceived Training Intensity and Performance Changes Quantification in Judo. *J. Strength Cond. Res.* **2015**, *29*, 1570–1577. [CrossRef]
46. Cavar, M.; Marsic, T.; Corluka, M.; Culjak, Z.; Cerkez Zovko, I.; Müller, A.; Tschakert, G.; Hofmann, P. Effects of 6 Weeks of Different High-Intensity Interval and Moderate Continuous Training on Aerobic and Anaerobic Performance. *J. Strength Cond. Res.* **2019**, *33*, 44–56. [CrossRef]

International Journal of
Environmental Research and Public Health

MDPI

Review

Sports Performance Tests for Amputee Football Players: A Scoping Review

Agnieszka Magdalena Nowak *, Jolanta Marszalek and Bartosz Molik

Faculty of Rehabilitation, Jozef Pilsudski University of Physical Education in Warsaw, 00-968 Warsaw, Poland; jolanta.marszalek@awf.edu.pl (J.M.); bartosz.molik@awf.edu.pl (B.M.)
* Correspondence: agnieszka.nowak@awf.edu.pl

Abstract: Background: This scoping review aims to identify sports performance tests for amputee football players and to critically analyze the methodological quality, validation data, reliability, and standardization of sport-specific tests to indicate the best-fitting tests. Methods: Electronic database searches were conducted between January 2019 and October 2021. Twelve articles met the inclusion criteria. Qualitative assessment of each study was conducted by STROBE checklist. Results: Twenty-nine sports performance tests were identified. No sports performance test fully met all three criteria associated with the qualitative assessment of tests. The critical appraisal of the articles demonstrates a gap in study design, settings, and main results description. Some inconsistencies were found in the methodological descriptions of tests assessing the same motor skill. A STROBE score of 13 points was considered a satisfactory score for the article (it was obtained by 8 of the 12 studies). The weakest point of the analyzed studies was the description of how the test group size was accessed and later obtained. Conclusions: No test was found that was simultaneously presented as valid, reliable, and standardized. The authors can recommend the use of the two-sports performance tests that are the closest to ideal: the L test and the YYIRT1.

Keywords: field-based tests; amputee soccer; assessment; disability; impairment; athletes; adapted sport

Citation: Nowak, A.M.; Marszalek, J.; Molik, B. Sports Performance Tests for Amputee Football Players: A Scoping Review. *Int. J. Environ. Res. Public Health* **2022**, *19*, 4386. https://doi.org/10.3390/ijerph19074386

Academic Editors: Catarina Nunes Matias, Stefania Toselli, Cristina Monteiro and Francesco Campa

Received: 21 February 2022
Accepted: 5 April 2022
Published: 6 April 2022

1. Introduction

Amputee football (amputee soccer; AF) is an impairment-specific football for people with an amputation or limb deficiency (US Soccer Federation). The major part of AF rules is based on regular soccer rules, while the few paragraphs consider the physical impairment of players [1]. Accordingly, two halves are being played (2 × 25 min) on a smaller pitch (from 60 × 30 to 70 × 55 m) by seven players (six field players, one goalkeeper). Single-leg amputees (either above or below the knee) play without prosthesis on aluminum wrist crutches (field players). Goalkeepers must be single-arm amputees [2]. AF is still developing and has become a point of interest for many researchers since it is a non-Paralympic sport discipline that is applying to enter the Paralympic Games. AF has become greatly popular as a recreational and elite sport. It is also recommended as a continuation of the rehabilitation process for amputees to improve the level of functional fitness, as well as a form of physical activity that allows people to realize themselves as athletes. What is more, AF has a positive impact on body composition and quality of life, and it gives a sense of belonging to society [3–5].

It is assumed that AF is classified as a high-intermittent sport with periods of high-intensity activity [6,7]. AF requires from its players a high level of many physical attributes, such as power, speed, strength, balance, agility as well as endurance [8,9]. Short bursts of high-intensity power production and aerobic capacity play a major role in AF performance [2,9]. Some studies have confirmed this, indicating that athletes spend the majority of their playing time in a heart rate zone above 80% of their maximum heart rate

(HRmax) [6,7]. Given the high intensity of the game, it is important to emphasize that AF players should be in excellent condition to easily perform the entire spectrum of activities with and without the ball and while moving on crutches [9]. Studies underline the fact that using crutches is quite exhausting for AF players [10]. Therefore, it can be assumed that players should not only be well prepared technically and tactically but also, most importantly, physically for the game, as is the case with able-bodied soccer players [11]. Coaches should be obliged to evaluate the motor performance of players to notice progress or weaknesses in the training process.

In the literature, many different sports performance tests have been reported [2,3,9,12–20]. Moreover, by reviewing the literature and observing the various nomenclature of motor abilities used and the different descriptions of the same tests, we decided to organize the sports performance tests for assessing the motor performance of amputee football players to make them transparent and understandable for researchers in the field, coaches, and people interested in this type of sport. Considering how important the periodic assessment of athletes' motor performance is to both sport-specific and non-sport-specific tests related to AF, the fundamental aim of this study is to identify sports performance tests for amputee football players in a literature review and to critically analyze the methodological quality, validation data, reliability, and standardization of sport-specific tests to indicate the best-fitting tests. Furthermore, the quality of the reviewed articles is checked to indicate the quality of the studies' descriptions.

2. Materials and Methods

2.1. Search Strategy, Study Selection, and Data Extraction

Reporting of this scoping review was guided by the Preferred Reporting Items for Systematic Review and Meta-Analysis (PRISMA) statement standards. The review protocol was registered with PROSPERO (CRD42021286911), and the review itself was conducted in January–October 2019 with no restrictions on the date of study publication. It was then regularly updated until November 2021. Electronic databases (EBSCO (SPORTDiscus with Full Text, Academic Search Ultimate, Teacher Reference Center, Health Source: Nursing/Academic Edition, MasterFILE Premier), Web of Science, and PubMed (Medline)) were searched. Database settings were customized for each database (option to search all fields, scientific journals, peer-reviewed articles). The keywords used in the search were divided into three groups: amputee OR amputation AND physical AND soccer OR football and were conducted by the Boolean AND/OR. More specific keywords were not needed due to the small number of publications in the field. The keyword combinations were used according to the databases' capabilities and were presented in an online repository.

In the examination process, the title and the abstract were first checked for compatibility with at least one keyword. If an article met the inclusion criteria, it was carefully selected for this review by making sure that it: was available in an online database in full text (1), was written in English (2), was an original study (cohort, case–control, cross-sectional) (3), involved amputee football players (4), and used sports performance tests as research tools (5). The criteria according to which an article could not be included in the examination were as follows: no keyword in the title and/or abstract, papers of other types (reviews, case reports, conference reports, chapters in books), written in a language other than English, not related to amputee football players, and did not include sports performance tests. We used Microsoft Excel to collect the data and uploaded them to an online repository. The PRISMA flowchart was used to describe the review process (Figure 1). Two researchers (A.M.N., J.M.) independently conducted the process.

2.2. Studies Description

First, the studies included in the review are described in a table pointing out the type of research conducted, the purpose of the study, and the characteristics of the study group. A summary description of the included studies is presented in Table 1.

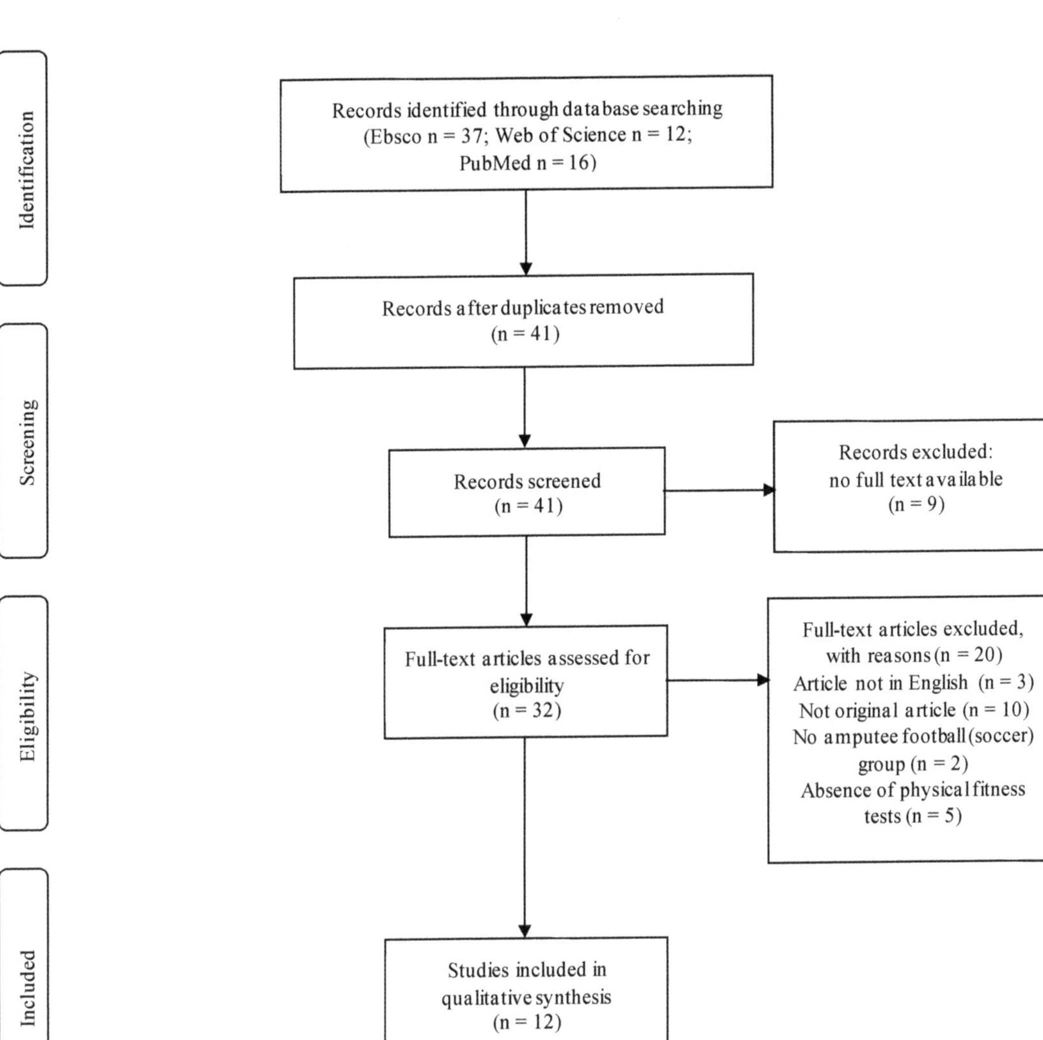

Figure 1. Study selection flow diagram.

2.3. Sports Performance Tests Description

The sports performance tests identified in the literature were analyzed in terms of the type of test and the entire procedure for conducting the test, including athlete preparation, warm-up, how to do the test, number of repetitions, intervals between repetitions of the test or between tests, and variables that are test results. The methodology of the identified sports performance tests is described in Table 2.

Table 1. General description of included studies (n = 12; studies arranged in chronological order).

Authors	Type of Study	Purpose	SG/CG	Training Experience of SG
Buckley et al., 2002	CC	To determine the balance performance of active lower limb amputees during quiet standing and under dynamic conditions.	n = 6 (AF) 25.7 ± 5.8 yrs./n = 6 (AB) 24.7 ± 2.7 yrs.	ND
Yazicioglu et al., 2007	CS-C	To investigate the effect of playing football on balance, muscle strength, locomotor capabilities, and health-related quality of life in subjects with unilateral below-knee amputation.	n = 12 AF, 28.3 ± 4.6 yrs./n =12 (AMP) 29.8 ± 1.4 yrs.	≥6 mths
Ozkan et al., 2012	C	To investigate the relationship between body composition, anaerobic performance, and sprint performance of AF.	n = 15 (AF) 25.5 ± 5.8 yrs.	3.3 ± 2.9 yrs.
Simim et al., 2013	C	To describe anthropometric and physical characteristics of AF and to compare these results, taking into consideration the players' tactical function, and to verify if there are differences between HR after maximum test and the employment of six equations for prediction of HRmax.	n = 12 (AF) 29.3 ± 8.6 yrs.	≥5 yrs.
Mine et al., 2014	C	To examine relationships between quickness and speed performance in AF.	n = 10 (AF) 25.8 ± 4.32 yrs.	ND
Wieczorek et al., 2015	C	To find the relationship between handgrip strength and sprint time in AF.	n = 13 (AF) 26.1 ± 7.7 yrs.	30.8 ± 14.3 mths
Guchan et al., 2017	CC	To determine the effects of playing soccer on various components of physical performance such as body composition, muscular endurance, anaerobic power, flexibility, balance, and speed of individuals with transtibial amputation.	n = 12 (AF) 26.67 ± 7.76 yrs./n = 12 (AMP, sedentary) 33 ± 6.7 yrs.	≥1 yr.
Simim et al., 2017	CC	To quantify the degree of game-induced muscular fatigue in AF.	n = 33 (AF) 31 ± 7 yrs./n = 5 (AF, not playing all matches)	≥4 yrs.
Simim et al., 2018	C	To investigate the match demands of amputee football and its impact on muscular endurance and power.	n = 16 (AF) 32 ± 5 yrs.	≥5 yrs.
Mikami et al., 2018	CC	To examine the difference in measured CPX values among two-legged, one-legged, and two-armed exercises in AB, and to preliminarily evaluate the endurance of AF through CPX with two-armed exercise.	n = 20 (AB) 28.3 ± 5.6 yrs./n = 8 (AF) 36.4 ± 5.7 yrs.	0.4–5 yrs.
Miyamoto et al., 2019	C	To analyze sprint motion in outfield positioned AF using crutches and to clarify the relationship between sprint speed and sprint motion.	n = 12 (AF) 42.3 ± 4.6 yrs.	3.58 ± 2.48 yrs.
Zwierko et al., 2020	CC	To examine postural control during single-leg stance test with progressively increased balance-task difficulty in soccer players with unilateral transfemoral amputation compared to AB soccer players.	n = 11 (AF) 27.45 ± 5.2 yrs./n = 11 (AB football players) 21.91 ± 3.11 yrs.	8.27 ± 3.63 yrs.

AB—able-bodied; AF—amputee football players; AMP—individuals with amputation; C—cohort; CC—case-control; CG—control group; CS—cross-sectional; CS-C—cross-sectional control; mths—months; ND—no data; SG—study group; yrs.—years.

Table 2. A detailed description of the sports performance tests in included studies.

I	II	III					IV
		Test Descriptions					
Authors	Methods	Participants' Preparation	Warm-Up	Procedures	Repetitions	Breaks	Outcomes
Buckley et al., 2002	Static balance test (Kistler force platform)	sportswear; prosthesis	ND	comfortable position on the force platform surface with feet equidistant from a central dividing line, hands on the hips; a large visual target at eye level on a wall 5 m from the force platform; stand stationary, look straight at the visual target for 30 s	ND	ND	CP excursion range, sum of the square's deviations from the mean CP location in the AP and ML directions
	Dynamic balance test (modified dynamic stabilimeter)		approx. 5 min; standing on the stabilimeter	ML—standing on the board pivoted side-to-side in the frontal plane; AP—pivoted forwards and backwards in the sagittal plane; trials in a random order; place hands on hips, focus on a large visual target positioned at eye level on a wall 5 m in front of them	3 trials of 20 s	step down from the stabilimeter	time spent in and out of balance; number of times the board contacted the ground; mean number of contacts per limb (prosthetic or intact, dominant, or non-dominant) or per direction (forwards or backwards)
Yazicioglu et al., 2007	Berg Balance Scale	ND	ND	14 tasks common in everyday life (sitting, standing, reaching, leaning over, turning, and looking over each shoulder, turning in a complete circle, and stepping)	ND	ND	each item is scored on a scale from 0 to 4, max. 56 points
	One-leg static balance test (KAT 2000; Kinesthetic Ability Trainer; Breg, Vista, CA)	ND	5 min.	standing on the intact limb; arms folded across the chest, the knee kept in approx. 10 degrees in flexion; unable to maintain a balance and touch the railing—test discontinued and restart performed in the first difficulty level according to subject's body mass	once a day, for 3 days	15 min.	distance from the central point to the reference position of each trial; balance index
	Dynamic balance test (KAT 2000; Kinesthetic Ability Trainer; Breg, Vista, CA)	prosthesis		standing on both legs; as above			
	Isokinetic muscle strength test (Cybex dynamometer)	ND	ND	peak torques of nonamputee side knee in extension and flexion; perform maximal concentric-concentric motion at angular velocities of 60, 120, and 180 degrees/s	once	20 s	Nm
Özkan et al., 2012	CMJ, SJ (force plate; Sport Expert TM, MPS-501 multi-purpose measuring system, Tumer Electronic LDT, Turkey; centimeter)	no crutches; no prosthesis	Test familiarization	jump as high as possible; SJ: starting position with knees flexed to 90°, hands fixed on the hips, and no allowance for preparatory counter movement; CMJ: performed from an upright standing position, hands fixed on the hips, and with a counter movement preparatory phase, with end position as SJ starting position	3 CMJs; 3 SJs	2 min.	jump height; total work produced in each jump (the Genuario and Dolgener formulas)
	T10, T20, T30 (light gates with timing system; Prosport, Tumer Electronics, Ankara, Turkey)	crutches; no prosthesis	ND	indoor court; light gates placed at the start and at the finish of each sprint test	2 times each distance	1 min.	time

Table 2. *Cont.*

I	II	III					IV
		Test Descriptions					
Authors	Methods	Participants' Preparation	Warm-Up	Procedures	Repetitions	Breaks	Outcomes
Simim et al., 2013	T20 (stopwatch)	crutches	10 min, test familiarization	official field with natural grass; player initiated a movement	ND	at least 24 h of recovery between testing sessions; 5 min. between the two first tests on day one; 3 min. rest between each T-square trial	time; mean speed
	T-square (stopwatch)				3 times		time
	The YYIRT1; Polar F5 to HR_{max}; Six equations to predict HR_{max}			official field with natural grass; 20 m shuttle run test with increasing velocities, 10 s of active recovery between runs until exhaustion; test end if: participant fails twice to reach the front line within the time limit or is unable to complete another run at the imposed speed; HR_{max} record immediately after the test	ND		total distance; HR_{max} result compared with 6 equations used to predict HR_{max}
Mine et al., 2014	T30 (electronic timing system)	rubber-soled track shoes	ND	photocells set on 0, 5, 30 m; standardized starting position; players started the approx. 30 cm back from the starting line; quickness on 5 m, speed on 30 m	3 times	3 min. intervals	time
Wieczorek et al., 2015	Handgrip test (SEHAN hydraulic hand dynamometer, Jamar)	ND	ND	sit with arms along the body, elbow joint in 90° flexion, forearm, wrist in a neutral position; grip the handle of the dynamometer	twice by each hand	ND	the highest value
	T30 (Fusion Smart Speed System; Fusion Sport, Coopers Plains, QLD, Australia)	crutches	ND	8 infrared working gates; 3 m distance between photocells and mirrors; splits recorded: 1, 5, 10, 15, 20, 25, 30 m; standing start; deciding themselves when to start; stretched crutches, no crossing the starting line	2 times	ND	time; mean running velocity
Güçhan et al., 2017	Sit-ups isotonic	prosthesis; shoes	5 min.	lie back with bent knees and sit up until the scapula is no longer in contact with the surface	ND	2 min. between tests	repetitions; time
	Isotonic PUT			trunk and head to the floor and push the body up			
	Back extensors isometric			lie on a table face down, with inguinal points and lower body on the end table, upper body over the table; cross arms in front of the shoulders, raise the trunk; assessor fixed participant's leg			time sustaining in the position
	Trunk flexors			lie back with knees flexed; arms straight toward knees, raise head, neck, shoulders stay in the position			
	Vertical jump test			stand up, fix amputated limb next to the wall, extend arm above; the end of the longest finger was marked before and after a jump; jump vertically; repetition with intact limb near the wall; distance between two marked heights	3 times for both sides		the best result; Lewis' formula (anaerobic power)

Table 2. *Cont.*

I	II	III					IV
Authors	**Methods**	**Test Descriptions**					**Outcomes**
		Participants' Preparation	**Warm-Up**	**Procedures**	**Repetitions**	**Breaks**	
	Modified Thomas test			sit on the end of a table and lie down, with hip joint fixed 28 cm away from the end of the table; flex the contralateral lower limb maximally with arms; tested both sides	ND		distance between the table and the tested knee
	Sit-and-reach test	no prosthesis; no shoes		sit on the floor, knees straight, feet resting vertically; reach forward with straight arms as far as possible; distance between the toes and the longest finger	3 trials		the best result
	Berg Balance Scale			14 items			each item scored 0–4
	L test	prosthesis; shoes		walk an "L" shaped path (7 × 3 m); starting position: sit on a chair without armrests; get up from a chair, walk with usual pace to the end and come back to the sitting position	ND		time
	F8W test			stand in the middle between 2 cones and complete the 8-shaped path with preferred speed walk, come back to the same point; 1.22 × 1.52 m			time; steps
Simim et al., 2017	PUT	ND	tests familiarization; dynamic warm-up and stretching; 1–3 repetitions of each test	max. number of repetitions in 60 s; result divided by body mass	ND	5 min. between tests	repetitions; relative measure
	CMJ (accelerometer Myotest, Sion, Switzerland; centimeter)			use preliminary movement by rapidly flexing the knee, before launching the body vertically	3 trials	as above; 30 s between jumps; 1 min. between throws	jump height; power
	MBT (medicine ball 3 kg)			sit with your back against the wall; lower back stays in contact with the wall during the test; hold a medicine ball with both hands against a chest and throw it on command as far as possible			distance
Simim et al., 2018	PUT	ND	tests familiarization; dynamic warm-up and stretching; 3 repetitions of each test	max. number of repetitions in 60 s; result divided by body mass	ND	one day; pre-and post-match; 5 min. rest between tests	repetitions; relative measure
	CMJ (accelerometer Myotest, Sion, Switzerland)			use preliminary movement by rapidly flexing the knee, before launching the body vertically		as above; 30 s between jumps; 1 min. rest between throws	jump height; power
	MBT (medicine ball 3 kg)			sit with your back against the wall; lower back stays in contact with the wall during the test; hold a medicine ball with both hands against a chest and throw it on command as far as possible			distance

Table 2. *Cont.*

I	II	III					IV
		Test Descriptions					
Authors	Methods	Participants' Preparation	Warm-Up	Procedures	Repetitions	Breaks	Outcomes
Mikami et al., 2018	CPX exercise test (Strength Ergometer; Strength Ergo 8, Mitsubishi Electric Engineering Co., Ltd., Tokyo, Japan); Expired gas monitoring breath-by-breath (cardiorespiratory exercise monitoring system, AE-310s; Minato Medical Science Co., Tokyo, Japan); Fatigue (Modified Borg Scale)	ND	ND	a multistage, Ramp-wise upgrading continuous load; two-legged exercise: Ramp 25 M/min.; one-legged exercise: used leg on the dominant hand side; Ramp 15 W/min.; two-armed exercise: Ramp 15 W/min.; Modified Borg Scale after exercise	ND	ND	anaerobic threshold value of oxygen uptake/weight; peak value of oxygen uptake/weight; HR; VE; WR; c-RPE; p-RPE
Miyamoto et al., 2019	T30 (electronic timing gates, TC Timing System, Brower Timing System, USA)	crutches; no prosthesis	ND	splits recorded: 10, 20, 30 m; sprint; running style by supporting both crutches together for 2 steps; the 30-m sprint test (10 m intervals); speed (between 10 to 20 m)	2 trials	ND	time
Zwierko et al., 2020	Static balance test with open eyes (Biodex Balance System Inc., Shirley, NY, USA)	barefoot	3 trials of 20 s adaptation (in 12, 8, and 4 level of platform stability)	12 dynamic stability levels (12 is the most stable, 1 is the most unstable); single-leg stance on rigid platform; single-leg stance with decreasing platform stability—levels 8 to 4; single-leg stance with platform stability—level 4; 20 s each balance task; during the tests, look straight ahead with arms folded along the chest	3 trials	10 s	OSI; API; MLI

AP—anteroposterior; API—anterior–posterior index; % BF—relative body fat; BI—balance index; BMI—body mass index; CMJ—countermovement jump; CP—center of pressure; CPX—cardiopulmonary exercise test; F8W—figure-of-8 walk; HR - heart rate; HR_{max}—maximum heart rate; HRQOL—health-related quality of life; J-P method—Jackson and Pollock method; MBT—medicine ball throw; ML—mediolateral; MLI—medio-lateral index; ND—no data; OSI—overall stability index; PUT—push up test; RPE—rating of perceived exertion; c-RPE—central rate of perceived exertion; p-RPE—peripheral rate of perceived exertion; SJ—squat jump; T10, T20, T30—10, 20, 30 m sprint test; VE—ventilation equivalent; WR—work rate; YYIRT1—the Yo-Yo intermittent recovery test—level 1.

2.4. Sports Performance Tests' Quality Assessment

In this phase, we divided the sports performance tests according to their characteristics (motor abilities), which assess balance, aerobic capacity, strength, endurance, power, sprint performance, agility, and flexibility. Two researchers (A.M.N., J.M.) independently assessed all the papers and then consulted the results among themselves.

All found sports performance tests were analyzed for reliability, validity, and standardization based on the authors' descriptions in the methods section of the articles. Test reliability and validity were recognized based on information about the reliability and validity of the test in the study and whether test references or expert validity were used. Expert validity implies that the researcher, based on their knowledge and experience, selected a sports performance test to assess specific motor abilities, e.g., the 30 m sprint test was used to assess sprint performance. Standardization means that the researchers have written down all the information necessary to repeat the test (participant preparation, environment, methodology, number of repetitions, intervals, outcomes). A description of this assessment is provided below, and points were allocated for each parameter:

- validity, reliability and/or standardization information present and/or cited references that have confirmed validity, reliability and/or standardization and/or expert validity: "1";
- cited references present but not available or in a language other than English or in unavailable books; no information on validity, reliability and/or standardization or insufficient standardization: "0".

Table 3 presents the qualitative assessment of the sports performance tests identified through the literature review process.

Table 3. Quality assessment of sports performance tests in the review (n = 12).

I	II	III	IV		
Physical Attribute Tested	Test Name (and Tools)	Authors	Sports Performance Tests Assessment		
			R	V	S
Balance	Static balance test (Kistler force platform) *	Buckley et al., 2002	0	1	0
	Dynamic balance test (modified dynamic stabilimeter) *		0	1	1
	Static balance test (Biodex)	Zwierko et al., 2020	0	1	1
	One-leg static balance test (KAT 2000)	Yazicioglu et al., 2007	0	1	0
	Dynamic balance test (KAT 2000) *		0	1	1
	Berg Balance Scale	Yazicioglu et al., 2007, Güçhan et al., 2017	1	1	0
Muscle strength	Isokinetic trunk strength test (Cybex dynamometer)	Yazicioglu et al., 2007	0	1	0
	Handgrip test (hydraulic hand dynamometer)	Wieczorek et al., 2015	1	1	0
	PUT	Simim et al., 2017, 2018	0	1	0
	Isotonic PUT	Güçhan et al., 2017	0	1	0
	Isotonic sit-ups test		0	1	0
	Isometric back extension test *		0	1	0
	Isometric trunk flection test		0	1	0
Power	Vertical jump tests—CMJ, CMJs, SJ, SJs (force plate Sport Expert TM)	Özkan et al., 2012	0	1	1
	CMJ (accelerometer Myotest)	Simim et al., 2017, 2018	1	1	0
	MBT (medicine ball 3 kg)		0	1	0
	Vertical jump test (Lewis' formula) *	Güçhan et al., 2017	0	1	1

Table 3. *Cont.*

I	II	III	IV		
Physical Attribute Tested	Test Name (and Tools)	Authors	Sports Performance Tests Assessment		
			R	V	S
Anaerobic performance (sprint and movement speed [1])	T10, T20, T30	Özkan et al.2012	0	1	0
	T20	Simim et al., 2013	0	1	0
	T30 (5 m)	Mine et al., 2014	0	1	0
	T30 (1, 5, 10, 15, 20, 25 m)	Wieczorek et al., 2015	0	1	0
	T30 (10, 20 m)	Myiamoto et al., 2019	0	1	0
	L test *	Güçhan et al., 2017	1	1	0
	F8W test *		0	1	0
Aerobic capacity	YYIRT1	Simim et al., 2013	1	1	0
	CPX two-armed exercise	Mikami et al., 2018	0	1	0
Flexibility	Modified Thomas test *	Güçhan et al., 2017	0	1	1
	Sit-and-reach test		0	1	1
Agility	T-square	Simim et al., 2013	0	1	1

V—valid; R—reliable; S—standardization; CMJ—countermovement jump; CPX—cardiopulmonary exercise test; F8W—figure-of-8 walk; MBT—medicine ball throw; PUT—push up test; SJ—squat jump; T10, T20, T30—10, 20, 30 m sprint test; YYIRT1—the Yo-Yo intermittent recovery test—level 1; "1"—presence of validity, reliability, standardization; "0"—absence of validity, reliability, standardization; *—test performed with prosthesis; [1]—sprint refers to tests in which movement is as fast as possible in one line, movement speed refers to tests in which movement is as fast as possible with changing directions.

2.5. Studies' Quality Assessment

The studies included in the review were qualitatively assessed to highlight the value of the papers in terms of their methodological design. To accomplish this, the Strengthening the Reporting of Observational Studies in Epidemiology (STROBE) statement was used, which was created to improve the quality of reported observational studies, and such studies were included in our study. The STROBE statement allows the strengths and weaknesses of the observational studies to be identified and provides an opportunity to generalize the results of the report [21]. The STROBE checklist consists of a checklist of 22 items that relate to the title and abstract (1 item), introduction (2 items), methods (9 items), results (5 items), discussion section (4 items), and other information (1 item) in the articles. One point for each item in the paper was given [22]. Some of these items originally had sub-items. In that case, one point was awarded for more positive responses. The outcome was the score obtained when consensus was reached (A.M.N., J.M.). Discrepancies were resolved by consensus with a third researcher (B.M.).

3. Results

Twenty-nine sports performance tests were found in the 12 included studies to assess AF players. They assessed motor abilities such as balance, anaerobic performance (strength, power, sprint performance), aerobic performance (capacity), flexibility, and agility (speed performance) (see Table 3). Despite measuring the same motor ability, the identified tests had different methodologies. For example, the jump test was performed once with and once without a prosthesis, and, in the second case, there was no information about it.

Through 29 sports performance tests, no test met all three quality assessment criteria. In eight cases, participants performed tests with a prosthesis, as marked with an asterisk and presented in Table 3.

3.1. Qualitative Assessment of Sports Performance Tests

3.1.1. Reliability

Five out of twenty-nine tests had confirmed reliability in the cited publications (handgrip test, CMJ, L test, YYIRT1, and Berg Balance Scale [2,3,16,19]), and the L test was tested for reliability among amputees.

Two tests, despite the references provided (isometric test of back extensors and trunk flexor test), were not described as reliable or used in the cited publications [3].

3.1.2. Validity

In total, 18 sports performance tests were considered valid based on expert validity and 11 on literature reference; 4 of the 11 cited books were not available.

3.1.3. Standardization

Although all the identified tests had a description of the procedure, only 28% of them met the standardization criteria. Sports performance tests that had complete instructions (subject preparation, environment, methodology, number of repetition, intervals, outcomes) were the T-square test, modified Thomas test, sit-and-reach test, vertical jump test, static balance test, and dynamic balance test [3,9,12,16,20]; 8 tests lacked information about participants' preparation, 9 tests lacked information about the warm-up, 16 tests lacked information about the number of test repetitions, and 6 tests lacked information about intervals between test attempts. The qualitative assessment of sports performance tests is presented in Table 3.

3.2. *Qualitative Assessment of Articles*

In this scoping review, we included observational studies available in the field of amputee football (5 case–control studies, 6 cohort studies, and 1 cross-sectional study). In total, 10 out of 12 articles met eligibility criteria and were from the past 10 years; 50% of the studies had a study group and a control group. Participants were AF players aged 24–32 years from local [2,9,12,15] or national teams [3,16–20]. Training experience ranged from two months to eight or more years. Two studies did not provide information on players' training experience [12,14] (see Table 1 for details).

The qualitative assessment of the studies resulted in STROBE scores ranging from 5 to 17 (mean 12.9 points; 65%). Two studies had the highest score of 17 points [17,18], while two different studies had the lowest possible score [14,19]. Six of the twelve studies had an appropriately constructed abstract and title, with two studies indicating the study design in the title or abstract and four studies indicating the study design in the methods section. All studies stated specific objectives, and 11 of 12 studies sufficiently explained the background of the study. In most cases, the participant description was correct. Simim et al. (2018) obtained the highest and the maximum score in the methods section. Providing information on how the study size was obtained in the methods section was the weakest aspect of the evaluation (only 2 of 12 authors reported this data [12,18]). In the results section, two articles met the requirements of item 13 (participants), eight articles met the requirements of item 14 (descriptive data), six articles met the requirements of item 15 (outcome data), three articles met the requirements of item 16 (main results), and six articles met requirements of item 17 (other analysis). In summary, in four studies, the key results concerning the study objectives were presented in the discussion section [3,14,15,20]. The items on limitations, interpretation, and generalizability were met by most of the included studies. Four studies provided information on the source of funding (item 22). The qualitative assessment of the included studies is presented in Table 4.

Table 4. The STROBE qualitative assessment of included studies.

	Introduction				Methods								Results					Discussion				Other Information	STROBE Points [1]
Items / Authors	1	2	3	4	5	6	7	8	9	10	11	12	13	14	15	16	17	18	19	20	21	22	Total
Simim et al., 2018	0	1	1	1	1	1	1	1	1	1	1	1	0	1	1	0	1	0	1	1	1	0	17
Guchan et al., 2017	1	1	1	0	0	1	1	1	1	0	1	1	0	1	1	0	0	1	1	1	1	0	15
Simim et al., 2017	1	1	1	1	1	1	1	1	1	0	1	1	0	0	1	0	1	0	1	1	1	1	17
Wieczorek et al., 2015	0	1	1	0	0	1	0	1	0	0	0	0	0	0	0	0	0	0	0	1	0	0	5
Simim e al. 2013	0	1	1	1	1	1	1	1	1	0	1	1	0	1	0	1	1	0	0	1	1	0	15
Ozkan et al., 2012	0	1	1	0	0	1	1	1	0	0	1	0	1	0	0	0	0	0	0	1	1	0	9
Yazicioglu et al., 2007	1	1	1	1	0	1	1	1	1	0	1	1	1	1	1	0	0	0	1	1	1	0	16
Mikami et al., 2018	0	1	1	0	0	0	1	1	0	0	1	1	0	1	1	0	0	0	1	1	1	1	12
Mine et al., 2014	1	0	1	0	0	0	1	0	0	0	1	0	0	0	0	0	0	1	0	0	0	0	5
Buckley et al., 2002	0	1	1	0	1	1	1	1	1	1	1	1	0	1	0	1	1	0	0	1	1	0	15
Miyamoto et al., 2019	1	1	1	0	0	0	1	1	0	0	1	0	0	1	1	1	1	1	1	1	1	1	15
Zwierko et al., 2020	1	1	1	0	0	0	1	1	0	0	1	1	0	1	0	0	1	1	1	1	1	1	14

[1]—max. 22 items; 1—title, abstract; 2—background; 3—objective; 4—study design; 5—settings; 6—participants; 7—variables; 8—data sources, measurement; 9—bias; 10—study size; 11—quantitative variables; 12—statistical methods; 13—participants; 14—descriptive data; 15—outcome data; 16—main results; 17—other analysis; 18—key results; 19—limitations; 20—interpretations; 21—generalizability; 22—funding.

4. Discussion

The purpose of this scoping review was to identify sports performance tests for amputee football (AF) players in the scientific papers and to critically analyze these tests for reliability (i), validity (ii), and standardization (iii) to indicate the best-fitting tests. Along this line, 29 sports performance tests used in AF were found in the current literature (12 studies). We found no sports performance test that would fully meet all three criteria associated with a qualitative assessment of sports performance tests.

When discussing the first parameter (i), the authors of the included studies did not conduct a test reliability examination. The reliability of five tests (YYIRT1, L test, handgrip test, CMJ, and Berg Balance Scale) has been confirmed by the authors of the included studies based on the references [2,3,16,19]. The reliability of only one test, the L test, was verified on amputees, which is an advantage of the reported study [3,23] compared to other tests in which reliability was verified on able-bodied individuals. The authors of the analyzed studies used reliable tools to assess muscle strength [19], lower limb power [17], aerobic capacity [16], and balance [2,3].

The PUT, the isometric back extension test, and the isometric trunk flexion test had inappropriate references to prove the reliability of these tests because the works cited were off-topic [3,17,18]. Consequently, we suggest that researchers and coaches pay attention to the reliability of sports performance tests applied to their groups of athletes.

In terms of validity (ii), from one point of view, the indispensable information was obtained in seven sports performance tests, which included the static balance one-leg test, dynamic balance test, handgrip test, L test, F8W test, YYIRT1, and Berg Balance Scale [2,3,16,19]. Whereas, in the case of four tests, such as the modified Thomas test, the sit-and-reach test, the vertical jump test by the Lewis formula, and the isometric back extension test, we could not approve their validity due to the inability to find the reference cited by the authors [3]. Additionally, about the PUT, it was performed differently than reported in the original paper [24]. Consecutively, it also remains unknown whether the presented PUT is truly valid [16]. Moreover, in articles that used static (Kistler force platform) and dynamic balance tests, isokinetic trunk strength tests, PUT, isotonic sit-ups tests, isometric back extension and trunk flections tests, CMJ and SJ (force plate Sport Expert TM), MBT, CPX two-armed exercise tests, modified Thomas tests, sit-and-reach tests, T-square, and sprint tests, there was no information on validity and reliability verification [2,3,9,12–14,16–19]. It is probably the case that the authors of included studies, when selecting tests to assess the motor abilities of AF players, verified these tests based on their experience and general knowledge (e.g., sprint tests to assess speed or sprint performance); therefore, we decided to give them one point as an expert validation.

It must be admitted that in most sports performance tests, the standardization (iii) was clearly explained. Information regarding the starting and finishing positions, the number of repetitions and break times, and the type of movement (running, walking with or without prosthesis) was adequately introduced. This renders them easily repeatable and, thus, helpful for both researchers and coaches. When analyzed in detail, 8 of the 29 test descriptions met all standardization criteria (T-square test, modified Thomas test, sit-and-reach test, vertical jump test, static balance test, and dynamic balance test [3,9,12,16,20]). For a test such as the YYIRT1, the only information about the number of repetitions of the test performed was missing, but we believe that this information is not necessary in this case because this type of aerobic capacity test is usually performed only once due to the maximal stimulation of the aerobic system, after which a long recovery is necessary [25]. The lack of descriptions regarding the warm-up and intervals between repetitions in sprint tests [9,14,19], in which a maximal effort is performed, deserves significant criticism since all these elements are crucial in the assessment of anaerobic performance. In the case of balance tests, information about the use of a familiarization session is important in the context of repeating and comparing the test in the future, and the question of whether and how this session affects test performance (learning process) and the final result is still unknown [26].

On the other hand, some of the tests might be misleading, e.g., the PUT, sprint tests, CMJ (by Myotest), MBT, and CPX two-armed exercise tests, in which it was not explained why and how the procedures were followed and how they were adapted for amputees [9,13,16–18]. The PUT did not have information about the position and the type of movement included in the description, as well as whether a prosthesis was used during this test and other tests [13,17,18]. Because of these confusions, we expected that performing the MBT in a seated position with or without a prosthesis might influence the stability of the trunk position, and, consequently, the final results might be different (athlete sits close to the wall vs. athlete performs a full backward and forward movement to complete the task). In some locomotion tests, participants used a prosthesis (L test and F8W test), while in others, they performed the tests on crutches without a prosthesis (T-square). At this point, it is worth asking ourselves under which conditions we want to evaluate the AF players, as it must be remembered that the athlete is moving on crutches during the match. The same dilemma regarding the use of a prosthesis or not has been noted in vertical jump tests [3,9] and balance tests [2,12]. Consequently, the reader does not know if these tests were performed in a single-leg standing position or if the athletes had three or four points of support. Moreover, we noted several discrepancies concerning the start of the tests. For the T10, T20, and T30 procedures [9], there was no information about the starting position or whether the starting signal was given by the researcher or whether the athlete decided to start the test. Then, in the MBT, it was not clear where the starting point was for measurement. Without such information, it is difficult to compare the results obtained by different groups of participants and then repeat and compare the tests with each other. The differences in results are likely due to erroneous measurements rather than the athletes' skills, making the ratings unreliable. Therefore, it is recommended that in future papers, authors describe their tests accurately.

The studies included in this review have many limitations in the clarity of the names of the motor abilities assessed in sports performance tests because of various wording. In other words, three different groups of researchers used different terms to match tests to the physical attribute; for instance, T30 was used to assess anaerobic performance, sprint performance, or movement speed [9,16,19]. It would be clearer for readers to use only one term. Surprisingly, the L test and the F8W have been classified as sprint tests, together with the T10, T20, and T30, which are speed tests [3,14,16,19]. It becomes obvious that the sprint tests were performed as fast as possible in a straight line, while the L test and the F8W were performed with changes in direction, which may affect the change in running speed and is more to assess agility than speed. Moreover, the result of the L test and the F8W may consist of the route execution technique, which is unlikely for the sprint tests.

A similar observation was made for the vertical jump tests and the MBT. The latter has been used as a power test, a muscle test, a neuromuscular performance test, and an anaerobic performance test and has been positioned as a test focused on strength assessment [3,9,17,18]. Given these achievements, we suggest classifying the MBT as a power assessment because it is the same physical attribute that vertical jump tests assess. We believe that future manuscripts should pay more attention to the terms and expressions used in the sports performance tests and to the description of the physical attributes. Maintaining this level of vocabulary clarity will be beneficial to both coaches and athletes in understanding which motor abilities are being tested in each sports performance test.

The articles included in this review had large discrepancies in scoring in the qualitative assessment. The authors of the current study believe that the methods and results sections of the included studies need the most correction and attention. First, providing the study design in the abstract and/or methods section is important because it gives the reader an understanding of what type of research they will be dealing with. Most studies correctly described the participants. The reader can read about: eligibility criteria and how participants were selected, outcomes, exposures, predictors, potential confounders, and details of assessment methods. The above-mentioned description is important because it indicates whether the study group was homogeneous and whether there were confounding factors.

In this review, the authors of the included studies did not mention any possible confounding factors or description of the test location (whether the tests were performed in the same setting, such as a gym, laboratory, or outdoor soccer pitch). Different conditions and environments can affect the results: e.g., headwind, a slippery floor in sprint tests, and low temperatures can cause poorer results in sprint or flexibility tests. In addition, researchers and coaches should be cautious when interpreting their results concerning the already existing results of others, as there have been times when the results of the same test have depended on different variables. For example, in the PUT, the duration of the test or the number of repetitions performed within a specified time was evaluated; in the sprint tests, the time or speed of the distance covered was evaluated; in the jump tests (vertical jump, CMJ, SJ), the height of the jump or power was evaluated.

What is more, we were concerned about the lack of explanation of how the study group size was obtained (only two articles reported this [12,18]). This issue is particularly relevant when judging null results, which might indicate that there was no real difference between the study groups or that the power of the statistical analysis was too low to detect a real difference. It is worth noting that some studies on AF players included relatively few participants (6–33 people). In the result section, the items were quite complex, and a study had to meet most of the criteria for each item to receive one full point. If a sub-criterion did not apply to the study in some cases, we did not count it. It seems worrying that most articles do not state the key results at the beginning of the discussion section (item 18). Another important point to indicate is if the purpose of the study was achieved in order to lead the discussion section fluently.

Although the STROBE checklist was designed for observational studies, it is important to keep in mind when using this tool that not all criteria are mandatory for every subtype of study, e.g., cohort studies usually do not have any follow-ups or reduction in the number of participants because they have only one group and the study is conducted over one or two days. The STROBE statement is a particularly detailed tool; on one hand, it can help in the preparation of the manuscript, but, on the other hand, it can cause difficulties in the evaluation of the study due to its precision. Considering the presented conclusions and the fact that most of the studies were single-case studies and that we could not give a positive score for some criteria (not because there was an error in the article but because the criteria did not apply to the study), we judged that 60% (13 points) was a satisfactory score, and, thus, 8 of the 12 articles achieved it.

Limitations and Perspectives

This is the first review to bring together all the sports performance tests used in AF and organize them in detail in terms of motor abilities and test descriptions. The available literature lacks a "gold standard", a battery of sports performance tests, or a compilation of which tests are dedicated to AF players (sport-specific tests). Our study indicates that some tests, based on their standardization, may be suitable for assessing sports performance in AF, and coaches may use them in their practice. However, further research is needed to investigate the tests' validity and reliability and characterize them for AF players.

We understand that the literature search performed in this research field may be conducted differently in future studies. This manuscript presents a structured way of literature review (keywords, inclusion/exclusion criteria). Other authors may search the literature using different methodology and other guidelines for reporting the main types of studies, such as the STROBE guidelines that were used in our study. However, in our opinion, recommendations for future studies seeking sports performance tests in a specific sport should be structured as a research review and a presentation of the advantages and disadvantages of the tests and research, such as was done in this study (quality of paper, presence of validity and reliability of tests, and the completeness of description of selected tests). A well-planned research review and manuscript organization will be important for the next steps in AF development as a future Paralympic sport, considering the development of sports classification based on evidence (evidence-based classification system) [27]. The International Paralympic Committee (IPC) has outlined the steps in this process, and the identification of tests in this manuscript is relevant to step 2 (identifying key activities and determinants) and step 3 (identifying appropriate tests to assess key determinants) of the IPC classification process [28]. Authors of future research may consider this rationale and address the need for an evidence-based classification approach as a purpose of their work.

5. Conclusions

Our study constitutes a practical and detailed description of the sports performance tests identified in the literature and includes a qualitative assessment of sports performance tests and a qualitative evaluation of the included articles. The authors of the studies included in this review have verified the reliability and validity of sports performance tests based on results from others' studies. Considering the final conclusions of the reviewed studies and our evaluation of these studies, we conclude that none of the 29 tests from the 12 research papers included in this review were simultaneously reported as valid, reliable, and standardized. We found few tests for amputee football players, which were only partially verified for validity and reliability; thus, we recommend verifying those tests using, for instance, the test–retest method [29]. Despite the deficiencies in the test descriptions, we recommend using two sports performance tests: the L test and the YYIRT1, to assess agility and endurance, respectively.

Author Contributions: Conceptualization, A.M.N.; data curation, A.M.N. and J.M.; formal analysis A.M.N., B.M. and J.M.; investigation, A.M.N., B.M. and J.M.; methodology, A.M.N. and J.M.; project administration A.M.N. and J.M.; supervision, J.M. and B.M.; visualization, A.M.N., J.M. and B.M.; writing—original draft, A.M.N.; writing—review and editing, A.M.N., B.M. and J.M. All authors have read and agreed to the published version of the manuscript.

Funding: This work was supported by the Ministry of Science and Higher Education in the years 2021 and 2022 under Research Group No. 4 at Jozef Pilsudski University of Physical Education in Warsaw ("Physical activity and sports for people with special needs").

Institutional Review Board Statement: Not applicable.

Informed Consent Statement: Not applicable.

Data Availability Statement: The data presented in this study are openly available in FigShare at: https://doi.org/10.6084/m9.figshare.16850194.v1 (accessed on 15 February 2022) and https://doi.org/10.6084/m9.figshare.16850185.v1 (accessed on 15 February 2022).

Conflicts of Interest: The authors declare no conflict of interest.

References

1. Frere, J. The History of "modern" Amputee Football. In *Amputee Sports for Victims of Terrorism*; Centre of Excellence—Defence Against Terrorism, Ed.; IOS Press: Ankara, Turkey, 2008; pp. 5–13.
2. Yazicioglu, K.; Taskaynatan, M.; Guzelkucuk, U.; Tugcu, I. Effect of Playing Football (Soccer) on Balance, Strength, and Quality of Life in Unilateral Below-Knee Amputees. *Am. J. Phys. Med. Rehabil.* **2007**, *86*, 800–805. [CrossRef] [PubMed]
3. Güçhan, Z.; Bayraklar, K.; Ergun, N.; Ercan, Y. Comparison of mobility and quality of life levels in sedentary amputees and amputee soccer players. *J. Exerc. Ther. Rehabil.* **2017**, *4*, 47–53.
4. Ilkım, M.; Canpolat, B.; Akyol, B. The Effects of Eight-Week Regular Training in Amateur Amputee Football Team Athletes' Body Composition. *Turk. J. Sport Exerc.* **2018**, *20*, 199–206. [CrossRef]
5. Monteiro, R.; Pfeifer, L.; Santos, A.; Sousa, N. Soccer practice and functional and social performance of men with lower limb amputations. *J. Hum. Kinet.* **2014**, *43*, 33–41. [CrossRef] [PubMed]
6. Maehana, H.; Miyamoto, A.; Koshiyama, K.; Yanagiya, T.; Yoshimura, M. Profile of match performance and heart rate response in Japanese amputee soccer. *J. Sports Med. Phys. Fit.* **2018**, *58*, 816–824. [CrossRef] [PubMed]
7. Nowak, A.M. Match performance in Polish amputee soccer Extra Ligue—A pilot study. *Adv. Rehabil.* **2020**, *34*, 16–25. [CrossRef]
8. Chin, T.; Sawamura, S.; Fujita, H.; Nakajima, S.; Ojima, I.; Oyabu, H.; Nagakura, Y.; Otsuka, H.; Nakagawa, A. Physical fitness of lower limb amputees. *Am. J. Phys. Med. Rehabil.* **2002**, *81*, 321–325. [CrossRef]
9. Özkan, A.; Kayıhan, G.; Köklü, Y.; Ergun, N.; Koz, M.; Ersöz, G.; Dellal, A. The Relationship Between Body Composition, Anaerobic Performance and Sprint Ability of Amputee Soccer Players. *J. Hum. Kinet.* **2012**, *35*, 141–146. [CrossRef]

10. Tatar, Y.; Gercek, N.; Ramazanoglu, N.; Gulmez, I.; Uzun, S.; Sanli, G.; Karagozoglu, C.; Cotuk, H.B. Load distribution on the foot and lofstrand crutches of amputee football players. *Gait Posture* **2018**, *64*, 169–173. [CrossRef]
11. Hoff, J. Training and testing physical capacities for elite soccer players. *J. Sports Sci.* **2005**, *23*, 573–582. [CrossRef]
12. Buckley, J.G.; O'Driscoll, D.; Bennett, S.J. Postural sway and active balance performance in highly active lower-limb amputees. *Am. J. Phys. Med. Rehabil.* **2002**, *81*, 13–20. [CrossRef] [PubMed]
13. Mikami, Y.; Fukuhara, K.; Kawae, T.; Sakamitsu, T.; Kamijo, Y.; Tajima, H.; Kimura, H.; Adachi, N. Exercise loading for cardiopulmonary assessment and evaluation of endurance in amputee football players. *J. Phys. Ther. Sci.* **2018**, *30*, 960–965. [CrossRef] [PubMed]
14. Mine, T.; Cengiz, T.; Turgut, K.; Halil, T. Relationship between quickness and speed performance in amputee footballers. *Sci. Mov. Health* **2014**, *14*, 580–584.
15. Miyamoto, A.; Maehana, H.; Yanagiya, T. The relationship between sprint speed and sprint motion in amputee soccer players. *Eur. J. Adapt. Phys. Act.* **2019**, *12*, 9. [CrossRef]
16. Simim, M.A.M.; Silva, B.V.C.; Marocolo Jr, M.; Mendes, E.L.; de Mello, M.T.; da Mota, G.R.T. Anthropometric profile and physical performance characteristic of the Brazilian amputee football (soccer) team. *Mot. Rev. Educ. Física* **2013**, *19*, 641–648. [CrossRef]
17. Simim, M.A.; Bradley, P.S.; da Silva, B.V.; Mendes, E.L.; de Mello, M.T.; Marocolo, M.; da Mota, G.R. The quantification of game-induced muscle fatigue in amputee soccer players. *J. Sports Med. Phys. Fit.* **2017**, *57*, 766–772. [CrossRef]
18. Simim, M.A.M.; da Mota, G.R.; Marocolo, M.; da Silva, B.V.C.; de Mello, M.T.; Bradley, P.S. The Demands of Amputee Soccer Impair Muscular Endurance and Power Indices but Not Match Physical Performance. *Adapt. Phys. Act. Q.* **2018**, *35*, 76–92. [CrossRef]
19. Wieczorek, M.; Wiliński, W.; Struzik, A.; Rokita, A. Hand Grip Strength Vs. Sprint Effectiveness in Amputee Soccer Players. *J. Hum. Kinet.* **2015**, *48*, 133–139. [CrossRef]
20. Zwierko, M.; Lesiakowski, P.; Zwierko, T. Postural Control during Progressively Increased Balance-Task Difficulty in Athletes with Unilateral Transfemoral Amputation: Effect of Ocular Mobility and Visuomotor Processing. *Int. J. Environ. Res. Public Health* **2020**, *17*, 6242. [CrossRef]
21. Vandenbroucke, J.P.; von Elm, E.; Altman, D.G.; Gøtzsche, P.C.; Mulrow, C.D.; Pocock, S.J.; Poole, C.; Schlesselman, J.J.; Egger, M. Strengthening the Reporting of Observational Studies in Epidemiology (STROBE): Explanation and Elaboration. *PLoS Med.* **2007**, *4*, e297. [CrossRef]
22. Sorensen, A.A.; Wojahn, R.D.; Manske, M.C.; Calfee, R.P. Using the Strengthening the Reporting of Observational Studies in Epidemiology (STROBE) Statement to Assess Reporting of Observational Trials in Hand Surgery. *J. Hand Surg.* **2013**, *38*, 1584–1589. [CrossRef] [PubMed]
23. Deathe, A.B.; Miller, W.C. The L Test of Functional Mobility: Measurement Properties of a Modified Version of the Timed "Up & Go" Test Designed for People with Lower-Limb Amputations. *Phys. Ther.* **2005**, *85*, 626–635. [CrossRef] [PubMed]
24. Negrete, R.J.; Hanney, W.J.; Pabian, P.; Kolber, M.J. Upper body push and pull strength ratio in recreationally active adults. *Int. J. Sports Phys. Ther.* **2013**, *8*, 138–144. [PubMed]
25. Bangsbo, J.; Iaia, F.M.; Krustrup, P. The Yo-Yo Intermittent Recovery Test: A Useful Tool for Evaluation of Physical Performance in Intermittent Sports. *Sports Med.* **2008**, *38*, 37–51. [CrossRef]
26. Giboin, L.-S.; Gruber, M.; Kramer, A. Additional Intra- or Inter-session Balance Tasks Do Not Interfere with the Learning of a Novel Balance Task. *Front. Physiol.* **2018**, *9*, 1319. [CrossRef]
27. Tweedy, S.M.; Vanlandewijck, Y.C. International Paralympic Committee position stand—Background and scientific principles of classification in Paralympic sport. *Br. J. Sports Med.* **2009**, *45*, 259–269. [CrossRef]
28. Tweedy, S.M.; Mann, D.; Vanlandewijck, Y.C. Research needs for the development of evidence-based systems of classification for physical, vision, and intellectual impairments. In *Training and Coaching the Paralympic Athlete*; Vanlandewijck, Y.C., Thompson, W.R., Eds.; Wiley Online Library: Hoboken, NJ, USA, 2016; pp. 122–149. [CrossRef]
29. Marszałek, J.; Kosmol, A.; Morgulec-Adamowicz, N.; Mróz, A.; Gryko, K.; Molik, B. Test–retest reliability of the newly developed field-based tests focuses on short time efforts with maximal intensity for wheelchair basketball players. *Adv. Rehabil.* **2019**, *33*, 23–27. [CrossRef]

International Journal of
Environmental Research and Public Health

MDPI

Systematic Review

Biomechanical Performance Factors in the Track and Field Sprint Start: A Systematic Review

Maria João Valamatos [1,2,3,4,*], João M. Abrantes [1,3], Filomena Carnide [1,2,3], Maria-José Valamatos [1] and Cristina P. Monteiro [1,2,5]

[1] Sport and Health Department, Faculdade de Motricidade Humana, Universidade de Lisboa, Estrada da Costa, 1499-688 Cruz-Quebrada, Portugal; jabrantes@fpatletismo.pt (J.M.A.); fcarnide@fmh.ulisboa.pt (F.C.); mzvalamatos@fmh.ulisboa.pt (M.-J.V.); cmonteiro@fmh.ulisboa.pt (C.P.M.)
[2] Interdisciplinary Center for the Study of Human Performance (CIPER), Faculdade de Motricidade Humana, Universidade de Lisboa, Estrada da Costa, 1499-688 Cruz-Quebrada, Portugal
[3] Biomechanics and Functional Morphology Laboratory, Faculdade Motricidade Humana, Universidade Lisboa, Estrada da Costa, 1499-688 Cruz-Quebrada, Portugal
[4] Neuromuscular Research Laboratory, Faculdade Motricidade Humana, Universidade Lisboa, Estrada da Costa, 1499-688 Cruz-Quebrada, Portugal
[5] Laboratory of Physiology and Biochemistry of Exercise, Faculdade de Motricidade Humana, Universidade de Lisboa, 1499-688 Cruz-Quebrada, Portugal
* Correspondence: mjvalamatos@fmh.ulisboa.pt; Tel.: +351-214-149-207

Abstract: In athletics sprint events, the block start performance can be fundamental to the outcome of a race. This Systematic Review aims to identify biomechanical factors of critical importance to the block start and subsequent first two steps performance. A systematic search of relevant English-language articles was performed on three scientific databases (PubMed, SPORTDiscus, and Web of Science) to identify peer-reviewed articles published until June 2021. The keywords "Block Start", "Track and Field", "Sprint Running", and "Kinetics and Kinematics" were paired with all possible combinations. Studies reporting biomechanical analysis of the block start and/or first two steps, with track and field sprinters and reporting PB100m were sought for inclusion and analysis. Thirty-six full-text articles were reviewed. Several biomechanical determinants of sprinters have been identified. In the "Set" position, an anthropometry-driven block setting facilitating the hip extension and a rear leg contribution should be encouraged. At the push-off, a rapid extension of both hips and greater force production seems to be important. After block exiting, shorter flight times and greater propulsive forces are the main features of best sprinters. This systematic review emphasizes important findings and recommendations that may be relevant for researchers and coaches. Future research should focus on upper limbs behavior and on the analysis of the training drills used to improve starting performance.

Keywords: track and field; sprinters; sprint start; block start; block velocity; biomechanics; kinematics; kinetics; sprint running; initial acceleration; sprint first stance; sprint first two steps

Citation: Valamatos, M.J.; Abrantes, J.M.; Carnide, F.; Valamatos, M.-J.; Monteiro, C.P. Biomechanical Performance Factors in the Track and Field Sprint Start: A Systematic Review. *Int. J. Environ. Res. Public Health* **2022**, *19*, 4074. https://doi.org/10.3390/ijerph19074074

Academic Editor: Paul B. Tchounwou

Received: 14 February 2022
Accepted: 25 March 2022
Published: 29 March 2022

1. Introduction

The 100 m race is perhaps the highlight of the Olympic Games, as it defines who is the fastest man and woman in the world. In this type of event, the block start performance and the subsequent first two steps can be of critical importance since they have a direct influence on the overall 100 m time [1–8]. Given the importance of the sprint start, a new body of research has emerged in the past two decades that involved advanced technologies, high-precision methods, and sprinters of a higher performance level. For this reason, several technical (kinematic) and dynamic (kinetic) aspects are currently identified as determinant factors for starting block phase and initial sprint acceleration performances [1,4,6,9–25]. However, the concepts, outcomes, and findings between studies are sometimes inconsistent and difficult to interpret and conclude from. These inconsistencies may be accounted for by different study designs, methods, technologies of measure (e.g., external reaction forces under or on the blocks), statistical analyses, or more importantly, the ambiguity between samples of sprinters with different performance levels (e.g., elite, sub-elite, well-trained or trained) and/or between-group analyses based on the overall 100 m performance (i.e., personal best at 100 m—PB100m), and not on block performance. Although two important narrative reviews have already been published [26,27], to our knowledge, no previous review conducted a systematic search of literature exploring the inter-individual variability on block start performance across different performance levels. Thus, the main purposes of this systematic review were: (a) determine the biomechanical parameters of greatest influence on the sprint start, including the "set" position and push-off phase, and the first two steps of initial sprint acceleration and (b) identify the kinematic and kinetic biomechanical variables that best differentiate sprinters of different performance levels in each of those three phases of the sprint start. Considering the impact of the sprint in the sports field and the absence of systematic studies on the kinematics and kinetics factors that determine success in block starts and initial sprint acceleration, we hypothesized that this systematic review will have a relevant impact on researchers to better design experimental/intervention studies, as well as constituting relevant support for coaches and athletes in the definition of efficient strategies for performance in the 100 m race.

2. Materials and Methods

2.1. Article Search, Eligibility, Inclusion, and Exclusion Criteria

The systematic search of relevant articles was conducted based on PRISMA (Preferred Reporting Items for Systematic Reviews and Meta-analyses) guidelines [28]. PubMed, Web of Science, and SPORTDiscus databases were searched for the following mesh terms: "Block Start" OR "Track and Field" OR "Sprint Running" OR "Acceleration" AND "Kinetics and Kinematics" pairing them with all possible combinations. In addition, filters for 'English' and 'articles' have been applied. The last search took place on 30 June 2021.

The inclusion criteria were: publications in English; original observational and experimental studies published in peer-reviewed journals; studies mainly focused on the block phase and/or one or two of the subsequent stance phases concerning kinematic and kinetic variables; and studies that included track and field sprinters with the indication of their PB100m. The following types of records were excluded: conference abstracts; studies focused exclusively on the acceleration phase (beyond the first two stance phases) or mainly focused on limitations imposed by motor and neurological impairments; studies reporting data referring to samples evaluated in previously published papers; studies not mentioning the performance level of the sprinters through their PB100m; case reports; and studies without reference to biomechanical variables.

The records identified from the databases with the aforementioned mesh terms were exported to the reference manager software EndNote X8 that eliminated duplicates. All articles' eligibility was then assessed independently by two reviewers' authors (JMA and FC). The articles identified were first screened by title and abstract for relevance. Studies that raised any uncertainty in exclusion were conservatively retained for subsequent full-text review. The full text of the articles selected as relevant or having raised uncertainty in exclusion was read and further scrutinized for meeting the inclusion criteria and their quality was evaluated. Disagreements on final inclusion or exclusion of studies were resolved by consensus, and if disagreement persisted, a third reviewer (first author, MJV) was available for adjudication. Articles that did not meet the selection criteria or presented a quality score below 50% were excluded.

2.2. Quality of the Studies

The study quality of each publication was evaluated according to the guidelines of the Strengthening the Reporting of Observational Studies in Epidemiology (STROBE) Initiative [29]. This analysis was based on 22 items. Title and abstract. Introduction: background and rationale. Methods: study design, setting, participants, variables, data sources, bias, sample size, quantitative variables, and statistical methods. Results: participants, descriptive data, outcome data, main results, and other analyses. Discussion: key results, limitations, interpretation, and generalizability. Funding. These criteria were scored on a binary scale (1 = yes, 0 = no) independently by two of the authors, and a quality score was then calculated for each study by adding its binary scores and dividing the result by the maximum possible score the study could have achieved. This was then expressed as a percentage to reflect a measure of methodological quality. The quality scores were classified as follows (a) low methodological quality for scores < 50%; (b) good methodological quality for scores between 50% and 75%; and (c) excellent methodological quality for scores > 75%. The studies with a score lower than 50% [30] were excluded from the systematic review. The inter-rater reliability analysis was evaluated by the Cohen's Kappa for nominal variables (2 dimensions) [31]. Standards for strength of agreement for the kappa coefficient were: ≤0 = poor, 0.01–0.20 = slight, 0.21–0.40 = fair, 0.41–0.60 = moderate, 0.61–0.80 = substantial, and 0.81–1 = almost perfect [32].

2.3. Data Extraction

An Excel form was used for data extraction. Of each manuscript selected for review, the following information was extracted from each included study: (a) the primary focus of study, means the phase of sprint start, e.g., block phase, first stance, and study design; (b) the main purpose, e.g., associations between biomechanical variables of starting blocks and the sprint start performance, comparing athletes of different performance levels, comparing different footplate spacing and block angles; (c) type of kinematic and kinetic analyses systems used—two dimensional (2D) or three dimensional (3D) analysis and starting blocks instrumented or placed on force platforms; (d) study sample—the number per gender of participants, and per level of expertise of participants according with the authors, and their PB100m; (e) biomechanical measurement protocols—the variables used to characterize the biomechanical factors of sprint start, number and distance of repeated trials; and (f) key findings of sprint start kinematic and kinetic factors.

3. Results

3.1. Search Results

The initial search identified 756 titles in the described databases. With the reference manager software, 406 duplicates were eliminated automatically. The remaining 350 articles were then screened according to title and abstract for relevance, resulting in another 289 studies being eliminated from the database. The full text of the remaining 61 articles was read and another 22 were rejected for not meeting the inclusion criteria defined for the current study and 3 studies were excluded for not meeting the quality criteria (quality index < 50%). A total of 36 studies was fully reviewed.

Studies were excluded in the screening stage due to not including track and field athletes or sprint starts using starting blocks (n = 289). In the eligibility stage, there were several reasons for exclusion, namely studies with results focused exclusively on the acceleration phase (n = 8), case studies (n = 4), studies reporting data referring to samples of previously published papers (n = 3) or mainly focused on the limitations of disability (n = 3), lack of information about the PB100m (n = 2) and studies presenting only results for electromyography and reaction time data (n = 2). Figure 1 presents the complete flow diagram.

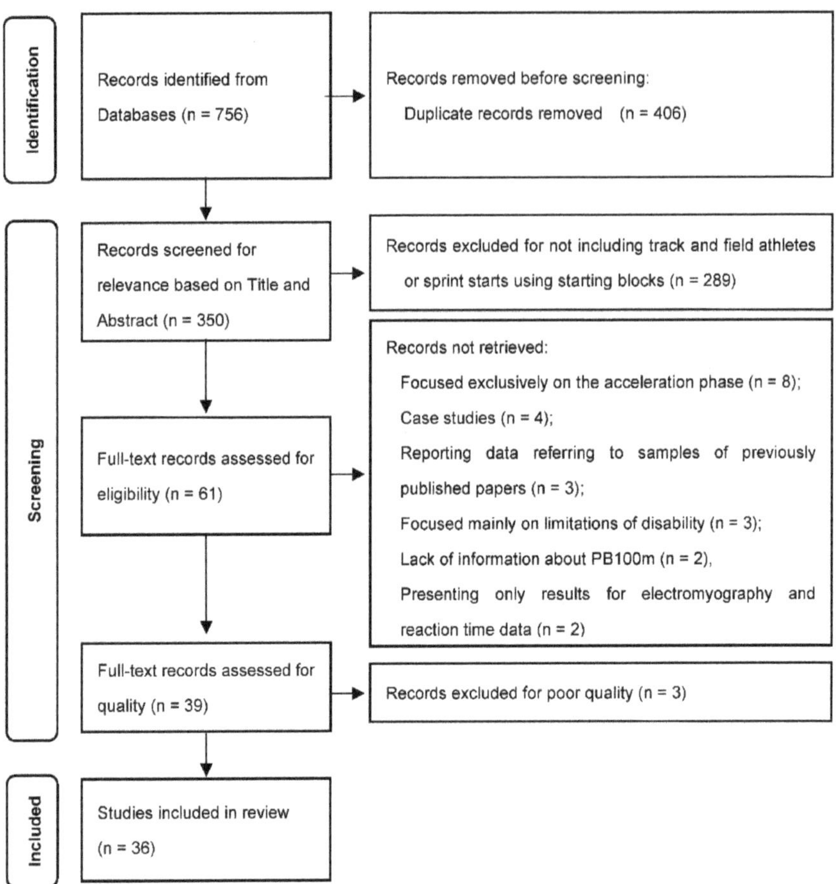

Figure 1. Flow diagram of the study selection process according to PRISMA guidelines.

3.2. Quality of Studies

In the evaluation of methodological quality, the inter-rater reliability analysis achieved a Kappa value of 0.91 (0.84–0.98), indicating almost perfect agreement between raters. The mean quality score of the included studies was 74.92%. None of the studies achieved the maximum score of 100% and 3 studies (excluded) scored below 50%. Sixteen studies were classified with good methodological quality (quality score between 50 and 75%), while 20 studies had excellent methodological quality (quality score > 75%). The main deficiencies in methodological quality were related to the estimation of sample size and study limitations discussion.

3.3. Basic Characteristics of Included Studies

Fifteen studies [2,3,10–12,17,20,21,23,25,33–37] focused specifically on the block phase, 18 studies [1,4–8,13–16,18,19,24,38–42] on the block phase and, at least one of the subsequent two flight and stance phases, and 3 studies [9,22,43] on the initial acceleration (the first and/or the second step). A summary of all the individual studies reviewed is presented in Table 1.

Table 1. Studies are listed in reverse-chronological order by year, followed by alphabetically for studies published in the same year. Samples (n) are restricted to total participating sprinters and are classified by performance level(s) according to the original authors.

Study Details				Sample				Main Findings	Quality
Reference	Primary Focus of the Study	Main Purpose	Biomechanics Analysis	Sex	n	Level	PB100m (s)		Mean Score (%)
Werkhausen, Willwacher [43]	First 2 steps. Two force platforms for the GRFs of the first 2 steps. Three-dimensional kinematic model (pelvis and lower limbs)	Investigate how plantar flexor muscle-tendon behavior is modulated during the first 2 steps	3D GRF of the first 2 steps	F	11	Germany national level	12.66 ± 0.49	Ankle and knee joint angles revealed no statistical differences at any time of both steps. Ankle joint power was negative after touchdown and positive during the rest of the stance phase, whereas net ankle joint work was positive during both steps. Knee joint power was positive during most of the stance phase.	67.78
Graham-Smith, Colyer [39]	Block phase and first 2 steps. An array of 6 force platforms.	Compare force production between elite senior and junior academy sprinters	3D block and first 2 steps GRF, and spatiotemporal data	M M	17 20	Elite Senior Junior Academy	8.2% worse than senior WR [(a)] 12.2% worse than junior WR [(b)]	Senior sprinters presented higher relative anteroposterior force and power during the initial block phase, higher forces during the transition from bilateral to unilateral pushing and lower (more horizontal) projection angle across the initial 2 steps of the sprint compared with junior athletes.	76.82
Nagahara, Gleadhill [35]	Block phase. Two force platforms with a coordinate transformation matrix to the coordinate block system.	Examine whether modulation of COP location on the starting block improves sprint start performance	3D GRF under each block and spatiotemporal data	M	20	National level	11.22 ± 0.41	The modulation of COP location did not show an effect on AHEP and 10 m time. However, instructing to push the calcaneus onto the block (posterior location) may improve the 10 m time and/or AHEP for some individuals and may be accomplished through a shorter reaction time.	82.50
Sado, Yoshioka [23]	Block phase. Separated starting blocks secured onto separate force plates.	Examine the 3D lumbo-pelvic-hip kinetics during block start	3D GRF under each block	M	12	University of Tokyo team	10.78 ± 0.19	The peak lumbosacral extension torque was larger than any other peak torque.	66.36

Table 1. *Cont.*

Study Details				Sample				Main Findings	Quality
Reference	Primary Focus of the Study	Main Purpose	Biomechanics Analysis	Sex	n	Level	PB100m (s)		Mean Score (%)
Bezodis, Walton [10]	Block phase. Four synchronized force platforms under each block and each hand.	Identify the continuous GRF features which contribute to blocking phase performance	3D GRF under each block and each hand	M	23	Trained	11.37 ± 0.37	The resultant magnitude of the GRFs on the rear block is the most important predictor of block phase performance, followed in importance by front block force magnitude features. Features related to the direction of application of these forces are not relevant predictors of performance	87.50
Cavedon, Sandri [12]	Block phase. Instrumented starting blocks and 2 high-speed video cameras.	Analyze the effect of 2 block setting conditions on block start performance	2D kinematic, horizontal, and vertical forces components, and spatiotemporal data	M (c) F	22 20	Regional and National	10.45–11.30 11.45–12.68	An anthropometry-driven block setting condition based on the sprinter's leg length was associated with several significant changes in postural parameters at the "set" position, as well as in kinetic and kinematic variables at the pushing and acceleration phases in comparison with the sprinter's usual block setting, leading to improved performance	88.64
Colyer, Graham-Smith [33]	Block phase. Four force platforms under each of the legs and arms separately.	Analyze the associations between block reaction forces and average horizontal external power	2D anteroposterior and vertical block reaction forces, and spatiotemporal data	M	5 32 20	Elite National (d) Academy	<10.15 — —	Both higher magnitudes of force and more horizontally orientated force vectors were associated with higher performance levels. The ability to sustain high forces during the transition from bilateral to unilateral pushing was a performance-differentiating factor. Faster sprinters produced less negative horizontal impulses under hands compared with their slower counterparts	78.60

Table 1. *Cont.*

Study Details				Sample				Main Findings	Quality
Reference	Primary Focus of the Study	Main Purpose	Biomechanics Analysis	Sex	n	Level	PB100m (s)		Mean Score (%)
Nagahara and Ohshima [20]	Block phase. Two force platforms under each block.	Examine the association of block clearance performance with COP location on the starting block surface	3D GRF under each block	M	21	Sprinters	11.24 ± 0.41	The COP location was related to sprint start performance (AHEP). Better sprint start performance was achieved with a higher and more to the rear COP on the starting block surface during the force production phase	75.45
Sandamas, Gutierrez-Farewik [24]	Block phase and 1st stance. Three-dimensional kinematic full-body model. Instrumented blocks and a force platform. Natural technique (Skating); 1st step inside a 0.3 m lane (Narrow).	Analyze the block reaction forces when 1st step width is manipulated	3D kinematic data and external block and 1st step reaction forces	M F (e)	8 2	Competitive, Including international championships finalists	11.03 ± 0.36 11.60 ± 0.45	The mediolateral impulses decreased with reduced step width; The propulsive component of the net anteroposterior impulse is significantly smaller for the narrow step width in the 1st stance; restricting step width, vertical block impulses increased while the mediolateral motion of the CM from Start to 1st stance toe-off decreased; reducing step width does not lead to any immediate improvement in performance. On the contrary, the skating style has a greater propulsive impulse during the 1st stance	79.77

Table 1. *Cont.*

Study Details				Sample				Main Findings	Quality
Reference	Primary Focus of the Study	Main Purpose	Biomechanics Analysis	Sex	n	Level	PB100m (s)		Mean Score (%)
Aeles, Jonkers [9]	First stance phase. 3D kinematic full-body model. Force platform to measure the GRFs of the 1st step.	Compare young and adult sprinters in kinematic and kinetic parameters during the 1st stance phase	3D kinematics and 3D GRF of 1st step	M F (f) M F	7 9 11 10	Adult Well-trained Young Well-trained	10.67 ± 0.14 12.12 ± 0.41 11.47 ± 0.34 12.75 ± 0.36	Well-trained young and adult sprinters have no differences in ankle joint stiffness, range of dorsiflexion or plantar flexor moment. Surprisingly, the young sprinters show a greater maximal and mean ratio of horizontal to total GRF. Adult sprinters have more MTU shortening and higher maximal MTU shortening velocities in all plantar flexors and in the rectus femoris.	80.68
Brazil, Exell [11]	Block phase. Force instrumented starting blocks. Three-dimensional kinematic lower limb model.	Explore the relationships between lower limb joint kinetics, external force production and starting block performance	3D block reaction forces and 3D kinematics	M	17	Sprinters	10.67 ± 0.32	86% of the variation in block performance is explained by the horizontal force applied to the front and rear blocks, and at the joint level 55% of the variation in block performance is explained by average rear ankle extensor moment, front hip extensor moment and front knee positive extensor power.	87.73
Brazil, Exell [4]	Block phase and 1st stance. Three-dimensional kinematic lower limb model. Force platform to the GRFs—1st step.	Examine lower limb joint kinetics during the block and 1st stance phases	3D kinematics and 3D block and 1st step reaction forces	M	10	Sprinters	10.50 ± 0.27	The asymmetrical nature of the block phase is most pertinent at the knee joint, and the leg extensor energy is predominantly generated at the hip joint in both the front and the rear block whereas during 1st stance, energy generation favors the ankle joint as a result of a significant reduction in relative hip work.	83.18

Table 1. *Cont.*

Study Details				Sample				Main Findings	Quality
Reference	Primary Focus of the Study	Main Purpose	Biomechanics Analysis	Sex	n	Level	PB100m (s)		Mean Score (%)
Ciacci, Merni [38]	Block phase and first 2 steps. 3D kinematic full-body model.	Compare kinematic differences between sexes	3D kinematic data	M F M F	6 6 4 4	Elite World-Class	10.74 ± 0.21 11.95 ± 0.24 10.03 ± 0.14 11.10 ± 0.17	The start kinematics is only partially affected by sex (men have shorter pushing phase, higher block horizontal velocity, and shorter contact times of first 2 steps), whereas a bigger role is played by the performance level (faster sprinters have CM closer to the ground and a more flexed front knee in the "set" position, longer pushing phase, lower block vertical velocity, and shorter contact times/longer flight times in first 2 steps.	85.23
Coh, Peharec [5]	Block phase and first 2 steps. Two independent force platforms for 2 independent starting block pads. 3D kinematic full-body model.	Compare the kinematic and kinetic factors between faster and slower high-level sprinters	3D GRF under each block and spatiotemporal data	M	6 6	Faster Slower	10.66 ± 0.18 11.00 ± 0.06	Faster sprinters show motor patterns of greater force development (rear block total force, rear block vertical maximal force, and the rate of force development) than their slower counterparts; The importance of other indicators as block clearance time, block velocity, and block acceleration was not confirmed in this study.	78.41
Debaere, Vanwanseele [15]	Block phase through until the start of 2nd touchdown. 3D kinematic full-body model and 2 force platforms for the first 2 steps.	Compare joint power generation between well-trained adult and young sprinters	3D Kinematics and 3D GRF of the first 2 steps	M F M F M F	8 6 8 10 5 6	Well-Trained Adult Under 18 Under 16	10.65 ± 0.07 11.87 ± 0.14 11.21 ± 0.11 12.42 ± 0.25 11.56 ± 0.08 12.86 ± 0.30	Adult sprinters generated more joint power at the knee during the 1st step compared to young sprinters, inducing longer step length and therefore higher velocity. Younger athletes employed a different technique: the hip contributes more to total power generation, whereas the contribution of the knee is far less.	82.95

Table 1. *Cont.*

Study Details				Sample				Main Findings	Quality
Reference	Primary Focus of the Study	Main Purpose	Biomechanics Analysis	Sex	n	Level	PB100m (s)		Mean Score (%)
Schrodter, Bruggemann [25]	Block phase. 2D ankle kinematic data and 3D block reaction force from instrumented blocks.	Describe the stretch-shortening behavior of ankle plantarflexion MTU during the push-off phase	2D kinematics and 3D block GRF	M (e) F	54 30	World-class National	10.98 ± 0.58 12.12 ± 0.68	This study provided the 1st systematic observation of ankle joint stretch-shortening behavior during sprint start for sprinters of a wide range of performance levels. It showed clear signs of a dorsi-flexion in the front and rear ankle joints preceding plantarflexion, seeming that the stretch-shortening cycle like the motion of the soleus muscle-tendon unit has an enhancing influence on push-off force generation.	68.86
Chen, Wu [37]	Block phase. Two-dimensional kinematic full-body model (15 segmented model).	Identifies optimal crouched position (bunched, medium, or elongated) from push-off through the first 2 steps	2D kinematic data—sagittal plane and spatiotemporal data	M	7	Skilled sprinters	10.94 ± 0.20	The medium starting position was the ideal starting position.	60.19
Bezodis, Salo [3]	Block phase. Two-dimensional kinematic full-body model and kinetic data calculated from consequent data procedures.	Identify the key characteristics of the lower-limb kinematic patterns during the block phase	2D kinematic data—(kinetic energy calculated from CM horizontal velocity)	M	16	World-class to university level	10.95 ± 0.51	Describes the lower limb joint kinematics patterns explicative of high levels of sprint start performance. The rear hip angle at block exit was highly related to block power, and there were moderate positive relationships between block power and rear hip ROM and peak angular velocity.	80.68

Table 1. *Cont.*

Study Details				Sample				Main Findings	Quality
Reference	Primary Focus of the Study	Main Purpose	Biomechanics Analysis	Sex	n	Level	PB100m (s)		Mean Score (%)
Debaere, Delecluse [14]	Block phase and first 2 steps. Three-dimensional kinematic full-body model and 2 force platforms for first 2 steps for inverse dynamics analysis.	Analyze the contribution of joint moments and muscle forces to the CM acceleration	3D kinematics and two 1st steps GRF	M (e) F	2 5	Well-Trained	11.10 to 11.77 12.05 to 12.36	Relates the specific joint and muscle contribution to the horizontal acceleration (propulsion) and vertical acceleration (body lift) of the CM during the initial two steps after block clearance. Torque-driven simulations identify the ankle joint as the major contributor to propulsion and body lift.	68.41
Otsuka, Kurihara [21]	Block phase. Separated starting blocks secured onto separate force platforms. Three-dimensional kinematic 7-segment model of the lower limb.	Clarify the effect of widened stance width at the "set" position during the block start phase	3D kinematic data and 2D GRF under each block	M	14	3 international 11 national	10.99 ± 0.40	A widened stance width at the "set" position affects the hip-joint kinematics in both legs and enhanced the hip power generation in the rear leg during the block start phase. However, when considering sprinting performance during the whole block start phase, there was no significant effect of the widened stance width on block-induced power and the subsequent sprint time.	68.86

Table 1. *Cont.*

Study Details				Sample				Main Findings	Quality
Reference	Primary Focus of the Study	Main Purpose	Biomechanics Analysis	Sex	n	Level	PB100m (s)		Mean Score (%)
Rabita, Dorel [22]	Initial sprint acceleration. Six individual force platforms connected in series.	Describe the sprint acceleration mechanics in elite and sub-elite sprinters	3D GRF of initial steps and spatiotemporal data	M	4 5	Elite Sub-Elite	9.95 to 10.29 10.40 to 10.60	Describes for the 1st time the mechanical characteristics of the acceleration phase in elite and sub-elite sprinters: (i) while step length increases regularly during the acceleration phase, step frequency is almost instantaneously leveled at the maximal possibility of elite athletes; (ii) F-V and P-V relationships during sprints were well described by linear and quadratic models, respectively; and (iii) the effectiveness of force application greatly accounts for the differences in performance among highly trained athletes.	74.31
Milanese, Bertucco [41]	Block phase and first 2 stance phases. 3D kinematic full-body model.	Investigate the rear knee angle associated with the impulse and the horizontal velocity in the starting block and acceleration phases	3D kinematics	M (e) F	6 5	University sprinters	12.0 ± 0.1 13.1 ± 0.9	Horizontal CM velocity increased significantly at the block clearance and along the first 2 strides when witching from 135° to 115° and then to 90° the rear knee angle. The horizontal velocity was directly determined by force impulse which tended to be greater at 90° rear knee angle.	79.55
Otsuka, Shim [42]	Block phase and first 2 steps. Ten individual force platforms connected in series.	Compare 3D force application in the blocks between 3 sprinting groups	3D GRF under each block and first 2 steps	M	9 9 11	Well-Trained Trained Non-Trained	10.87 ± 0.41 11.31 ± 0.42 ——	The greater anterior acceleration of well-trained sprinters during the starting block phase may be accompanied, not by a greater GRF magnitude, but by a more forward-leaning sagittal GRF vector.	72.50

Table 1. *Cont.*

Study Details				Sample				Main Findings	Quality
Reference	Primary Focus of the Study	Main Purpose	Biomechanics Analysis	Sex	n	Level	PB100m (s)		Mean Score (%)
Debaere, Delecluse [6]	Block phase and first 2 steps. Three-dimensional kinema-tic full-body model. Two force platforms for first 2 steps.	Characterize the sprint technique during the transition from start block into sprint running	3D kinematics and 3D GRF of the first 2 steps	M (e) F	11 10	Elite/Well-Trained	10.62 + 0.18 11.89 + 0.30	During the 1st step, maximal power was predominately generated by the hip (54%) followed by the knee (31%) and the ankle (15%). The importance of power generation at the knee decreased at second stance since it only accounted for 9% of total power generation and the importance of the ankle increased up to 38%.	64.77
Aerenhouts, Delecluse [1]	Block phase and initial acceleration (first 5 steps). Instrumented start blocks and a universal laser velocity sensor.	Compare starting performance between adults and juniors sprinters having reached their adult height	Horizontal block forces and spatiotemporal data	M F (g) M F	16 9 23 19	Elite Adult Elite Junior	10.81 ± 0.40 11.29 ± 0.29 11.85 ± 0.24 12.54 ± 0.26	The higher muscularity of senior athletes did not result in significantly higher forces against the starting blocks nor block velocity compared with the junior athletes. The more muscular senior athletes had a better running acceleration than the junior athletes. In female athletes, a higher body fat percentage negatively correlated with 1st step length.	79.32
Slawinski, Dumas [8]	Block phase and 1st step. Three-dimensional kinematic full-body model.	Compare the influence of bunched, medium, and elongated start on start performance	3D kinematics and spatiotemporal data	M (e) F	6 3	National sprinters	10.58 ± 0.27 11.61 ± 0.42	Head and trunk limb movements were important to create a high CM velocity during the starting block phase. The elongated start, compared to the bunched or medium start, induced an increase in block velocity and a decrease in the time at 5 and 10 m.	72.73

Table 1. *Cont.*

Study Details				Sample				Main Findings	Quality
Reference	Primary Focus of the Study	Main Purpose	Biomechanics Analysis	Sex	n	Level	PB100m (s)		Mean Score (%)
Bezodis, Salo [2]	Block Phase. High-speed camera and a laser distance measurement device.	Choose the measure that best describes sprint start performance	Spatiotemporal data and horizontal block forces derivations	M	12	University sprinters	11.30 ± 0.42	For the 1st time, normalized average horizontal external power was identified as the most appropriate measure of performance. One single measure reflects how much a sprinter is able to increase velocity and the time taken to achieve this, whilst accounting for variations in morphologies between sprinters.	79.55
Slawinski, Bonnefoy [7]	Block phase and first 2 steps. Three-dimensional kinematic full-body model.	Identify the most relevant kinematic and kinetic parameters differentiators of elite and well-trained sprinters	3D kinematics and spatiotemporal data	M	6 6	Elite Well-Trained	10.27 ± 0.14 11.31 ± 0.28	Anterior and vertical components of CM, rate of force development and force impulse were significantly greater in elite sprinters. The muscular strength and arm coordination appear to characterize the efficiency of the sprint start.	67.73
Slawinski, Bonnefoy [36]	Block phase. Three-dimensional kinematic full-body model.	Measure the joint angular velocity and the kinetic energy of the different segments in elite sprinters	3D kinematics and 3D Euler angular velocities	M	8	Elite	10.30 ± 0.14	Highlights the importance of a 3D analysis of a sprint start. Joints such as shoulders, thoracic, or hips did not reach their maximal angular velocity with a movement of flexion-extension, but with a combination of flexion–extension, abduction–adduction and internal–external rotation.	67.73

Table 1. *Cont.*

Study Details				Sample				Main Findings	Quality
Reference	Primary Focus of the Study	Main Purpose	Biomechanics Analysis	Sex	n	Level	PB100m (s)		Mean Score (%)
Maulder, Bradshaw [40]	Block phase and first 3 steps. Two-dimensional kinematic full-body model.	Examine the changes in block start and early sprint acceleration kinematics with resisted sled loading	2D kinematics—Sagittal plane	M	10	National and Regional	10.87 ± 0.36	A load of approximately 10% BM had no "negative" effect on sprint start technique or step kinematic variables. The kinematic changes produced by the 10% BM load may be more beneficial than those of the 20% BM load.	76.14
Gutierrez-Davilla, Dapena [17]	Block phase. Two synchronized force platforms under blocks (1) and hands (2)	Compare the CM velocities and positions between pre-tensed and conventional starts	Horizontal forces and spatiotemporal data	M	19	Experienced competitive sprinters	11.09 ± 0.30	The pre-tensed and conventional starts produced similar performance. The increased propulsive force exerted through the legs in the early part of the block acceleration phase in the pre-tensed starts was counteracted by an increased backward force exerted through the hands during the same period.	72.95
Mero, Kuitunen [19]	"Set" position (block phase and 1st step). Sixteen individual force platforms connected in series.	Examine the effects of muscle-tendon length on joint moment and power	2D kinematics and horizontal and vertical GRF under blocks, hands and 1st step	M	9	Sprinters	10.86 + 0.34	Lower block angles (40° vs. 65°) were associated with enhanced starting performance by increasing the final block velocity. The inverse association between block angles and muscle-tendon lengths of the gastrocnemius and soleus in both legs, which may generate higher peak joint moments and powers, especially in the ankle joint, may explain this result.	68.86

Table 1. *Cont.*

Study Details						Sample		Main Findings	Quality
Reference	Primary Focus of the Study	Main Purpose	Biomechanics Analysis	Sex	n	Level	PB100m (s)		Mean Score (%)
Fortier, Basset [16]	Block phase and first 2 steps. Three-dimensional full-body kinematic model. Instrumented blocks.	Examine if kinetic and kinematic parameters could differentiate elite from sub-elite sprinters	3D kinematics and horizontal block forces	M M	6 6	Elite Sub-Elite	<10.70 10.70 to 11.40	Four kinetic parameters differentiating elite from sub-elite sprinters: delay between the end of rear and front force offset, rear peak force, total block time, and time to rear peak force.	72.73
Čoh, Jost [13]	Block phase and first 2 steps. Two-dimensional kinematic full-body model. Instrumented blocks.	Determine the most important kinematic and kinetic parameters of the "set" position and push-off	Horizontal block forces, 2D kinematic and spatiotemporal data	M F	13 11	Slovene national team	10.73 ± 0.2 11.97 ± 2.6	Identification of three parameters that best define an efficient start for both male and female sprinters: horizontal start velocity, start reaction time and impulse of push-off force from the front starting block.	65.22
Guissard, Duchateau [34]	Block phase. Strain gauges mounted on each footplate and behind the starting block. Two-dimensional kinematic front leg model.	Analyze the mechanical parameters about EMG activity at different front block inclinations	EMG, 2D kinematics and horizontal GRF behind blocks	M F	14 3	Trained	10.4 to 11.9	Decreasing front block obliquity induced neural and mechanical modifications that contribute to increasing the block start velocity without any increase in the duration of the push-off phase.	76.36

Table 1. *Cont.*

Study Details				Sample				Main Findings	Quality
Mero [18]	Block phase and 1st stance. Starting blocks over a force platform. Two-dimensional kinematic full-body model (14 points).	Analyze the force-time characteristics during the 1st stance and the relationships between force and run velocity	2D kinematics and horizontal and vertical GRF under blocks and 1st step	M	8	Trained	10.79 ± 0.21	In the 1st contact after leaving the blocks there was a significant braking phase and the force produced in the propulsion phase was associated with running velocity; Muscle strength strongly affected running velocity in sprint start.	57.27

2D—two-dimensional analysis; 3D—three-dimensional analysis; AHEP—average horizontal external power; BM—body mass; CM—center of mass; COP—center of pressure; EMG—electromyography; F—female sample; F-V—force-velocity; GRF—ground reaction forces; MTU—muscle-tendon unit; M—male sample; P-V—power-velocity; ROM—range of motion; WR—world record; (a) 100 m world record at the study time was 9.58 s; (b) 100 m U20 world record at the study time was 9.97 s; (c) all sample was divided into 3 groups according to the Cormic Index (12 brachycormic, 19 metricormic, and 11 macrocormic); (d) sample divided into two groups: 5 elite sprinters and remaining 52 sprinters; (e) all subjects included in a single experimental group; (f) sample divided into 2 experimental groups: adult/senior vs. junior sprinters; (g) sample divided into 4 experimental groups.

Study purposes included evaluation of specific block start and initial acceleration variables and their influence on block performance (14 studies) [2–4,6,10,11,14,18,23,24,33,36,40,43]; analysis of different "set" position or block configurations (11 studies): location [20] and modulation [35] of center of pressure (COP) on the starting block surface, different block spacing [8,12,37] and widened conditions [21], different block plate obliquities [19,25,34], changed "set" position knee angles [41] and block pre-tension [17]; and comparisons between sprinters of different performance levels, despite the subjectivity associated with the descriptor of the performance level of the athletes (11 studies) [1,5,7,9,13,15,16,22,38,39,42]. The ambiguity in the performance level descriptors includes categories such as: elite vs. sub-elite or well-trained [7,16,22], world-class vs. elite [38], faster vs. slower [5], adult well-trained vs. trained [9,15,42]; elite or well-trained senior vs. junior academy, elite junior, U18 or young well-trained [1,39]; and top sprinters [13]. All studies comparing groups of athletes included male sprinters, but only 4 [1,9,15,38] included women of different performance levels. The studies included in the systematic review presented a cross-sectional study design, except for one study that presented a follow-up design [16].

Twenty-one studies evaluated kinetic variables from blocks start placed on force platforms (12 studies) [5,10,17–21,23,33,35,39,42] or instrumented starting blocks sensors (9 studies) [1,4,11–13,16,24,25,34]. Twelve studies [4,6,9,14,15,18,19,22,24,39,42,43] used a large variety of force platforms arrangements to analyze the dynamic characteristics of the first steps of the initial acceleration.

Concerning kinematic variables, a bi-dimensional analysis, including one or two high-speed digital cameras, was applied in 9 studies [3,12,13,18,19,25,34,37,40], and a 3D kinematic analysis, including 3 [38], 6 [16], or 8 or more cameras [5–9,21,24,36,41] was applied in 11 studies.

Total participants are 766 track and field sprinters, including 179 women and 587 men, and 11 non-trained male subjects [42]. Regarding the sample size of the individual studies selected, Chen, Wu [37] and Debaere, Delecluse [14] are those with the smallest number, 7 participants, and Schrodter, Bruggemann [25] conducted the study with 84 subjects (the largest sample size). The sample sizes from the other studies ranged from 8 [18,36] to 67 [1] subjects, with a mean sample size of 20 participants per study. The mean age of the participants in the selected studies ranged from 15.3 years (under 16) to 28 years. For women, PB100m ranged from 11.10 s (world-class) to 13.10 s (university level), with more classification terms being used, such as "elite" (11.29 to 11.95 s), "well-trained" (11.87 to 12.20 s), "trained" (<11.90 s), or "national level" (11.45 to 12.66) sprinters. Men were classified as "world-class" (10.03 to 10.98 s), "elite" (9.95 to 10.81 s), "sub-elite" (10.40 to 10.95 s), "well-trained" (10.65 to 11.77 s), "trained" (10.40 to 11.37 s), "national level" (10.58 to 11.22 s), "university level" (10.78 to 12.00 s), or just "sprinters" (10.50 to 11.24 s). Among studies, male PB100m ranged from 9.95 s to 12.00 s.

Through the analysis of the research setup protocols, it was possible to identify a "standard experimental setup". Sixty-nine percent of the studies used distances between 10 and 30 m, with distances shorter than 10 m used only in 4 studies [5,24,41,43] and distances greater than 30 m used in 7 studies [10,20,22,33,37–39]. The number of trials performed ranged between 3 and 10 in 86% of the studies, but in 3 studies [10,20,38] the participants performed 1 or 2 trials, and in 2 studies [40,41] more than 10 trials. Fifty-eight percent of the studies were carried out on an indoor track, 4 studies [12,37,38,40] on an outdoor track, 2 studies [24,41] in a laboratory context, and 9 studies [1,8,10,16,20,23,25,42,44] did not mention the measurement location.

3.4. Data Organization and Analysis

There was a very large diversity of kinematic and kinetic variables reported among selected studies. Since it is impossible to discuss them all, we will highlight those reported as explicative of high levels of the sprint start performance and that best differentiate faster from slower sprinters. Based on the main findings highlighted in Table 1, the explanatory variables of superior performance levels were identified and systematized in a sequence of tables in Appendixes A–C, related to the "Set" position (Appendix A Table A1), block phase (Appendix B Tables A2 and A3), and first two steps of the initial acceleration (Appendix C Tables A4 and A5). With this strategy of results presentation, it is expected that readers will have access to the primary data extracted from all the studies included in the systematic review. Therefore, Appendix A Table A1 summarizes the kinematic variables in the "Set" position, showing that anthropometry-driven block setting and muscle-tendon unit (MTU) length have an important role in the block start performance. Furthermore, faster sprinters tend to move their center of mass (CM) closer to the starting line and closer to the ground. Concerning joint angles, the knee angular position seems to be a greater performance predictor than any other lower limb joint. At the push-off phase (Appendix B Tables A2 and A3, for kinematic and kinetic variables, respectively) a rear hip extension range of motion (ROM) and a rapid extension of both hips appear to be positively associated with block performance. Moreover, greater average force production during the push against the blocks, especially from the rear leg and particularly the hip, appears to be important for performance. A posterior COP location on block surfaces can also improve sprint performance. Immediately after exiting the blocks, shorter first flight durations and longer first stance durations (allowing more time to generate propulsive force) are the kinematic features of best sprinters (Table A4). During the first two steps of initial acceleration, higher levels of performance seem to be associated with shorter flight times, longer contact times, and the ability to extend the knee throughout both stance phases (Table A5).

4. Discussion

This paper systematically reviews the kinematic and kinetic biomechanical variables of the block start and initial sprint acceleration phase that influence performance and best differentiate sprinters of different levels. Despite the large number of variables reported in the reviewed studies it was possible to identify some that effectively best describe the influential factors of these events as they are associated with better performance outcomes or best differentiate sprinters of different performance levels. However, notice should be made to the difficulty in analyzing data between studies as there are still no standards for reporting the data, such as measurement units (e.g., m vs. cm) [12,17,18,35], joint angular measurement norms and conventions [3,4,6,12,13,36,38] and/or data normalization methodologies (e.g., for full-height/lower limb length, body mass/body weight) [2,4,17,22,24,25]. Additionally, there is some subjectivity associated with inconsistent descriptors of performance level [26], confirmed by the variability of the sprinter's classifications used (e.g., from just sprinters to well-trained sprinters, elite sprinters, world-class sprinters, or high-level sprinters) [5,7,16,22,36,38,42]. Another critical factor that somehow may influence data variability between studies is the period of the season in which the data collection took place (e.g., prior to the competition phase of the indoor season vs. during the competitive indoor season or beginning of the summer season) [18].

To better understand the determinant factors of sprint start, the findings from the reviewed studies have been organized into three focuses: (i) the "set" position, (ii) the push-off phase, and (iii) the first two steps of initial acceleration, according to the data presented in Appendixs A–C.

4.1. The "Set" Position

The "Set" position is the first performance key factor in the block start performance because it depends on block settings and the body posture assumed by sprinters. For the question: "Is there one optimal "Set" position which should be adopted by sprinters?" the answer seems to be no. The researched studies [3,38] showed that it is not an important differentiating factor of performance, since it does not present any correlation with PB100m or normalized block power [3]. However, there are some interesting aspects that sprinters should look out for in a more effective "Set" position [5,12]. The ideal "Set" position depends on the individual anthropometric features [12], strength [38], and morphologic characteristics and motor abilities [13].

4.1.1. Block Settings

The "Set" position depends largely on the anteroposterior block distance, which defines the type of start used. There are three types of block starts based on inter-block spacing: bunched—less than 0.30 m; medium—0.30 to 0.50 m; and elongated—greater than 0.50 m [27,37].

Studies that reported block spacing based on the individual sprinter's preferences [5,12,13,18,35] reported distances between 23.5 ± 1.9 cm (for female sprinters; PB100: 11.97 ± 2.6 s) [13] and 32 ± 5 cm (for male sprinters; PB100m: 10.79 ± 0.21) [18]. This suggests that most sprinters adopt distances within or very close to the bunched start type, favoring CM positioning closer to the starting line [7,38]. Slawinski, Dumas [8] have demonstrated that elongated start settings increase the block velocity (i.e., horizontal CM velocity at the block clearing [7]), but linked to an increase in the pushing time on the blocks which implies a significantly worse performance at 5 and 10 m compared to the bunched start. The same authors showed that the medium start offers the best compromise between the pushing time and the force exerted on the blocks, allowing better times at 10 m [8]. Additionally, more recently, Cavedon, Sandri [12] have demonstrated that the anthropometry-driven block setting based on the sprinter's leg length has an important role in the block start performance leading to a postural adaptation that promotes several kinematic and kinetic advantages [12]. Adjusting inter-block spacing to the relative lengths of the sprinter's trunk and lower limbs (increasing 25.02% the usually bunched start inter-block spacing), allows greater force and impulse on the rear leg and greater total normalized average horizontal external power (NAHEP) [12], the latter one identified as the best descriptor of starting block performance [2].

Other blocks setting features that should be considered in the "set" position are the feet plate obliquity and the amount of pre-tension exerted on the blocks prior to the gunshot. The block inclination (relative to the track) affects the plantar flexor muscle-tendon units' (MTU) initial lengths and determines the muscle mechanics and the external force parameters during the block phase [19,25,34]. Faster sprinters presumably produce the peak torque at longer MTU lengths and adopting a more crouched position would allow them to produce a higher force on the block phase [38]. Research data shows that reductions in both footplates' inclinations (from 65 to 40°), meaning more muscle-tendon pre-stretch, lead to acute increases in block velocity and higher peak joint moments and powers, especially in the ankle [19]. Reductions in front block inclination alone (from 70 to 30°) also acutely increase block velocity without affecting push-off phase duration [34]. In another study [25], however, a greater mean rear block horizontal force was achieved by switching the rear foot to a steeper position (to 65°). This potential conflict between evidence might have arisen from differences in the location of the COP and the length of the footplates' surface between studies since a better sprint start performance is accomplished with a higher and more to the rear COP on the starting block surface [20,35]. Conversely, a pre-tensioned start does not seem to yield a performance advantage over a conventional start, because the increase in the propulsive force of the lower limbs is reversed by an increase in the back force exerted through the hands during the same period [17].

4.1.2. Sprinter Body Posture

Apart from block configuration, the choice of the sprinter's body posture also determines the effectiveness of the "Set" position on the subsequent block push-off phase. The horizontal distance between starting line and the vertical projection of the CM to the ground in the "Set" position (XCM) [7] is a factor that differentiates sprinters with different performance levels. As said before, faster sprinters tend to move their CM closer to the starting line [7,38] and closer to the ground [38]. Elite (PB100: 10.27 ± 0.14 s) and well-trained (PB100: 11.31 ± 0.28 s) male sprinters showed XCM of 22.9 and 27.8 cm, respectively [7]. Likewise, world-class (PB100: 11.10 ± 0.17 s) and elite (PB100: 11.95 ± 0.24 s) female sprinters presented XCM of 16.2 and 24.8 cm, respectively [38]. This more crouched position is only possible due to the high explosive strength of best sprinters, which allows them to produce higher levels of strength in the blocks [38] and reduce the horizontal travel distance of the CM. This body position is complemented by a more advanced shoulder position, putting more tension on the arms, allowing greater blocking speed during the subsequent phase [7].

Related to sprinter joint angles configuration in the "set" position, Milanese and Bertucco [41] have shown that horizontal CM velocity at the block take-off and along the first two steps increases significantly when the rear knee angle is set to 90° instead of 135° or 115°. A 90° rear knee angle allows for a better push-off of the rear leg than larger angles, showing such condition may be a strategy that allows some elite sprinters to maximize their strength capacity [41]. A more flexed front knee may facilitate the optimal joint moment production, but only in sprinters with exceptionally high levels of explosive strength [38].

4.2. *The Push-Off Phase*

The "block-phase" or "push-off phase" in the starting blocks initiates immediately after the gunshot and is considered a complex motor task that helps to determine sprint start performance [1]. Reaction time is the first factor in the time sequence of the block phase and it is the period from the gun signal to the first measurable change of pressure detected in the instrumented blocks [16]. While a sprinter's ability to react is undeniably important, it is related to the information-processing mechanisms that do not seem to correlate with the performance level [7,45] and, therefore, is beyond the scope of our review (for a review of factors that affect response times, see Milloz, Hayes [46]). Having reacted, the aim of the block phase is to maximize horizontal velocity in as little time as possible. The motion variables during the block phase are, therefore, the focus of this section.

4.2.1. Push-Off Kinematics Analysis

The efficiency of the starting action depends mainly on the compromise between horizontal start velocity (or block velocity) and the block time (referring to the time elapsing from the first movement at the "set" position to the exiting from the block [7]), resulting in the horizontal start acceleration [13]. Despite the horizontal block velocity could be considered the main parameter for an efficient sprint start [13], it cannot be used solely [2] because an increased block velocity could be due to either an increase in the net propulsion force generated or to an increased push-off duration [2,18]. Thus, best sprinters tend to present higher block velocity and greater block acceleration than slower sprinters [1,5,7,13,16,22,39,42], because they are able to produce a greater impulse in a shorter time [2,5,36] and optimize their force production on the blocks [16,19]. In fact, if sprinters increase their anteroposterior force impulse (FI = force × time) from a longer block time, they decrease their block acceleration [2,42] and the performance at 5 and 10 m [8]. Studies comparing data between sprinters of different performance levels mostly show higher block velocities (3.38 ± 0.10 vs. 3.19 ± 0.19 $m{\cdot}s^{-1}$; 3.48 ± 0.05 vs. 3.24 ± 0.18 $m{\cdot}s^{-1}$; 3.61 ± 0.08 vs. 3.17 ± 0.19 $m{\cdot}s^{-1}$; and 3.36 ± 0.15 vs. 3.16 ± 0.18 $m{\cdot}s^{-1}$) [5,7,22,33] and greater block accelerations (9.5 vs. 8.8 $m{\cdot}s^{-2}$; 8.2 vs. 7.9 $m{\cdot}s^{-2}$; 9.72 vs. 8.4 $m{\cdot}s^{-2}$; and 7.47 vs. 7.35 $m{\cdot}s^{-2}$) [1,5,7,42] for faster sprinters. Furthermore, higher performance levels also appear to be slightly related to lower block vertical velocities [38] and more horizontal CM projection angles (i.e., resultant direction from the CM horizontal and vertical block exit velocities) [33,39].

Lower limbs joints pattern during the pushing phase (i.e., from movement onset until block exit) is mostly associated with extension movements, especially on the hips and knees [3,4,6,25,36]. The front leg joints typically extend through a considerable ROM in a proximal-to-distal extension pattern [3], reaching their maximum at the beginning of the flight phase (e.g., hip: 183.2 ± 6.8°, knee: 177.4 ± 5.2°, and ankle: 133.1 ± 6.7°) [6]. Contrarily, the rear leg does not exhibit the same proximal-to-distal extension strategy, with the knee reaching its peak angular velocity before the hip and the ankle [3,36]. This happens perhaps due to considerably less ROM of the rear knee compared to the front knee [3], as it starts from a more extended angle in the "set" position (e.g., rear knee: 120.7 ± 9.7°; front knee: 91.0 ± 9.8°). The movement of the ankles is more complex because it involves first a dorsiflexion and after an extension resulting in a stretch-shortening cycle of the triceps surae muscle [3,6,25,36]. The duration of the ankle's flexion is greater for the rear ankle (50% of the block phase) than for the front ankle (20% of the block phase) [36]. Experimental manipulations on footplates' inclinations [19,34] have shown an inverse association between block angles and muscle-tendon lengths of the gastrocnemius and soleus, highlighting that block angles steeper than 65° could have disadvantageous effects on plantar flexor function [19]. Peak angular velocities at both hips are reached by a combination of flexion–extension, abduction–adduction, and internal–external rotation [23,36], reinforcing the importance of a 3D analysis of the sprint start [36]. Whilst there is a consistent trend among sprinters in the joint angular velocity sequence during the block phase, the lack of comparative data between sprinters of different performance levels does not allow to highlight the technical aspects critical to success. However, a rapid hip extension should be one of the first aspects to consider on a sprinter's technique during the start, as peak angular velocities at both hips and rear hip range of extension are positively associated with block power ($r = 0.49$) [3].

Although upper body kinematics in the push-off phase has been the focus of a small number of studies, some important findings are noteworthy. The action of the upper limbs is more variable between sprinters than that observed for the lower limbs [36]. Despite this, it is possible to recognize a 3D movement pattern for shoulders and trunk with a combination of flexion–extension, abduction–adduction, and internal–external rotation movements, while the elbows exhibit an extension and pronation movement [36]. The velocity of the rear shoulder tends to be slightly greater than that of the other joints, but the peak resultant angular velocities at the upper limb joints are comparable to those at lower limbs during the push-off phase, particularly that of both knees and front ankle [36]. However, there is no evidence linking different upper limb kinematic patterns with any block phase performance predictor, and further research is needed to compile relevant recommendations for athletes and coaches.

4.2.2. Push-Off Kinetic Analysis

According to Newton's second law of motion, horizontal CM acceleration requires net propulsive forces to be applied to the athlete's body in the sprinting direction. Therefore, as said before, the horizontal force impulse, made up by the mean horizontal force and push-off time, is the determining factor of the horizontal velocity at block exit [2,5,36,42]. The relationship between these factors (i.e., horizontal force and push-off time) shows that the application of a greater amount of horizontal force is a key performance factor [42], as an increase in the time action (block time) conflicts with the criterion for 100 m performance: 'shortest time possible'. Thus, best sprinters generate greater average forces [10,22], higher rates of force development [7,25], and larger net [7] and horizontal [5] block impulses than their slower counterparts. Likewise, Graham-Smith, Colyer [39] comparing senior to junior athletes also showed that sprinters with faster PB100m (senior athletes) exhibit higher relative horizontal force during the initial block phase and higher forces during the transition from bilateral to unilateral pushing [39]. The evident importance of the force generated against the blocks for proficient execution of the starting block phase has encouraged researchers to gain a deeper understanding of the kinetic determinants of such a crucial phase of sprinting. Bezodis, Salo [2] tried to find the push-off performance measure that was more adequate, objective, and possible to quantify in the field. From their analysis, the NAHEP was identified as the most appropriate measure of performance because it objectively reflects, in a single measure, how much sprinters are able to increase their velocities and the associated length of time taken to achieve this, whilst accounting for variations in morphologies between sprinters [2]. Later, the identification of the magnitude of the force applied to both blocks and their optimal orientation as major determinants of performance encouraged researchers to gain a deeper understanding of the push-off forces applied against each block separately. Consequently, some studies support the importance of the force generated by the front leg for forwards propulsion [6,42] and show that faster sprinters are able to produce higher force impulses in the front block than slower sprinters [5,33] (for example: 221.3 ± 15.8 N·s vs. 178.3 ± 13.1 N·s for faster and slower sprinters, respectively [5]). Colyer, Graham-Smith [33] reinforce this feature highlighting that higher front block force production during the transition (when the rear foot leaves the block, 54% of the block push) and a more horizontally orientated front block force vector in the block phase (81–92%) are important performance-differentiating factors. However, other evidence ensures that the rear block force magnitudes are the most predictive external kinetic features of block power [10,33] and sprint performance [5,7,12,16]. For example, Coh, Peharec [5] found that a faster group of sprinters (PB100m = 10.66 ± 0.18 s; 913 ± 89.23 N) produced greater total forces against the rear block than a group of slower sprinters (PB100m = 11.00 ± 0.06 s; 771 ± 55.09 N). A longer relative rear leg push (i.e., as a percentage of the total push-off phase) is also positively associated ($r = 0.53$ [3]) with greater block power [3,10] and is present in sprinters with faster PB100m [5,7,33]. Modulations of the COP on the starting block surface showed that COP location may also be related to initial sprint performance [20,35]. Better sprint start performance appears to be achieved with a higher and more to the rear COP during the force production phase [20]. Thus, athletes and coaches should keep in mind that pushing the calcaneus onto the block (posterior location) may improve the 10 m time and/or horizontal external power for some individuals [35].

Forces under the hands have been reported in relatively few studies [10,33,42], showing somewhat contradictory results. While some point to a primary support role [42], others point out that the best athletes produced less negative horizontal impulse under hands compared with their slower counterparts [33]. Therefore, the importance of the hands' kinetics during the push-off phase remains unclear and should be the subject of future research.

In addition to external kinetic analyses, which provide valuable insight into starting block performance, the analysis of internal kinetics (i.e., joint kinetics) helps to increase the understanding of the segment motions that are responsible for CM acceleration. Recent research of joint kinetics has shown that 55% of the variance in NAHEP of a group of sprinters with a PB100m of 10.67 s was mainly accounted for by rear ankle joint moment (23%), front hip joint moment (15%), and front knee joint power (15%). The remaining 2% was shared by the remaining lower limbs joint kinetic variables [11]. In the rear block, the magnitude of the horizontal force produced is determined by the rear hip extensor moment and the rear hip extensor power coupled with large ankle joint plantarflexion moment [4,11,19], without any significant knee joint contribution [4,11]. At the front block, a proximal–distal pattern of peak joint power is evident [4], highlighting a strategy often adopted in power demanding tasks, with the main periods of positive extensor power at the front ankle and knee occurring after the rear foot has left the block [4]. In a study with 12 sprinters from the University of Tokyo team (PB100m: 10.78 ± 0.19 s), Sado, Yoshioka [23] showed that the peak lumbosacral extension moment was significantly larger than any other lumbosacral and lower-limb moment, being positively correlated with the starting performance. This peak value appeared in the double-stance phase where both hip joints exerted extension moments. The aforementioned evidence supports the findings of Slawinski, Bonnefoy [36] who showed that the lower limbs and the head–trunk segments are the two main segments that contribute to the kinetic energy of the total body. Upper limbs contribute 22% to the total body kinetic energy, demonstrating that their actions in the pushing phase on the blocks are not negligible [36].

4.3. The First Two Steps

The primary goal of the first steps is to generate a high horizontal velocity [40]. However, the transition between block start and the first steps represents a specific biomechanical paradigm: integrate temporal and spatial acyclic movements into a cyclic action [5]. The efficiency of this transition depends on the biomechanical demands of the first stances after block clearance, which are very different from the other stances during acceleration [14]. The sprinter aims to generate maximal forward acceleration during the transition from start block into sprint running [2,14,22,42] while generating sufficient upward acceleration to erect itself from a flexed position in the start blocks to a more extended position [6,14]. Specific technical (kinematic) and dynamic (kinetic) skills are therefore needed to successfully achieve this transition, and they are the focus of this section.

4.3.1. First Two Steps Kinematic Analysis

The primary goal of the initial steps of a sprint running is to generate a high horizontal sprint velocity, which results from the product of the length and frequency of the sprinter's steps [22,40]. Spatiotemporal parameters have shown that the sprinter's step length increases regularly during the acceleration phase, while step frequency is almost instantaneously leveled to the maximum possible [22]. Typically, the step frequency reaches the maximal values very quickly (80% at the first step and about 90% after the third step) [22], achieving around 4 Hz immediately after block exit [26,40]. The length of the first steps is more variable between sprinters, ranging from 0.82 to 1.068 m (senior females) [1,38] or 0.85 to 1.371 m (senior males) [1,7] on the first step, and from 1.06 to 1.30 m (senior females) [1,13] or 1.053 to 2.10 m (senior males) [7,37] on the second step. Despite this variability, step length tends to be longer in faster sprinters, particularly in the first step (e.g., 1.371 ± 0.090 vs. 1.208 ± 0.087 m [7]; 1.30 ± 0.51 vs. 1.06 ± 0.60 m [5]; 1.135 ± 0.025 vs. 0.968 ± 0.162 m [38]), exhibiting an increase of about 14 cm for every 1 s less in PB100m [38]. This may be a consequence of the lower vertical velocity of the CM at the block clearing shown by faster sprinters, allowing them to travel a longer distance despite shorter flight times [38]. Indeed, the kinematics of faster sprinters is also characterized by a tendency to assume long ground contact times in the first two steps (e.g., mean first contact duration for Diamond League sprinters is 0.210 s for males and 0.225 s for females, which is greater than those of lower-level Italian junior sprinters: 0.176 and 0.166 s, respectively), associated to short flight times (0.045 and 0.064 s, for the first flight of world-class and elite male sprinters, respectively) [38]. This strategy allows the high-level sprinters to optimize the time during which propulsive force can be generated, minimizing the time spent in flight where force cannot be generated. Combined with this, best sprinters have their CM projected further forward [7] at the first touchdown, putting the foot behind the vertical projection of the CM [3], and minimizing the braking phase. At the takeoff of the first and second steps, the CM horizontal position is also greater in elite than well-trained sprinters [7]. This means that the CM resultant and horizontal velocity in the first two steps are generally greater in high-level sprinters [7,15]. Slawinski, Bonnefoy [7], for example, reported that elite sprinters have a CM resultant velocity 5.8% higher than well-trained sprinters, at the end of the first step (4.69 ± 0.15 vs. 4.42 ± 0.11 $m \cdot s^{-1}$ for elite and well-trained sprinters, respectively). Furthermore, high-level sprinters also show slightly lower vertical velocities [7,39] and more horizontal CM projection angles at the end of the first two support phases [39].

Lower limb joints pattern during the first two steps is associated with a proximal-to-distal sequence of the hip, knee, and ankle of the stance leg [4,9,43]. During both first and second steps, the ankle joint undergoes dorsiflexion during the first half of stance (e.g., 17 ± 3° and 18 ± 3° for the first and second steps, respectively [43]) and subsequently a plantarflexion movement (e.g., 45 ± 6° and 44 ± 5° for the first and second steps, respectively [43]).

The hip performs extension for the entire stances, the knee extends until the final 5% of stances, and the ankle is dorsi-flexed during the first half of stances before the plantar flexing action [6]. After leaving the rear block, there is a small increase in ankle joint dorsiflexion during the swing phase, preceding the plantarflexion that occurs just before touchdown [6]. Although the ankle plantar-flexes slightly at the end of the flight, the ankle is in a dorsi-flexed position at initial contact (e.g., first stance: 70.6 ± 5.8° and second stance: 72.4 ± 7.1° [6]). During both first and second steps, the ankle joint dorsi-flexes during the first half of stance (e.g., 17 ± 3° and 18 ± 3° for the first and second steps, respectively [43]) and subsequently performs a plantarflexion movement (e.g., 45 ± 6° and 44 ± 5° for the first and second stance, respectively [43]). Note that a reduction in the range of dorsiflexion during early stance, requiring high plantar flexor moments, has already been associated with increases in first stance power [47]. Maximal plantarflexion occurs immediately following takeoff reaching, for example, 111.3° at the first stance and 107.1° at the second stance [6]. The extension of both knees occurs just after the block exit and reaches its maximum at the beginning of the flight phase, with larger extension in the front compared with the rear leg (e.g., rear: 134.9 ± 11.2°; front: 177.4 ± 5.2°) [6]. From a flexed position at initial contact, the knee extensors generate power to induce extension throughout stance and to attain maximal extension at takeoff, achieving peak extension angles of around 160–170° (not full extension; e.g., first stance: 165.2 ± 20.6°; second stance: 163.6 ± 17.7° [6]). This extension action of the knee during stances on its own may play a role in the rise of the CM during early acceleration [26]. The hip joints extend during block clearance to reach maximal extension during the beginning of the flight phase. During stance, the hips are in a flexed position at initial contact and continue to extend throughout stance, achieving maximal extension immediately following takeoff (e.g., first stance: 180.6 ± 20.9°; second stance: 181.1 ± 20.0° [6]). There is also a considerable ROM in hip and pelvis rotation during stance as well as abduction. Although there are detailed descriptions of the lower limb angular kinematics during the first two stances and flight phases [3,6], there seems to be no clear evidence about the joint kinematic features that differentiate faster from slower sprinters. Furthermore, there is also a lack of experimental data on arm actions during early acceleration and its relationship to performance descriptors, making necessary future research in this area to help identify the most important performance features.

4.3.2. First Two Steps Kinetic Analysis

As said before, fast acceleration is a crucial determinant of performance in sprint running, where a high horizontal force impulse in a short time [13] is essential to reach high horizontal velocity [43]. Thus, as the highest CM acceleration during a sprint occurs during the first stances [7,9,14] (e.g., first stance: 0.36 ± 0.05 $m{\cdot}s^{-2}$; second stance: 0.23 ± 0.04 $m{\cdot}s^{-2}$ [14]), the ability to generate during this phase greater absolute impulse [7,18], maximal external power [39,42], and a forward-leaning force oriented in the sagittal plane [21,22,24,42] is linked to an overall higher sprint performance. Larger propulsive horizontal forces are particularly important during early acceleration, being a discriminating factor for superior levels of performance [48]. Experienced male sprinters (PB100m: 10.79 ± 0.21 s) can produce propulsive horizontal forces of around 1.1 bodyweight during the first stance [18]. However, a negative horizontal force has also been reported during the first contact after the block exit, even if the foot is properly placed behind the vertical projection of the CM [18]. During the first stance, for example, the braking phase represents about 13% of the total stance phase and the magnitude of the braking forces can reach up to 40% of the respective propulsive forces [18].

Furthermore, 3D analysis studies also highlight a lower body motion outside the sagittal plane during the first few ground contact phases [6,21,22,24,36,42]. In fact, during the first steps of a sprinter, a stance medial deviation is often observed that results from an impulse in the transverse plane. Although the medial impulse is the smallest of the three orthogonal stance impulses [21,22,42], the fact that it is non-zero can have an effect on the motion of the CM and on step width. However, it has been shown that well-trained sprinters present similar step widths in the early acceleration to those of the trained and non-trained sprinters [42]. Moreover, manipulations of both "set" position [21] and first step [24] widths have shown no effect on block-induced power nor braking force or net anteroposterior impulse, showing that smaller step width is not a discriminator factor of superior performance levels. Therefore, the perception that the adoption of a widened stance during initial acceleration (referred to as "skating style") is detrimental to performance is not

phase should be a priority in sprinter focus and coach feedback; (vii) the large role played by the hips on the push-off phase and by both the knee and ankle at the early stance must be acknowledged within physical and technical training to ensure strength and power are developed effectively for the nature of the sprint start.

Author Contributions: Conceptualization, M.J.V., J.M.A. and F.C.; methodology, M.J.V., J.M.A., F.C. and C.P.M.; formal analysis, M.J.V., J.M.A. and F.C.; writing—original draft preparation, M.J.V., J.M.A. and M.-J.V.; writing—review and editing, M.-J.V., F.C. and C.P.M.; visualization, M.J.V., M.-J.V., F.C. and C.P.M.; supervision, M.J.V., F.C. and C.P.M. All authors have read and agreed to the published version of the manuscript.

Funding: This work was partly supported by a national grant through the FCT—Fundação para a Ciência e Tecnologia within the unit I&D 447 (UIDB/00447/2020).

Institutional Review Board Statement: Not applicable.

Informed Consent Statement: Not applicable.

Data Availability Statement: Not applicable.

Conflicts of Interest: The authors declare no conflict of interest.

Appendix A

Table A1. Summary of the kinematic variables in the "Set" position. Data are the magnitude of the mean ± SD presented in the reviewed studies. Groups are male, female, and mixed (when authors joined data without discriminating by sex) sprinters. Studies are listed, in each variable, in reverse chronological order. Data, terms, conditions, and sprinters' performance levels are presented according to the original authors. Statistical differences between groups are marked with asterisks (* $p < 0.05$; *** $p < 0.001$).

"Set" Position Kinematics	Study	Male		Female	Mixed	
Inter-block spacing (cm) (m)	Cavedon, Sandri [12]				Usual condition 27.6 ± 2.4 cm	Anthropometric condition 36.8 ± 2.3 cm
	Čoh, Jost [13]	Slovene national sprinters 26.72 ± 2.33 cm		Slovene national sprinters 23.47 ± 1.88 cm ***		
	Mero [18]	Trained sprinters 0.32 ± 0.05 m				
Front block distance (to start line) (cm) (m)	Nagahara, Gleadhill [35]	National level sprinters 0.439 ± 0.045 m				
	Cavedon, Sandri [12]				Usual condition 52.3 ± 4.8 cm	Anthropometric condition 49.1 ± 3.0 cm
	Coh, Peharec [5]	Faster sprinters 0.54 ± 0.05 m	Slower sprinters 0.51 ± 0.04 m			
	Čoh, Jost [13]	Slovene national sprinters 55.15 ± 6.22 cm		Slovene national sprinters 45.49 ± 5.37 cm ***		
	Mero [18]	Trained sprinters 0.51 ± 0.05 m				

Table A1. *Cont.*

"Set" Position Kinematics	Study	Male		Female		Mixed		
Rear block distance (to start line) (cm) (m)	Nagahara, Gleadhill [35]	National level sprinters 0.686 ± 0.049 m						
	Coh, Peharec [5]	Faster sprinters 0.84 ± 0.09 m	Slower sprinters 0.79 ± 0.07 m					
	Čoh, Jost [13]	Slovene national sprinters 81.88 ± 7.47 cm		Slovene national sprinters 68.96 ± 5.91 cm ***				
	Mero [18]	Trained sprinters 0.83 ± 0.07 m						
Horizontal projection of the CM to the starting line (cm) (m)	Ciacci, Merni [38]	World-class 0.199 ± 0.054 m	Elite 0.202 ± 0.066 m	World-class 0.162 ± 0.037 m	Elite 0.248 ± 0.056 m			
	Ciacci, Merni [38]	Independent of category 0.201 ± 0.058 m		Independent of category 0.214 ± 0.066 m				
	Slawinski, Dumas [8]					Bunched start 21.7 ± 2.0 cm	Medium start 25.2 ± 1.9 cm	Elongated start 30.9 ± 3.0 cm
	Slawinski, Bonnefoy [7]	Elite 22.9 ± 1.5 cm	Well-trained 27.8 ± 2.8 cm *					
	Gutierrez-Davilla, Dapena [17]	Normal start 0.310 ± 0.057 m	Pre-tensed start 0.346 ± 0.068 m ***					
	Čoh, Jost [13]	Slovene national sprinters 18.77 ± 5.07 cm		Slovene national sprinters 15.03 ± 3.00 cm				
	Mero [18]	Trained sprinters 0.29 ± 0.05 m						

Table A1. *Cont.*

"Set" Position Kinematics	Study	Male			Female		Mixed		
Vertical height of CM (cm) (m)	Ciacci, Merni [38]	World-class 0.643 ± 0.025 m	Elite 0.655 ± 0.038 m		World-class 0.533 ± 0.032 m	Elite 0.587 ± 0.037 m			
		Independent of category 0.650 ± 0.033 m			Independent of category 0.565 ± 0.044 m *				
	Chen, Wu [37]	Bunched start 0.57 ± 0.04 m	Medium start 0.56 ± 0.03 m	Elongated start 0.57 ± 0.03 m					
	Slawinski, Dumas [8]						Bunched start 66.6 ± 2.4 cm	Medium start 66.5 ± 2.9 cm	Elongated start 65.5 ± 2.9 cm
	Slawinski, Bonnefoy [7]	Elite 65.7 ± 3.8 cm	Well-trained 62.6 ± 3.9 cm						
	Čoh, Jost [13]	Slovene national sprinters 54.38 ± 4.81 cm			Slovene national sprinters 53.18 ± 2.04 cm				
	Mero [18]	Trained sprinters 0.57 ± 0.04 m							

Table A1. *Cont.*

"Set" Position Kinematics	Study	Male		Female		Mixed	
Front leg hip angle (°)	Cavedon, Sandri [12]					Usual condition 47 ± 6 [(a)]	Anthropometric condition 43 ± 6 [(a)]
	Ciacci, Merni [38]	World-class 37.6 ± 0.6 [(a)]	Elite 44.9 ± 3.3 [(a)]	World-class 48.4 ± 14.6 [(a)]	Elite 46.7 ± 7.5 [(a)]		
		Independent of category 42.0 ± 4.5 [(a)]		Independent of category 47.4 ± 10.1 [(a)]			
	Bezodis, Salo [3]	World-class to university sprinters 47 ± 6 [(a)]					
	Debaere, Delecluse [6]					Elite sprinters 82.8 ± 10.1 [(b)]	
	Mero, Kuitunen [19]	Block angle 40° 52 ± 2 [(a)]	Block angle 60° 49 ± 2 [(a)]				
	Čoh, Jost [13]	Slovene national sprinters 44.78 ± 6.15 [(a)]		Slovene national sprinters 42.36 ± 9.43 [(a)]			
	Mero [18]	Trained sprinters 39 ± 7 [(a)]					

Table A1. *Cont.*

"Set" Position Kinematics	Study	Male		Female		Mixed	
Front leg knee angle (°)	Cavedon, Sandri [12]					Usual condition 92 ± 9 (c)	Anthropometric condition 98 ± 8 (c)
	Ciacci, Merni [38]	World-class 91.0 ± 9.8 (c)	Elite 99.3 ± 10.8 (c)	World-class 91.0 ± 10.1 (c)	Elite 100.1 ± 9.0 (c)		
	Ciacci, Merni [38]	Independent of category 95.9 ± 10.7 (c)		Independent of category 96.4 ± 10.0 (c)			
	Bezodis, Salo [3]	World-class to university sprinters 86 ± 5 (c)					
	Debaere, Delecluse [6]					Elite sprinters 94.5 ± 11.2 (c)	
	Slawinski, Bonnefoy [7]	Elite 110.7 ± 9.3 (c)	Well-trained 106.1 ± 13.7 (c)				
	Mero, Kuitunen [19]	Block angle 40° 103 ± 2 (c)	Block angle 60° 97 ± 2 (c)				
	Čoh, Jost [13]	Slovene national sprinters 93.75 ± 8.26 (c)		Slovene national sprinters 103.38 ± 6.97 (c) *			
	Mero [18]	Trained sprinters 96 ± 12 (c)					

Table A1. *Cont.*

"Set" Position Kinematics	Study	Male		Female		Mixed	
Front leg ankle angle (°)	Cavedon, Sandri [12]					Usual condition 92 ± 6 [(d)]	Anthropometric condition 93 ± 7 [(d)]
	Bezodis, Salo [3]	World-class to university sprinters 107 ± 2 [(d)]					
	Debaere, Delecluse [6]					Elite sprinters 82.3 ± 9.5 [(d)]	
	Mero, Kuitunen [19]	Block angle 40° 96 ± 2 [(d)]	Block angle 60° 111 ± 2 [(d)]				
	Čoh, Jost [13]	Slovene national sprinters 97.55 ± 10.55 [(d)]		Slovene national sprinters 102.65 ± 6.58 [(d)]			
	Mero [18]	Trained sprinters 94 ± 4 [(d)]					
Rear leg hip angle (°)	Cavedon, Sandri [12]					Usual condition 77 ± 8 [(a)]	Anthropometric condition 84 ± 8 [(a)]
	Ciacci, Merni [38]	World-class 71.2 ± 5.6 [(a)]	Elite 62.6 ± 3.7 [(a)]	World-class 75.2 ± 14.2 [(a)]	Elite 69.5 ± 5.1 [(a)]		
		Independent of category 66.0 ± 6.2 [(a)]		Independent of category 71.8 ± 9.5 [(a)]			
	Bezodis, Salo [3]	World-class to university sprinters 77 ± 9 [(a)]					
	Debaere, Delecluse [6]					Elite sprinters 107.1 ± 9 [(b)]	

Table A1. *Cont.*

<table>
<tr><th>"Set" Position Kinematics</th><th>Study</th><th colspan="2">Male</th><th colspan="2">Female</th><th colspan="2">Mixed</th></tr>
<tr><td></td><td>Mero, Kuitunen [19]</td><td>Block angle 40°
83 ± 2 (a)</td><td>Block angle 60°
79 ± 2 (a)</td><td colspan="2"></td><td colspan="2"></td></tr>
<tr><td></td><td>Čoh, Jost [13]</td><td colspan="2">Slovene national sprinters
24.91 ± 4.27 (e)</td><td colspan="2">Slovene national sprinters
19.25 ± 9.30 (e)</td><td colspan="2"></td></tr>
<tr><td></td><td>Mero [18]</td><td colspan="2">Trained sprinters
77 ± 9 (a)</td><td colspan="2"></td><td colspan="2"></td></tr>
<tr><td rowspan="10">Rear leg knee angle (°)</td><td>Cavedon, Sandri [12]</td><td colspan="2"></td><td colspan="2"></td><td>Usual condition
112 ± 11 (c)</td><td>Anthropometric condition
117 ± 11 (c)</td></tr>
<tr><td rowspan="2">Ciacci, Merni [38]</td><td>World-class
120.7 ± 9.7 (c)</td><td>Elite
116.1 ± 7.6 (c)</td><td>World-class
113.6 ± 20.9 (c)</td><td>Elite
118.4 ± 6.6 (c)</td><td colspan="2"></td></tr>
<tr><td colspan="2">Independent of category
118.0 ± 8.3 (c)</td><td colspan="2">Independent of category
116.5 ± 13.3 (c)</td><td colspan="2"></td></tr>
<tr><td>Bezodis, Salo [3]</td><td colspan="2">World-class to university sprinters
109 ± 9 (c)</td><td colspan="2"></td><td colspan="2"></td></tr>
<tr><td>Debaere, Delecluse [6]</td><td colspan="2"></td><td colspan="2"></td><td colspan="2">Elite sprinters
112.8 ± 15.1 (c)</td></tr>
<tr><td>Slawinski, Bonnefoy [7]</td><td>Elite
135.5 ± 11.4 (c)</td><td>Well-trained
117.3 ± 10.1 (c) *</td><td colspan="2"></td><td colspan="2"></td></tr>
<tr><td>Mero, Kuitunen [19]</td><td>Block angle 40°
131 ± 2 (c)</td><td>Block angle 60°
122 ± 2 (c)</td><td colspan="2"></td><td colspan="2"></td></tr>
<tr><td>Čoh, Jost [13]</td><td colspan="2">Slovene national sprinters
112.72 ± 13.31 (c)</td><td colspan="2">Slovene national sprinters
115.59 ± 13.86 (c)</td><td colspan="2"></td></tr>
<tr><td>Mero [18]</td><td colspan="2">Trained sprinters
126 ± 16 (c)</td><td colspan="2"></td><td colspan="2"></td></tr>
</table>

Table A1. *Cont.*

"Set" Position Kinematics	Study	Male	Female	Mixed
Rear leg ankle angle (°)	Cavedon, Sandri [12]			Usual condition 87 ± 6 (d); Anthropometric condition 85 ± 7 (d)
	Bezodis, Salo [3]	World-class to university sprinters 111 ± 12 (d)		
	Debaere, Delecluse [6]			Elite sprinters 82.5 ± 7.8 (d)
	Mero, Kuitunen [19]	Block angle 40° 95 ± 3 (d); Block angle 60° 109 ± 3 (d)		
	Čoh, Jost [13]	Slovene national sprinters 97.45 ± 10.28 (d); Slovene national sprinters 99.80 ± 6.44 (d)		
	Mero [18]	Trained sprinters 96 ± 8 (d)		
Trunk angle (°)	Chen, Wu [37]	Bunched start −20.4 ± 7.3 *; Medium start −14.9 ± 6.7 *; Elongated start −8.8 ± 10.8 (f) *		

CM—center of mass; (a) internal angle between the thigh and trunk in flexion/extension plane; (b) relative angle between the pelvis and the thigh according to the Biomechanical Convention [53]; (c) relative angle between the thigh and the shank according to the Medical Convention [53]; (d) relative angle between the shank and the foot according to the Biomechanical Convention [53]; (e) rear leg hip angle measured as front-rear leg angle; (f) relative angle between the vector from hip to shoulder and the horizontal plane.

Appendix B

Table A2. Summary of the kinematic variables in the "Block Phase". Data are the magnitude of the mean ± SD presented in the reviewed studies. Groups are male, female, and mixed (when authors joined data without discriminating by sex) sprinters. Studies are listed, in each variable, in reverse-chronological order, followed by alphabetically for studies published in the same year. Data, terms, conditions, and sprinters' performance levels are presented according to the original authors. Statistical differences between groups are marked with asterisks (* $p < 0.05$; ** $p < 0.01$; *** $p < 0.001$; # Cohen's d—large effect size (>0.8); § small effect size [0.2–0.6] of 90% confidence intervals; §§ moderate effect size [0.6–1.2] of 90% confidence intervals); ᵽ clearly associated with average horizontal power produced across the block phase—$p < 0.05$; ¥ significantly greater compared to the bunched start.

Block Phase Kinematics	Study	Male		Female	Mixed	
Block time (ms) (s)	Graham-Smith, Colyer [39]	Seniors 365 ± 18 ms	Juniors 412 ± 49 ms			
	Nagahara, Gleadhill [35]	National-level sprinters 0.369 s [(a)]				
	Sado, Yoshioka [23]	University-level sprinters 0.36 ± 0.03 s				
	Bezodis, Walton [10]	Sprint start-trained athletes 0.391 ± 0.038 s				
	Cavedon, Sandri [12]				Usual condition 0.421 ± 0.047 s	Anthropometric condition 0.427 ± 0.038 s
	Colyer, Graham-Smith [33]	Elite 0.360 ± 0.010 s [(ᵽ)]	All sample 0.390 ± 0.039 s			
	Sandamas, Gutierrez-Farewik [24]				Skating condition 0.37 ± 0.03 s	Narrow condition 0.38 ± 0.03 s
	Brazil, Exell [4]	Athletic sprinters 0.359 ± 0.014 s				

Table A2. *Cont.*

Block Phase Kinematics	Study	Male		Female		Mixed		
	Ciacci, Merni [38]	World-class 0.356 ± 0.011 s	Elite 0.323 ± 0.024 s	World-class 0.356 ± 0.018 s	Elite 0.323 ± 0.024 s			
	Ciacci, Merni [38]	Independent of category 0.336 ± 0.025 s		Independent of category 0.336 ± 0.027 s				
	Coh, Peharec [5]	Faster sprinters 332 ± 28.73 ms	Slower sprinters 305 ± 24.35 ms					
	Bezodis, Salo [3]	World-class to university sprinters 0.358 ± 0.022 s						
	Otsuka, Kurihara [21]	Normal condition 0.334 ± 0.031 s	Widened condition 0.330 ± 0.025 s					
	Rabita, Dorel [22]	Elite 376 ± 24 ms	Sub-elite 394 ± 13 ms #					
	Milanese, Bertucco [41]					Rear knee angle @ 90° 0.354 ± 0.015 s	@ 115° 0.348 ± 0.016 s	@ 135° 0.355 ± 0.014 s
	Otsuka, Shim [42]	Well-trained 0.349 ± 0.019 s	Trained 0.379 ± 0.022 s *					
	Aerenhouts, Delecluse [1]	Elite Seniors 357 ± 29 ms	Elite Juniors 367 ± 28 ms	Elite Seniors 380 ± 16 ms	Elite Juniors 383 ± 19 ms			
	Slawinski, Dumas [8]					Bunched start 0.371 ± 0.016 s	Medium start 0.377 ± 0.017 s	Elongated start 0.427 ± 0.056 s
	Slawinski, Bonnefoy [7]	Elite 0.352 ± 0.018 s	Well-trained 0.351 ± 0.020 s					

Table A2. *Cont.*

Block Phase Kinematics	Study	Male		Female	Mixed		
	Maulder, Bradshaw [40]	National and regional level sprinters 0.31 s [(a)]					
	Gutierrez-Davilla, Dapena [17]	Conventional start 0.375 ± 0.028 s	Pre-tensed start 0.386 ± 0.036 s *				
	Mero, Kuitunen [19]	Block angle 40° 0.343 ± 0.036 s	Block angle 65° 0.333 ± 0.027 s				
	Fortier, Basset [16]	Elite 399 ± 21 ms	Sub-elite 422 ± 33 ms *				
	Čoh, Jost [13]	Slovene national sprinters 0.30 ± 0.03 s		Slovene national sprinters 0.34 ± 0.02 s			
	Guissard, Duchateau [34]				Block angle 30° 0.321 ± 0.023 s	Block angle 50° 0.325 ± 0.035 s	Block angle 70° 0.317 ± 0.039 s
	Mero [18]	Trained sprinters 0.342 ± 0.022 s					
Rear leg block time (ms) (s)	Nagahara, Gleadhill [35]	National-level sprinters 0.212 ± 0.029 s					
	Sado, Yoshioka [23]	University-level sprinters 0.18 ± 0.02 s					
	Cavedon, Sandri [12]				Usual condition 0.211 ± 0.041 s		Anthropometric condition 0.212 ± 0.041 s
	Brazil, Exell [4]	Athletic sprinters 0.193 ± 0.012 s					
	Coh, Peharec [5]	Faster sprinters 162 ± 9.47 ms	Slower sprinters 149 ± 12.40 ms *				

Table A2. *Cont.*

Block Phase Kinematics	Study	Male		Female	Mixed		
	Otsuka, Kurihara [21]	Normal Condition 0.175 ± 0.034 s	Widened condition 0.180 ± 0.023 s				
	Milanese, Bertucco [41]				Rear knee angle @ 90° 0.12 ± 0.01 s	@115° 0.11 ± 0.01 s	@ 135° 0.09 ± 0.02 s
	Otsuka, Shim [42]	Well-trained 0.188 ± 0.022 s	Trained 0.187 ± 0.029 s				
	Slawinski, Bonnefoy [7]	Elite 0.154 ± 0.017 s	Well-trained 0.140 ± 0.026 s				
	Mero, Kuitunen [19]	Block angle 40° 0.188 ± 0.008 s	Block angle 65° 0.172 ± 0.015 s				
	Fortier, Basset [16]	Elite 370 ± 18 ms (b)	Sub-elite 268 ± 58 ms				
	Čoh, Jost [13]	Slovene national sprinters 0.20 ± 0.02 s		Slovene national sprinters 0.18 ± 0.03 s			
Ratio rear leg time/block time (%)	Sado, Yoshioka [23]	University-level sprinters 49.7 ± 5.1					
	Bezodis, Salo [3]	World-class to university sprinters 53 ± 5					
	Milanese, Bertucco [41]				Rear knee angle @ 90° 34.62 ± 3.60	@115° 31.30 ± 3.52	@135° 28.65 ± 3.57
	Slawinski, Bonnefoy [7]	Elite 43.5 ± 3.8	Well-trained 39.8 ± 8.1				

Table A2. *Cont.*

Block Phase Kinematics	Study	Male			Female	Mixed		
Block resultant velocity ($m \cdot s^{-1}$)	Chen, Wu [37]	Bunched start 3.32 ± 0.14	Medium start 3.36 ± 0.15	Elongated start 3.45 ± 0.22				
	Slawinski, Dumas [8]					Bunched start 2.76 ± 0.11	Medium start 2.84 ± 0.14 ¥	Elongated start 2.89 ± 0.13 ¥
	Slawinski, Bonnefoy [7]	Elite 3.48 ± 0.05		Well-trained 3.24 ± 0.18 *				
	Fortier, Basset [16]	Elite 3.28 ± 0.19		Sub-elite 3.12 ± 0.30				
	Čoh, Jost [13]	Slovene national sprinters 3.37 ± 0.35			Slovene national sprinters 3.09 ± 0.21 *			
	Mero [18]	Trained sprinters 3.46 ± 0.32						
Block horizontal velocity ($m \cdot s^{-1}$)	Graham-Smith, Colyer [39]	Seniors 3.36 ± 0.15		Juniors 3.16 ± 0.18 §§				
	Sado, Yoshioka [23]	University-level sprinters 3.31 ± 0.13						
	Bezodis, Walton [10]	Sprint start-trained athletes 3.12 ± 0.21						
	Cavedon, Sandri [12]					Usual condition 3.36 ± 0.35		Anthropometric condition 3.50 ± 0.39
	Colyer, Graham-Smith [33]	Elite 3.36 ± 0.13 (þ)		All sample 3.30 ± 0.20				

Table A2. *Cont.*

Block Phase Kinematics	Study	Male		Female		Mixed		
	Ciacci, Merni [38]	World-class 4.16 ± 0.39	Elite 4.08 ± 0.08	World-class 3.11 ± 0.39	Elite 3.48 ± 0.23			
		Independent of category 4.11 ± 0.24		Independent of category 3.33 ± 0.34 *				
	Coh, Peharec [5]	Faster sprinters 3.38 ± 0.10	Slower sprinters 3.19 ± 0.19 *					
	Rabita, Dorel [22]	Elite 3.61 ± 0.08	Sub-elite 3.17 ± 0.19 #					
	Milanese, Bertucco [41]					Rear knee angle @ 90° 2.67 ± 0.26	@ 115° 2.62 ± 0.23	@ 135° 2.56 ± 0.24
	Debaere, Delecluse [6]						Elite Sprinters 3.10 ± 0.25	
	Aerenhouts, Delecluse [1]	Elite Seniors 2.9 ± 0.3	Elite Juniors 2.9 ± 0.3	Elite Seniors 2.8 ± 0.2	Elite Juniors 2.7 ± 0.3			
	Bezodis, Salo [2]	University-level sprinters 3.28 ± 0.24						
	Maulder, Bradshaw [40]	National and regional level sprinters 3.40 ± 0.20						
	Gutierrez-Davilla, Dapena [17]	Conventional start 3.21 ± 0.22	Pre-tensed start 3.22 ± 0.24					
	Mero, Kuitunen [19]	Block angle 40° 3.39 ± 0.23	Block angle 65° 3.30 ± 0.21 **					
	Čoh, Jost [13]	Slovene national sprinters 3.20 ± 0.19		Slovene national sprinters 2.99 ± 0.23 *				

Table A2. *Cont.*

Block Phase Kinematics	Study	Male	Female	Mixed
	Guissard, Duchateau [34]			Block angle 30° 2.94 ± 0.20; Block angle 50° 2.80 ± 0.23; Block angle 70° 2.37 ± 0.31
Block vertical velocity ($m \cdot s^{-1}$)	Graham-Smith, Colyer [39]	Seniors 0.60 ± 0.12; Juniors 0.61 ± 0.13		
	Sado, Yoshioka [23]	University-level sprinters 0.58 ± 0.08		
	Colyer, Graham-Smith [33]	Elite 0.58 ± 0.06; All sample 0.60 ± 0.11		
	Ciacci, Merni [38]	World-class −0.21 ± 0.27 (c); Elite 0.59 ± 0.32	World-class 0.38 ± 0.06; Elite 0.52 ± 0.30	
		Independent of category 0.27 ± 0.50	Independent of category 0.47 ± 0.24	
	Chen, Wu [37]	Bunched start 0.49 ± 0.19; Medium start 0.40 ± 0.15; Elongated start 0.42 ± 0.33		
	Debaere, Delecluse [6]			Elite sprinters 0.84 ± 0.13
	Slawinski, Bonnefoy [7]	Elite 0.52 ± 0.06; Well-trained 0.51 ± 0.14		
	Čoh, Jost [13]	Slovene national sprinters 0.69 ± 0.21	Slovene national sprinters 0.76 ± 0.19	

Table A2. *Cont.*

Block Phase Kinematics	Study	Male		Female		Mixed		
Block acceleration ($m \cdot s^{-2}$)	Coh, Peharec [5]	Faster sprinters 7.47 ± 1.34	Slower sprinters 7.35 ± 0.90					
	Otsuka, Kurihara [21]	Normal condition 9.65 ± 0.72	Widened condition 9.73 ± 0.59					
	Otsuka, Shim [42]	Well-trained 9.72 ± 0.36	Trained 8.41 ± 0.49 *					
	Aerenhouts, Delecluse [1]	Elite Seniors 8.2 ± 0.9	Elite Juniors 7.9 ±0.7	Elite Seniors 7.3 ± 0.7	Elite Juniors 7.0 ± 0.8			
	Bezodis, Salo [2]	University-level sprinters 9.14 ± 0.99						
	Slawinski, Bonnefoy [7]	Elite 9.5 ± 0.4	Well-trained 8.8 ± 0.8					
	Maulder, Bradshaw [40]	National and regional level sprinters 8.00 ± 0.80						
	Guissard, Duchateau [34]					Block angle 30° 9.03 ± 0.91	Block angle 50° 8.36 ± 1.17	Block angle 70° 7.46 ± 1.42
Take-off angle (°) [(d)]	Milanese, Bertucco [41]					Rear knee angle @ 90° 40.42 ± 2.74	@ 115° 40.23 ± 2.13	@ 135° 39.77 ± 2.50
	Slawinski, Bonnefoy [7]	Elite 34.7 ± 1.4	Well-trained 34.3 ± 2.0					
	Maulder, Bradshaw [40]	National and regional level sprinters 42 ± 4						
	Čoh, Jost [13]	Slovene national sprinters 49.54 ± 2.91		Slovene national sprinters 53.20 ± 3.20 *				
CM projection angle (°) [(e)]	Graham-Smith, Colyer [39]	Seniors 10.2 ± 2.0	Juniors 11.0 ± 2.1 §					

Table A2. *Cont.*

Block Phase Kinematics	Study	Male	Female	Mixed
	Colyer, Graham-Smith [33]	Elite 9.8 ± 0.8 (h); All sample 10.3 ± 2.0		
Horizontal CM ROM (m)	Gutierrez-Davilla, Dapena [17]	Conventional start 0.600 ± 0.046; Pre-tensed start 0.619 ± 0.059 *		
Angular displacement (°)				
Trunk	Bezodis, Salo [3]	World-class to university sprinters 46 ± 8		
Front hip		World-class to university sprinters 113 ± 9		
Front knee		World-class to university sprinters 73 ± 7		
Front ankle		World-class to university sprinters 36 ± 10		
Rear hip		World-class to university sprinters 31 ± 13 (f)		
Rear knee		World-class to university sprinters 18 ± 6 (f)		
Rear ankle		World-class to university sprinters 19 ± 9 (f)		
Ankle joint dorsiflexion (°)	Schrodter, Bruggemann [25]	Front block (g) 15.8 ± 7.4; Rear block (g) 8.0 ± 5.7 ***		
Trunk angle at takeoff (°) (h)	Chen, Wu [37]	Bunched start 25.7 ± 6.1; Medium start 29.1 ± 4.5 *; Elongated start 28.9 ± 4.5		
	Maulder, Bradshaw [40]	National and regional level sprinters 22 ± 7		

Table A2. *Cont.*

Block Phase Kinematics	Study	Male	Female	Mixed	
Hip angle at takeoff (°)	Debaere, Delecluse [6]			Elite sprinters—Rear block 146.8 ± 9.4 [(i)]	Elite sprinters—Front block 183.2 ± 6.8 [(i)]
Knee angle at takeoff (°)				Elite sprinters—Rear block 134.9 ± 11.2 [(j)]	Elite sprinters—Front block 177.4 ± 5.2 * [(j)]
Ankle angle at takeoff (°)				Elite sprinters—Rear block 139.2 ±7.0 [(k)]	Elite sprinters—Front block 133.1 ± 6.7 [(k)]
Joint angular velocity (°.s^{-1})					
Trunk	Slawinski, Bonnefoy [36]	Elite sprinters 220.2 ± 57.5			
Front hip		Elite sprinters 456.3 ± 17.7			
Front knee		Elite sprinters 660.2 ± 40.5			
Front ankle		Elite sprinters 641.5 ± 44.9			
Rear hip		Elite sprinters 425.7 ± 61.0			
Rear knee		Elite sprinters 651.4 ± 112.3			
Rear ankle		Elite sprinters 462.9 ± 74.7			

CM—center of mass; ROM—range of motion; [(a)] block time calculated from the difference between the average data of total block time and reaction time data; [(b)] probably an incorrect data from the original paper; [(c)] presumably the negative signal is a gap in the data reported in the original paper; [(d)] the take-off or push-off angle is the angle between the horizontal and the line passing through the most front part of the contact foot and the center of mass at block clearance; [(e)] center of mass projection angle is calculated as the resultant direction from the horizontal and vertical block exit velocities of the center of mass; [(f)] angular displacement during rear block contact only; [(g)] higher magnitude of dorsiflexion was correlated to a faster stretch velocity, which was related to increased force generation (maximal rate of force development, maximal resultant and horizontal push force, and also normalized average horizontal block power); [(h)] the angle, measured relative to the horizontal, between the line passing through the hip and shoulder (trunk segment) of the side of the body in which the athlete's front foot at the block take off instant; [(i)] relative angle between the pelvis and the thigh according to the Biomechanical Convention [53]; [(j)] relative angle between the thigh and the shank according to the Medical Convention [53]; [(k)] relative angle between the shank and the foot according to the Biomechanical Convention [53].

Table A3. Summary of the kinetic variables in the "Block Phase". Data are the magnitude of the mean ± SD presented in the reviewed studies. Groups are male and mixed (when authors joined data without discriminating by sex) sprinters. Studies are listed, in each variable, in reverse-chronological order, followed by alphabetically for studies published in the same year. Data, terms, conditions, and sprinters' performance levels are presented according to the original authors. Statistical differences between groups are marked with asterisks (* $p < 0.05$; ** $p < 0.01$; *** $p < 0.001$; # Cohen's d—large effect size (>0.8); §§§ large effect size [1.2–1.6] of 90% confidence intervals; ᵬ clearly associated with average horizontal power produced across the block phase—$p < 0.05$; ᵬᵬ moderately associated (moderate effect size: >0.3) with average horizontal power produced across the block phase).

Block Phase Kinetics			Male		Mixed	
Block force	Initial force on blocks—"Set" position ($N.N^{-1}$)	Gutierrez-Davilla, Dapena [17]	Normal start 0.113 ± 0.04	Pre-tensed start 0.186 ± 0.053 ***		
	Relative average total force (NAF) ($N{\cdot}kg^{-1}$)	Sandamas, Gutierrez-Farewik [24]			Skating conditions 1.44 ± 0.07 BW (a)	Narrow condition 1.44 ± 0.07 BW (a)
		Cavedon, Sandri [12]			Usual condition 11.37 ± 1.19	Anthropometric condition 11.55 ± 1.12
		Otsuka, Shim [42]	Well-trained sprinters 15.03 ± 0.32	Trained sprinters 13.99 ± 0.65		
	Average horizontal force (AHF) (N)	Rabita, Dorel [22]	Elite sprinters 783 ± 59	Sub-elite sprinters 596 ± 47 #		
		Mero [18]	Trained sprinters 655 ± 76			
	Relative average horizontal force (NAHF) ($N{\cdot}kg^{-1}$) (BW)	Colyer, Graham-Smith [33]	Elite sprinters 9.4 ± 0.1 $N{\cdot}kg^{-1}$ (ᵬ)	All sample 8.7 ± 1.1 $N{\cdot}kg^{-1}$		
		Sandamas, Gutierrez-Farewik [24]			Skating condition 0.87 ± 0.10 BW (a) (b)	Narrow condition 0.86 ± 0.10 BW (a) (b)
		Rabita, Dorel [22]	Elite sprinters 9.59 ± 0.53 $N{\cdot}kg^{-1}$	Sub-elite sprinters 7.74 ± 0.82 $N{\cdot}kg^{-1}$ #		

Table A3. *Cont.*

Block Phase Kinetics		Male			Mixed	
	Otsuka, Shim [42]	Well-trained sprinters 9.72 ± 0.36 N·kg^{-1}	Trained sprinters 8.41 ± 0.49 N·kg^{-1} *			
Peak-to-minimum horizontal force average change (N·kg^{-1}) (transition from bilateral to unilateral pushing)	Colyer, Graham-Smith [33]	Elite sprinters −10.3 ± 3.1 (þ)	All sample −10.6 ± 2.5			
Resultant force front block resultant force (N)	Coh, Peharec [5]	Faster sprinters 1104 ± 82.53	Slower sprinters 1073 ± 56.21			
Relative front block resultant mean force (N·kg^{-1})	Nagahara, Gleadhill [35]	Normal condition 9.25 ± 0.39	Anterior condition 9.03 ± 0.63	Posterior condition 9.44 ± 0.84		
	Otsuka, Shim [42]	Well-trained sprinters 10.03 ± 1.07	Trained sprinters 9.62 ± 0.94			
Rear block resultant force (N)	Coh, Peharec [5]	Faster sprinters 913 ± 89.23	Slower sprinters 771 ± 55.09 **			
Relative rear block resultant mean force (N·kg^{-1})	Nagahara, Gleadhill [35]	Normal condition 7.20 ± 0.52	Anterior condition 6.05 ± 1.55	Posterior condition 8.23 ± 1.13 * (c)		
	Otsuka, Shim [42]	Well-trained sprinters 7.71 ± 1.24	Trained sprinters 7.46 ± 1.04			
Horizontal force						
Front block horizontal maximal force (N)	Coh, Peharec [5]	Faster sprinters 461 ± 51.05	Slower sprinters 398 ± 56.73			
	Aerenhouts, Delecluse [1]	Elite Seniors 686 ± 110	Elite Juniors 623 ± 105		Elite Seniors (d) 482 ± 98	Elite Juniors (d) 454 ± 65
Relative front block horizontal maximal force (N·kg^{-1})	Cavedon, Sandri [12]				Usual condition 6.02 ± 0.71	Anthropometric condition 5.91 ± 0.65
Relative front block horizontal mean force (N·kg^{-1})	Nagahara, Gleadhill [35]	Normal condition 5.87 ± 0.38	Anterior condition 5.93 ± 0.56	Posterior condition 6.25 ± 0.64		

Table A3. *Cont.*

Block Phase Kinetics		**Male**			**Mixed**	
	Otsuka, Shim [42]	Well-trained sprinters 6.70 ± 0.58		Trained sprinters 5.99 ± 0.67		
Rear block horizontal maximal force (N)	Coh, Peharec [5]	Faster sprinters 460 ± 58.12		Slower sprinters 423 ± 45.50		
	Aerenhouts, Delecluse [1]	Elite Seniors 785 ± 220		Elite Juniors 697 ± 143	Elite Seniors (d) 485 ± 986	Elite Juniors (d) 435 ± 115
Relative rear block horizontal maximal force ($N \cdot kg^{-1}$)	Cavedon, Sandri [12]				Usual condition 4.52 ± 1.09	Anthropometric condition 4.95 ± 1.34 *
Relative rear block horizontal mean force ($N \cdot kg^{-1}$)	Nagahara, Gleadhill [35]	Normal condition 5.18 ± 0.38	Anterior condition 3.97 ± 1.17	Posterior condition 6.14 ± 0.86 ** (c)		
	Otsuka, Shim [42]	Well-trained sprinters 5.82 ± 0.71		Trained sprinters 5.41 ± 0.88		
Vertical force						
Front block vertical maximal force (N)	Coh, Peharec [5]	Faster sprinters 1019 ± 69.99		Slower sprinters 978 ± 43.12		
Relative front block vertical maximal force ($N \cdot kg^{-1}$)	Cavedon, Sandri [12]				Usual condition 6.13 ± 0.92	Anthropometric condition 6.12 ± 0.90
Relative front block vertical mean force ($N \cdot kg^{-1}$)	Nagahara, Gleadhill [35]	Normal condition 7.15 ± 0.29	Anterior condition 6.81 ± 0.40	Posterior condition 7.07 ± 0.64		
	Otsuka, Shim [42]	Well-trained sprinters 7.43 ± 1.01		Trained sprinters 7.50 ± 0.78		
Rear block vertical maximal force (N)	Coh, Peharec [5]	Faster sprinters 795 ± 91.29		Slower sprinters 645 ± 41.55 **		
Relative rear block vertical maximal force ($N \cdot kg^{-1}$)	Cavedon, Sandri [12]				Usual condition 3.78 ± 1.12	Anthropometric condition 3.96 ± 1.20

Table A3. *Cont.*

Block Phase Kinetics			Male			Mixed	
Relative rear block vertical mean force (N·kg^{-1})		Nagahara, Gleadhill [35]	Normal condition 4.99 ± 0.57	Anterior condition 4.53 ± 1.17	Posterior condition 5.47 ± 0.83		
		Otsuka, Shim [42]	Well-trained sprinters 5.03 ± 1.15		Trained sprinters 5.12 ± 0.68		
Maximal rate of force development (N·s^{-1}) (N·kg^{-1}·s^{-1})		Schrodter, Bruggemann [25]				World-class sprinters 259 ± 79 N·kg^{-1}.s^{-1}	Well-trained sprinters 175 ± 86 N·kg^{-1}.s^{-1} **
		Slawinski, Bonnefoy [7]	Elite 15505 ± 5397 N·s^{-1}		Well-trained 8459 ± 3811 N·s^{-1} *		
Block power	Average horizontal block power (W) (e)	Bezodis, Walton [10]		Sprint start-trained athletes 832 ± 113			
		Bezodis, Salo [3] (f)		World-class to university sprinters 1171 ± 268			
		Rabita, Dorel [22]	Elite sprinters 1415 ± 118		Sub-elite sprinters 949 ± 124 #		
		Bezodis, Salo [2] (f)		University-level sprinters 1094 ± 264			
	Relative average horizontal external power (W·kg^{-1})	Graham-Smith, Colyer [39]	Seniors sprinters 15.5 ± 1.5		Juniors sprinters 12.4 ± 2.2 §§§		
		Nagahara, Gleadhill [35]	Normal condition 14.8 ± 1.0	Anterior condition 13.2 ± 1.3	Posterior condition 16.2 ± 2.1 ** (c)		
		Colyer, Graham-Smith [33]		Mix of elite, senior and junior sprinters 14.3 ± 2.3			
		Nagahara and Ohshima [20]		Sprinters 14.7 ± 1.4			
		Rabita, Dorel [22]	Elite sprinters 17.3 ± 1.3		Sub-elite sprinters 12.3 ± 1.9 #		

Table A3. *Cont.*

Block Phase Kinetics			Male		Mixed		
		Sado, Yoshioka [23]	University-level sprinters 0.55 ± 0.05				
		Bezodis, Walton [10]	Sprint start-trained athletes 0.43 ± 0.06 (associated with block velocity)				
		Cavedon, Sandri [12]			Usual condition 0.47 ± 0.90	Anthropometric condition 0.50 ± 0.10 *	
	Normalized average horizontal external power (g)	Sandamas, Gutierrez-Farewik [24]			Skating condition 0.46 ± 0.07	Narrow condition 0.45 ± 0.07	
		Schrodter, Bruggemann [25]			World-class sprinters 0.360 ± 0.098	Well-trained sprinters 0.305 ± 0.056 ** (h)	
		Bezodis, Salo [3]	World-class to university sprinters 0.53 ± 0.08 (associated with PB100m)				
		Otsuka, Kurihara [21]	Normal condition 0.539 ± 0.053	Widened condition 0.543 ± 0.051			
		Bezodis, Salo [2]	University-level sprinters 0.51 ± 0.09 (associated with block velocity and acceleration data)				
Force impulse	Absolute force impulse (N·s)	Coh, Peharec [5]	Faster sprinters 294.3 ± 21.1	Slower sprinters 269.5 ± 17.9 *			
		Milanese, Bertucco [41]			Rear knee angle @ 90° 175.00 ± 26.49	@ 115° 172.00 ± 25.49	@ 135° 168.35 ± 25.61
		Slawinski, Bonnefoy [7]	Elite sprinters 276.2 ± 36.0	Well-trained sprinters 215.4 ± 28.5 *			
		Mero, Kuitunen [19]	Block angle 40° 249.0 ± 21.5	Block angle 65° 240.3 ± 22.9			

Table A3. *Cont.*

Block Phase Kinetics			Male		Mixed	
	Relative force impulse ($N·s·kg^{-1}$) ($m·s^{-1}$)	Cavedon, Sandri [12]			Usual condition 4.76 ± 0.55 $N·s·kg^{-1}$	Anthropometric condition 4.93 ± 0.56 $N·s·kg^{-1}$
		Sandamas, Gutierrez-Farewik [24]			Skating conditions 3.27 ± 0.15 $m·s^{-1}$	Narrow conditions 3.25 ± 0.16 $m·s^{-1}$
	Horizontal force impulse (N·s)	Coh, Peharec [5]	Faster sprinters 140.7 ± 11.5	Slower sprinters 112.8 ± 10.4 ***		
		Mero [18]	Trained sprinters 223 ± 18			
	Relative horizontal force impulse ($m·s^{-1}$) ($N·s·kg^{-1}$)	Sandamas, Gutierrez-Farewik [24]			Skating condition 3.21 ± 0.16 $m·s^{-1}$	Narrow condition 3.19 ± 0.16 $m·s^{-1}$
		Otsuka, Kurihara [21]	Normal condition 3.20 ± 0.18 $N·s·kg^{-1}$	Widened condition 3.20 ± 0.20 $N·s·kg^{-1}$		
		Otsuka, Shim [42]	Well-trained sprinters 3.407 ± 0.149 $N·s·kg^{-1}$	Trained sprinters 3.179 ± 0.163 $N·s·kg^{-1}$		
	Vertical force impulse (N·s)	Coh, Peharec [5]	Faster sprinters 256.1± 9.7	Slower sprinters 209.8 ± 8.9 ***		
		Mero [18]	Trained sprinters 173 ± 30			
	Normalized vertical force impulse ($m·s^{-1}$)	Sandamas, Gutierrez-Farewik [24]			Skating condition 0.54 ± 0.07	Narrow condition 0.59 ± 0.08 *
	Normalized me-diolateral force impulse ($m·s^{-1}$)	Sandamas, Gutierrez-Farewik [24]			Skating conditions 0.23 ± 0.10	Narrow condition 0.08 ± 0.05 *

Table A3. *Cont.*

Block Phase Kinetics			Male		Mixed
	Force impulse of front block (N·s)	Coh, Peharec [5]	Faster sprinters 221.3 ± 15.8	Slower sprinters 178.3 ± 13.1 ***	
	Force impulse of rear block (N·s)	Coh, Peharec [5]	Faster sprinters 76.7 ± 8.8	Slower sprinters 71.1 ± 6.7	
COP location (m)	Front block anteroposterior location	Nagahara and Ohshima [20]	Sprinters −0.080 ± 0.024 (þþ)		
	Front block vertical location		Sprinters 0.061 ± 0.022 (þ)		
	Rear block anteroposterior location		Sprinters −0.082 ± 0.018		
	Rear block vertical location		Sprinters 0.064 ± 0.018 (þ)		
	Front block location		Sprinters −0.45 ± 0.05 (þþ)		
	Rear block location		Sprinters −0.69 ± 0.06		
Peak joint moments	Peak ankle ex-tension moment	Brazil, Exell [4]	Rear block 0.236 ± 0.044 (i)	Front block 0.172 ± 0.032 (i) *	
	Peak knee exten-sion moment		Rear block 0.054 ± 0.020 (i)	Front block 0.199 ± 0.067 (i) *	
	Peak hip exten-sion moment		Rear block 0.315 ± 0.086 (i)	Front block 0.349 ± 0.035 (i)	
	Peak lumbosa-cral extension moment ($N \cdot s^{-1}$)	Sado, Yoshioka [23]	University-level sprinters 3.64 ± 0.39 (j) (þ)		

Table A3. *Cont.*

Block Phase Kinetics			Male		Mixed
Peak joint powers	Peak positive ankle power	Brazil, Exell [4]	Rear block 0.236 ± 0.066 [i]	Front block 0.388 ± 0.084 [i] *	
	Peak positive knee power		Rear block 0.047 ± 0.026 [i]	Front block 0.440 ± 0.177 [i] *	
	Peak positive hip power		Rear block 0.408 ± 0.152 [i]	Front block 0.576 ± 0.071 [i] *	

[a] Normalized to body mass, gravity constant and sprinter's leg length; [b] units as reported in the original article; [c] significantly different from the anterior condition; [d] only elite female data; [e] average horizontal external power is calculated as the product of anteroposterior force and horizontal velocity; [f] average horizontal external power was calculated based on the rate of change of mechanical energy in a horizontal direction (i.e., change in kinetic energy divided by time) [2]; [g] normalized average horizontal external power is the average horizontal external power normalized to the mass and the leg length of the sprinter [2]; [h] for normalization, the body height was used instead of the sprinter's leg length [25]; [i] joint data normalized to the mass and the leg length of the sprinter; [j] significantly larger ($p < 0.05$, Cohen's $d = 2.02$–11.09) than any other lower-limb and lumbosacral torques, although quantitative data for the remaining joint torques are not available.

Appendix C

Table A4. Summary of the kinematic variables in the "first two steps". Data are the magnitude of the mean ± SD presented in the reviewed studies. Groups are male, female, and mixed (when authors joined data without discriminating by sex) sprinters. Studies are listed, in each variable, in reverse-chronological order, followed by alphabetically for studies published in the same year. Data, terms, conditions, and sprinters' performance levels are presented according to the original authors. Statistical differences between groups are marked with asterisks (* $p < 0.05$; ** $p < 0.01$; *** $p < 0.001$; # significant different from adults; § small effect size [0.2–0.6] of 90% confidence intervals; §§ moderate effect size [0.6–1.2] of 90% confidence intervals).

First and Second Steps Kinematics	Study	Male	Female	Mixed
First Step				
First step length (m) (cm)	Cavedon, Sandri [12]			Usual condition 1.09 ± 0.11 nor. to leg length; Anthropometric condition 1.12 ± 0.12 nor. to leg length
	Ciacci, Merni [38]	World-class 1.135 ± 0.025 m; Elite 0.968 ± 0.162 m	World-class 1.068 ± 0.032 m; Elite 0.950 ± 0.099 m	
		Independent of category 1.035 ± 0.149 m	Independent of category 0.997 ± 0.097 m	
	Coh, Peharec [5]	Faster sprinters 1.30 ± 0.51 m; Slower sprinters 1.06 ± 0.60 m §		
	Debaere, Vanwanseele [15]			Adult sprinters 1.00 ± 0.07 m; U18 sprinters 0.94 ± 0.11 m; U16 sprinters 0.94 ± 0.10 m
	Chen, Wu [37]	Bunched start 0.97 ± 0.10 m; Medium start 1.00 ± 0.12 m *; Elongated start 1.03 ± 0.10 m *		
	Bezodis, Salo [3]	World-class to university sprinters 1.10 ± 0.07 normalized to step length		
	Rabita, Dorel [22]	Elite 0.96 ± 0.16 m; Sub-elite 1.01 ± 0.06 m		

Table A4. *Cont.*

First and Second Steps Kinematics	Study	Male		Female		Mixed		
	Milanese, Bertucco [41]					Rear knee angle @ 90° 1.23 ± 0.12 m	@ 115° 1.22 ± 0.11 m	@ 135° 1.21 ± 0.13 m
	Aerenhouts, Delecluse [1]	Elite Seniors 85 ± 33 cm	Elite Juniors 63 ±27 cm *	Elite Seniors 82 ± 19 cm	Elite Juniors 61 ± 20 cm *			
	Slawinski, Bonnefoy [7]	Elite 137.1 ± 9.0 cm	Well-trained 120.8 ± 8.7 cm *					
	Maulder, Bradshaw [40]	National and regional level sprinters 1.04 ± 0.03 m						
	Mero, Kuitunen [19]	Block angle 40° 1.09 ± 0.06 m	Block angle 65° 1.06 ± 0.06 m					
	Čoh, Jost [13]	Slovene national sprinters 100.85 ± 9.79 cm		Slovene national sprinters 98.64 ± 6.74 cm				
First step contact time (ms) (s)	Werkhausen, Willwacher [43]			Germany national sprinters 0.20 ± 0.02 s				
	Graham-Smith, Colyer [39]	Seniors 0.195 ± 0.022 s	Juniors 0.202 ± 0.024 s					
	Sandamas, Gutierrez-Farewik [24]					Skating condition 0.21 ± 0.01 s		Narrow condition 0.20 ± 0.01 s
	Aeles, Jonkers [9]					Adult sprinters 0.191 ± 0.024 s		Young sprinters 0.199 ± 0.023 s

Table A4. *Cont.*

First and Second Steps Kinematics	Study	Male		Female		Mixed
	Ciacci, Merni [38]	World-class 0.210 ± 0.035 s	Elite 0.176 ± 0.008 s	World-class 0.225 ± 0.034 s	Elite 0.166 ± 0.017 s	
		Independent of category 0.189 ± 0.027 s		Independent of category 0.190 ± 0.038 s		
	Coh, Peharec [5]	Faster sprinters 170 ± 18.17 ms	Slower sprinters 174 ± 16.94 ms			
	Aerenhouts, Delecluse [1]	Elite Seniors 173 ± 67 ms	Elite Juniors 199 ± 24 ms	Elite Seniors 196 ± 62 ms	Elite Juniors 210 ± 17 ms	
	Slawinski, Bonnefoy [7]	Elite 0.173 ± 0.010 s	Well-trained 0.167 ± 0.011 s			
	Maulder, Bradshaw [40]	National and regional level sprinters 0.20 ± 0.02 s				
	Mero, Kuitunen [19]	Block angle 40° 0.185 ± 0.020 s	Block angle 65° 0.197 ± 0.019 s			
First flight time (ms) (s)	Ciacci, Merni [38]	World-class 0.045 ± 0.025 s	Elite 0.064 ± 0.009 s	World-class 0.045 ± 0.025 s	Elite 0.085 ± 0.011 s	
		Independent of category 0.056 ± 0.019 s		Independent of category 0.069 ± 0.027 s		
	Bezodis, Salo [3]	World-class to university sprinters 0.073 ± 0.022 s				
	Rabita, Dorel [22]	Elite 81 ± 13 ms	Sub-elite 70 ± 25 ms			

Table A4. *Cont.*

First and Second Steps Kinematics	Study	Male		Female	Mixed		
	Slawinski, Bonnefoy [7]	Elite 0.093 ± 0.009 s	Well-trained 0.087 ± 0.021 s				
	Maulder, Bradshaw [40]	National and regional level sprinters 0.07 ± 0.01 s					
Horizontal CM position—first step touchdown (cm) [(a)]	Slawinski, Bonnefoy [7]	Elite 68.5 ± 4.7	Well-trained 58.0 ± 8.1 *				
Normalized first step touchdown distance [(b)]	Bezodis, Salo [3]	World-class to university sprinters −0.20 ± 0.07					
Horizontal CM position—first step takeoff (cm) [(a)]	Slawinski, Bonnefoy [7]	Elite 137.1 ± 9.0	Well-trained 120.8 ± 8.7 *				
First step resultant velocity (m·s^{-1})	Debaere, Vanwanseele [15]				Adult sprinters 4.34 ± 0.25	U18 sprinters 4.06 ± 0.24 #	U16 sprinters 4.01 ± 0.25 #
	Slawinski, Dumas [8]				Bunched start 3.81 ± 0.18	Medium start 3.85 ± 0.16	Elongated start 3.90 ± 0.15
	Slawinski, Bonnefoy [7]	Elite 4.69 ± 0.15	Well-trained 4.42 ± 0.11 *				
	Čoh, Jost [13]	Slovene national sprinters 4.48 ± 0.29		Slovene national sprinters 4.29 ± 0.18			
	Mero [18]	Trained sprinters 4.65 ± 0.28					
First step horizontal velocity (touchdown) (m·s^{-1})	Sandamas, Gutierrez-Farewik [24]				Skating condition 3.10 ± 0.16		Narrow condition 3.08 ± 0.16

Table A4. *Cont.*

First and Second Steps Kinematics	Study	Male			Female	Mixed	
First step horizontal velocity (takeoff) (m·s^{-1})	Graham-Smith, Colyer [39]	Seniors 4.60 ± 0.23		Juniors 4.39 ± 0.21			
	Sandamas, Gutierrez-Farewik [24]					Skating condition 4.37 ± 0.18	Narrow condition 4.32 ± 0.15 *
	Debaere, Delecluse [6]					Elite sprinters 4.28 ± 0.27	
	Čoh, Jost [13]	Slovene national sprinters 4.47 ± 0.29			Slovene national sprinters 4.25 ± 0.18 *		
First step change in horizontal velocity (m·s^{-1})	Werkhausen, Willwacher [43]				Germany national sprinters 1.09 ± 0.06		
	Aeles, Jonkers [9]					Adult sprinters 0.82 ± 0.39	Young sprinters 1.09 ± 0.25 *
First step vertical velocity (m·s^{-1})	Graham-Smith, Colyer [39]	Seniors 0.46 ± 0.15		Juniors 0.54 ± 0.10			
	Chen, Wu [37]	Bunched start 0.27 ± 0.12	Medium start 0.28 ± 0.10	Elongated start 0.39 ± 0.13 *			
	Debaere, Delecluse [6]					Elite sprinters 0.67 ± 0.12	
	Slawinski, Bonnefoy [7]	Elite 0.35 ± 0.03		Well-trained 0.42 ± 0.09			
	Čoh, Jost [13]	Slovene national sprinters 0.37 ± 0.19 *			Slovene national sprinters 0.52 ± 0.10 *		

Table A4. *Cont.*

First and Second Steps Kinematics	Study	Male	Female	Mixed
First step CM projection angle (°) (c)	Graham-Smith, Colyer [39]	Seniors 5.7 ± 1.9; Juniors 7.1 ± 1.4 §§		
First step takeoff angle (°) (d)	Maulder, Bradshaw [40]	National and regional level sprinters 43 ± 2		
Trunk angle at touchdown—first step (°) (e)	Chen, Wu [37]	Bunched start 27.2 ± 5.4; Medium start 30.9 ± 4.1 *; Elongated start 29.9 ± 4.7		
	Maulder, Bradshaw [40]	National and regional level sprinters 32 ± 8		
Hip angle at touchdown—first step (°)	Bezodis, Salo [3]	World-class to university sprinters 95 ± 9 (f)		
	Debaere, Delecluse [6]			Elite sprinters 121.2 ± 11.3 (g)
Knee angle at touchdown—first step (°)	Bezodis, Salo [3]	World-class to university sprinters 101 ± 7 (h)		
	Debaere, Delecluse [6]			Elite sprinters 111.6 ± 9.1 (h)
Ankle angle at touchdown—first step (°)	Bezodis, Salo [3]	World-class to university sprinters 96 ± 7 (i)		
	Debaere, Delecluse [6]			Elite sprinters 70.6 ± 5.8 (i)
Maximal plantar-flexion—first step (°)	Debaere, Delecluse [6]			Elite sprinters 111.3 ± 11.2 (i)
Knee angle at takeoff—first step (°)	Debaere, Delecluse [6]			Elite sprinters 165.2 ± 20.6 (h)
Hip angle at takeoff—first step (°)	Debaere, Delecluse [6]			Elite sprinters 180.6 ± 20.9 (g)

Table A4. *Cont.*

First and Second Steps Kinematics	Study	Male			Female	Mixed	
Trunk angle at takeoff—first step (°) [(c)]	Chen, Wu [37]	Bunched start 30.5 ± 6.9	Medium start 31.9 ± 6.2	Elongated start 32.5 ± 6.0			
	Maulder, Bradshaw [40]	National and regional level sprinters 32 ± 8					
Hip ROM extension—first step (°)	Aeles, Jonkers [9]					Adult sprinters 64.50 ± 13.08	Young sprinters 69.45 ± 9.53
Knee ROM extension—first step (°)						Adult sprinters 60.09 ± 7.24	Young sprinters 58.24 ± 6.10
Ankle ROM dorsiflexion—first step (°)	Werkhausen, Willwacher [43]				Germany national sprinters 17 ± 3		
Ankle ROM plantar flexion—first step (°)					Germany national sprinters 45 ± 6		
	Aeles, Jonkers [9]					Adult sprinters 59.05 ± 7.40	Young sprinters 50.96 ± 9.39 *
Peak foot linear velocity (from the start to the first step) ($m \cdot s^{-1}$)	Chen, Wu [37]	Bunched start 6.31 ± 0.48	Medium start 6.66 ± 0.55 *	Elongated start 6.79 ± 0.99			

Table A4. *Cont.*

First and Second Steps Kinematics	Study	Male			Female		Mixed		
Second Step									
Second step length (m) (cm)	Cavedon, Sandri [12]						Usual condition 1.15 ± 0.14 nor. to leg length		Anthropometric condition 1.19 ± 0.12 nor. to leg length
	Ciacci, Merni [38]	World-class 1.143 ± 0.105 m		Elite 1.057 ± 0.150 m	World-class 1.098 ± 0.104 m	Elite 1.078 ± 0.181 m			
		Independent of category 1.091 ± 0.135 m			Independent of category 1.086 ± 0.148 m				
	Coh, Peharec [5]	Faster sprinters 1.03 ± 0.12 m		Slower sprinters 0.98 ± 0.33 m					
	Debaere, Vanwanseele [15]						Adult sprinters 1.09 ± 0.06 m	U18 sprinters 1.01 ± 0.08 m #	U16 sprinters 1.02 ± 0.08 m #
	Chen, Wu [37]	Bunched start 2.02 ± 0.18 m	Medium start 2.08 ± 0.18 m	Elongated start 2.10 ± 0.19 m					
	Milanese, Bertucco [41]						Rear knee angle @ 90° 1.96 ± 0.17 m	@ 115° 1.94 ± 0.12 m	@ 135° 1.93 ± 0.17 m
	Aerenhouts, Delecluse [1]	Elite Seniors 148 ± 25 cm		Elite Juniors 130 ± 20 cm	Elite Seniors 130 ± 14 cm	Elite Juniors 127 ± 11 cm			
	Slawinski, Bonnefoy [7]	Elite 106.6 ± 5.9 cm		Well-trained 105.3 ± 6.3 cm					

Table A4. *Cont.*

First and Second Steps Kinematics	Study	Male		Female		Mixed
	Maulder, Bradshaw [40]	National and regional level sprinters 1.08 ± 0.13 m				
	Čoh, Jost [13]	Slovene national sprinters 1.30 ± 0.51 m		Slovene national sprinters 1.06 ± 0.60 m		
Second step contact time (ms) (s)	Werkhausen, Willwacher [43]			Germany national sprinters 0.17 ±0.02 s		
	Graham-Smith, Colyer [39]	Seniors 0.173 ± 0.018 s	Juniors 0.173 ± 0.020 s			
	Ciacci, Merni [38]	World-class 0.170 ± 0.026 s	Elite 0.148 ± 0.008 s	World-class 0.180 ± 0.016 s	Elite 0.148 ± 0.013 s	
		Independent of category 0.157 ± 0.020 s		Independent of category 0.161 ± 0.021 s		
	Coh, Peharec [5]	Faster sprinters 157 ± 15.42 ms	Slower sprinters 149 ± 18.87 ms			
	Aerenhouts, Delecluse [1]	Elite Seniors 173 ± 28 ms	Elite Juniors 169 ± 20 ms	Elite Seniors 173 ± 19 ms	Elite Juniors 283 ± 23 ms	
	Slawinski, Bonnefoy [7]	Elite 0.138 ± 0.031 s	Well-trained 0.145 ± 0.016 s			
	Maulder, Bradshaw [40]	National and regional level sprinters 0.18 ± 0.03 s				

Table A4. *Cont.*

First and Second Steps Kinematics	Study	Male		Female	Mixed
Horizontal CM position—second step touchdown (cm) [(a)]	Slawinski, Bonnefoy [7]	Elite 168.2 ± 11.3	Well-trained 156.9 ± 12.4		
Horizontal CM position—second step takeoff (cm) [(a)]	Slawinski, Bonnefoy [7]	Elite 243.6 ± 13.9	Well-trained 224.9 ± 12.0 *		
Second step velocity $(m \cdot s^{-1})$	Slawinski, Bonnefoy [7]	Elite 5.50 ± 0.26	Well-trained 5.25 ± 0.13 *		
	Čoh, Jost [13]	Slovene national sprinters 5.40 ± 0.24		Slovene national sprinters 5.01 ± 0.29 **	
Second step horizontal velocity $(m \cdot s^{-1})$	Graham-Smith, Colyer [39]	Seniors 5.48 ± 0.26	Juniors 5.27 ± 0.26		
	Debaere, Delecluse [6]				Elite sprinters 5.19 ± 0.30
	Čoh, Jost [13]	Slovene national sprinters 5.38 ± 0.24		Slovene national sprinters 4.99 ± 0.29 **	
Second step change in horizontal velocity $(m \cdot s^{-1})$	Werkhausen, Willwacher [43]			Germany national sprinters 1.12 ± 0.07	
Second step vertical velocity $(m \cdot s^{-1})$	Graham-Smith, Colyer [39]	Seniors 0.54 ± 0.10	Juniors 0.62 ± 0.12		
	Debaere, Delecluse [6]				Elite sprinters 0.70 ± 0.17
	Slawinski, Bonnefoy [7]	Elite 0.35 ± 0.05	Well-trained 0.45 ± 0.07 *		
	Čoh, Jost [13]	Slovene national sprinters 0.45 ± 0.18		Slovene national sprinters 0.50 ± 0.10	

Table A4. *Cont.*

First and Second Steps Kinematics	Study	Male			Female	Mixed
Second step CM projection angle (°) [a]	Graham-Smith, Colyer [39]	Seniors 4.9 ± 1.3		Juniors 6.8 ± 1.4 §§		
Second step take off angle (°) [b]	Maulder, Bradshaw [40]	National and regional level sprinters 46 ± 2				
Ankle angle at touch-down—second step (°)	Debaere, Delecluse [6]					Elite sprinters 72.4 ± 7.1 [i]
Ankle ROM dorsiflexion—second step (°)	Werkhausen, Willwacher [43]				Germany national sprinters 18 ± 3	
Ankle ROM plantarflexion—second step (°)	Werkhausen, Willwacher [43]				Germany national sprinters 44 ± 5	
Maximal plantarflexion—second step (°)	Debaere, Delecluse [6]					Elite sprinters 107.1 ± 15.0 [i]
Knee angle at touch-down—second step (°)	Debaere, Delecluse [6]					Elite sprinters 115.6 ± 6.2 [h]
Knee angle at takeoff—second step (°)	Debaere, Delecluse [6]					Elite sprinters 163.6 ± 17.7 [h]
Hip angle at touch-down—second step (°)	Debaere, Delecluse [6]					Elite sprinters 124.48 ± 11.3 [g]
Hip angle at takeoff—second step (°)	Debaere, Delecluse [6]					Elite sprinters 181.1 + 20.0 [g]
Trunk angle at touch-down—second step (°) [c]	Chen, Wu [37]	Bunched start 33.4 ± 7.0	Medium start 34.9 ± 5.9	Elongated start 36.7 ± 5.9 *		
	Maulder, Bradshaw [40]	National and regional level sprinters 44 ± 8				

Table A4. *Cont.*

First and Second Steps Kinematics	Study	Male			Female	Mixed	
Peak foot linear velocity (from start to second step) $(m \cdot s^{-1})$	Chen, Wu [37]	Bunched start 8.51 ± 0.58	Medium start 8.72 ± 0.40	Elongated start 8.68 ± 0.61 *			
First and Second Steps							
Step frequency First to second step (Hz)	Maulder, Bradshaw [40]	National and regional level sprinters 4.2 ± 0.3					
Minimal step frequency (Hz) [j]	Rabita, Dorel [22]	Elite 3.94 ± 0.44		Sub-elite 3.90 ± 0.44			
Maximal step frequency (Hz) [j]	Rabita, Dorel [22]	Elite 4.95 ± 0.12		Sub-elite 4.80 ± 0.30 §§			
Maximal CM horizontal acceleration $(m \cdot s^{-2})$	Debaere, Delecluse [14]					First stance 0.36 ± 0.05	Second stance 0.23 ± 0.04 ***
Maximal CM vertical acceleration $(m \cdot s^{-2})$	Debaere, Delecluse [14]					First stance 0.28 ± 0.08	Second stance 0.25 ± 0.05 *
Net induced acceleration $(m \cdot s^{-2})$	Debaere, Delecluse [14]					First stance 501.4 ± 164.4 (33.2% horizontal/66.8% vertical)	Second stance 367.7 ± 36.7 *** (36.3% horizontal/63.7% vertical)

CM—center of mass; [a] horizontal distance relative to stat line: [b] represents the horizontal distance (divided by leg length) between the CM and the stance leg metatarsal-phalangeal joint (negative value means that foot is behind the CM; [c] center of mass projection angle is calculated as the resultant direction from the horizontal and vertical block exit velocities of the center of mass; [d] the angle, measured relative to the horizontal, between the line passing through the most front part of the contact foot and the CG during takeoff; [e] the angle, measured relative to the horizontal, between the line passing through the hip and shoulder (trunk segment) of the support leg; [f] internal angle between the thigh and trunk in flexion/extension plane; [g] relative angle between the pelvis and the thigh according the Biomechanical Convention [53]; [h] relative angle between the thigh and the shank according the Medical Convention [53]; [i] relative angle between the shank and the foot according the Biomechanical Convention [53]; [j] data referring to the values recorded in the entire acceleration phase (0–40 m) excluding the block phase.

Table A5. Summary of the kinetic variables in the "first two steps". Data are the magnitude of the mean ± SD presented in the reviewed studies. Groups are male and mixed (when authors joined data without discriminating by sex) sprinters. Studies are listed, in each variable, in reverse chronological order, followed by alphabetically for studies published in the same year. Data, terms, conditions, and sprinters' performance levels are presented according to the original authors. Statistical differences between groups are marked with asterisks (* $p < 0.05$; *** $p < 0.001$; # significant different from adults; þ different from descending phase—$p < 0.05$; § small effect size [0.2–0.6] of 90% confidence intervals; † significantly greater compared with either leg in the block phase).

First and Second Steps Kinetics			Male		Mixed	
First Step						
GRFs	Relative resultant GRF (N·kg^{-1})	Sandamas, Gutierrez-Farewik [24]			Skating condition 1.51 ± 0.10 BW [(a)]	Narrow condition 1.49 ± 0.12 BW [(a)]
		Otsuka, Shim [42]	Well-trained sprinters 14.93 ± 0.79	Trained sprinters 14.62 ± 1.44		
	Maximal horizontal force (N)	Aeles, Jonkers [9]			Adult sprinters 488.47 ± 268.16	Young sprinters 552.91 ± 147.40
	Average horizontal force (N)	Aeles, Jonkers [9]			Adult sprinters 289.63 ± 163.32	Young sprinters 333.15 ± 94.26
	Relative maximal horizontal force (N·kg^{-1})	Aeles, Jonkers [9]			Adult sprinters 7.09 ± 3.28	Young sprinters 9.02 ± 2.00 *
	Relative average horizontal force (N·kg^{-1})	Sandamas, Gutierrez-Farewik [24]			Skating condition 0.64 ± 0.06 BW [(a)]	Narrow condition 0.63 ± 0.04 BW [(a)]
		Aeles, Jonkers [9]			Adult sprinters 4.21 ± 2.04	Young sprinters 5.43 ± 1.28 *
		Otsuka, Shim [42]	Well-trained sprinters 5.87 ± 0.35	Trained sprinters 5.48 ± 0.77		
	Maximal ratio of hori-zontal to total GRF (%)	Aeles, Jonkers [9]			Adult sprinters 61.31 ± 20.75	Young sprinters 78.00 ± 12.37 *
	Mean ratio of horizontal force to total GRF (%)	Aeles, Jonkers [9]			Adult sprinters 28.49 ± 12.16	Young sprinters 37.04 ± 6.37 *

Table A5. *Cont.*

First and Second Steps Kinetics			Male		Mixed	
	Relative average vertical force ($N{\cdot}kg^{-1}$)	Otsuka, Shim [42]	Well-trained sprinters 13.59 ± 0.82	Trained sprinters 13.43 ± 1.35		
Power	Relative average hori-zontal external power—first step ($W{\cdot}kg^{-1}$)	Graham-Smith, Colyer [39]	Seniors 25.1 ± 3.6	Juniors 23.1 ± 6 §		
Impulses	Absolute Impulse (N·s)	Slawinski, Bonnefoy [7]	Elite 104.8 ± 16.5	Well-trained 78.6 ± 6.3 *		
	Net normalized horizontal impulse ($m{\cdot}s^{-1}$)	Sandamas, Gutierrez-Farewik [24]			Skating condition 1.29 ± 0.06	Narrow condition 1.26 ± 0.04
	Normalized horizontal braking impulse ($m{\cdot}s^{-1}$)				Skating condition 0.04 ± 0.04	Narrow condition 0.03 ± 0.02
	Normalized horizontal propulsive impulse ($m{\cdot}s^{-1}$)				Skating condition 1.33 ± 0.06	Narrow condition 1.29 ± 0.05 *
	Normalized vertical impulse ($m{\cdot}s^{-1}$)				Skating condition 0.71 ± 0.18	Narrow condition 0.71 ± 0.28
	Normalized mediolateral impulse ($m{\cdot}s^{-1}$)				Skating condition 0.33 ± 0.10	Narrow condition 0.17 ± 0.10 *

Table A5. *Cont.*

First and Second Steps Kinetics			Male	Mixed		
Joint moments	Relative ankle joint moment (Plantar Flexion) ($N \cdot m \cdot kg^{-1}$)	Aeles, Jonkers [9]		Adult sprinters 0.19 ± 0.05		Young sprinters 0.22 ± 0.07
		Debaere, Vanwanseele [15]		Adult sprinters 0.19 ± 0.07	U18 sprinters 0.24 ± 0.06	U16 sprinters 0.21 ± 0.12
		Debaere, Delecluse [6]			Elite sprinters 0.20 ± 0.03 $N \cdot m \cdot N^{-1}$	
	Relative knee joint moment (extension) ($N \cdot m \cdot kg^{-1}$)	Aeles, Jonkers [9]		Adult sprinters 0.29 ± 0.10		Young sprinters 0.21 ± 0.09 *
		Debaere, Vanwanseele [15]		Adult sprinters 0.30 ± 0.11	U18 sprinters 0.18 ± 0.08 #	U16 sprinters 0.18 ± 0.09 #
		Debaere, Delecluse [6]			Elite sprinters 0.20 ± 0.04 $N \cdot m \cdot N^{-1}$	
	Relative hip joint moment (extension) ($N \cdot m \cdot kg^{-1}$)	Aeles, Jonkers [9]		Adult sprinters 0.24 ± 0.08		Young sprinters 0.22 ± 0.07
		Debaere, Vanwanseele [15]		Adult sprinters 0.50 ± 0.22	U18 sprinters 0.34 ± 0.10 #	U16 sprinters 0.42 ± 0.09
		Debaere, Delecluse [6]			Elite sprinters 0.33 ± 0.15 $N \cdot m \cdot N^{-1}$	
	Relative hip joint moment (flexion) ($N \cdot m \cdot kg^{-1}$)	Aeles, Jonkers [9]		Adult sprinters 0.41 ± 0.22		Young sprinters 0.26 ± 0.12 *
		Debaere, Delecluse [6]			Elite sprinters 0.42 ± 0.16 $N \cdot m \cdot N^{-1}$	
	Normalized peak ankle joint moment (extension)	Brazil, Exell [4]	Athletic sprinters 0.388 ± 0.035 (b) †			
	Normalized peak knee joint moment (extension)		Athletic sprinters 0.242 ± 0.068 (b)			
	Normalized peak hip joint moment (extension)		Athletic sprinters 0.330 ± 0.071 (b)			

Table A5. *Cont.*

First and Second Steps Kinetics			Male		Mixed		
	Joint moments contributi-on to body propulsion (%)	Debaere, Delecluse [14]			Hip joint 10.3	Knee joint 9.6	Ankle joint 67.1 ***
	Joint moments contribu—tion to body lift (%)	Debaere, Delecluse [14]			Hip joint 12.3	Knee joint 38.1	Ankle joint 49.6
	Ankle joint stiffness as—cending phase (N·m/°) (c)	Aeles, Jonkers [9]			Adult sprinters 6.64 ± 2.01		Young sprinters 7.35 ± 3.12 þ
	Ankle joint stiffness des—cending phase (N·m/°) (d)	Aeles, Jonkers [9]			Adult sprinters 2.27 ± 0.62		Young sprinters 2.85 ± 1.23 þ
Joint powers	Relative ankle peak power (W·N^{-1})	Brazil, Exell [4]	Athletic sprinters 1.093 ± 0.069 (b) †				
		Debaere, Vanwanseele [15]			Adult sprinters 1.79 ± 0.96	U18 sprinters 2.30 ± 1.02	U16 sprinters 2.19 ± 1.46
	Relative knee peak power (W·N^{-1})	Brazil, Exell [4]	Athletic sprinters 0.468 ± 0.145 (b)				
		Debaere, Vanwanseele [15]			Adult sprinters 3.03 ± 1.24	U18 sprinters 1.31 ± 0.66 #	U16 sprinters 1.12 ± 1.2 #
	Relative hip peak power (W·N^{-1})	Brazil, Exell [4]	Athletic sprinters 0.908 ± 0.185 (b) †				
		Debaere, Vanwanseele [15]			Adult sprinters 3.79 ± 0.95	U18 sprinters 4.56 ± 1.42	U16 sprinters 4.33 ± 0.96
	Relative average hori-zontal external power (W·kg^{-1})	Graham-Smith, Colyer [39]	Seniors 25.1 ± 3.6	Juniors 23.1 ± 6 §			

Table A5. *Cont.*

First and Second Steps Kinetics			Male		Mixed		
	Joint moments contributi-on to body propulsion (%)	Debaere, Delecluse [14]			Hip joint 10.3	Knee joint 9.6	Ankle joint 67.1 ***
	Joint moments contributi-on to body lift (%)	Debaere, Delecluse [14]			Hip joint 12.3	Knee joint 38.1	Ankle joint 49.6 ***
Joint work	Ankle negative work ($J \cdot kg^{-1}$)	Werkhausen, Willwacher [43]				Germany national sprinters -0.32 ± 0.14 (e)	
Joint work	Ankle positive work ($J \cdot kg^{-1}$)	Werkhausen, Willwacher [43]				Germany national sprinters 1.58 ± 0.17 (e)	
Second Step							
GRFs	Relative resultant GRF ($N \cdot kg^{-1}$)	Otsuka, Shim [42]	Well-trained sprinters 14.93 ± 0.79	Trained sprinters 14.62 ± 1.44			
GRFs	Relative average horizontal force ($N \cdot kg^{-1}$)	Otsuka, Shim [42]	Well-trained sprinters 4.83 ± 0.70	Trained sprinters 4.79 ± 0.47			
GRFs	Relative average vertical force ($N \cdot kg^{-1}$)	Otsuka, Shim [42]	Well-trained sprinters 13.22 ± 1.15	Trained sprinters 13.72 ± 1.18			
Power	Relative average horizontal external power—second step ($W \cdot kg^{-1}$)	Graham-Smith, Colyer [39]	Seniors 26.7 ± 3.6	Juniors 24.9 ± 4.5 §			
Impulses	Absolute impulse ($N \cdot s$)	Slawinski, Bonnefoy [7]	Elite 75.0 ± 15.8	Well-trained 55.9 ± 9.4 *			

Table A5. *Cont.*

First and Second Steps Kinetics			Male	Mixed	
Joint moments	Relative ankle joint moment (plantar flexion) ($N{\cdot}m{\cdot}N^{-1}$)	Debaere, Delecluse [6]		Elite sprinters 0.23 ± 0.05	
	Relative knee joint moment (extension)($N{\cdot}m{\cdot}N^{-1}$)			Elite sprinters 0.10 ± 0.04	
	Relative hip joint moment (extension) ($N{\cdot}m{\cdot}N^{-1}$)			Elite sprinters 0.43 ± 0.01	
	Relative hip joint moment (flexion) ($N{\cdot}m{\cdot}N^{-1}$)			Elite sprinters 0.20 ± 0.06	
	Joint moments contributi-on to body propulsion (%)	Debaere, Delecluse [14]	Hip joint 0	Knee joint 7.1	Ankle joint 92.9 ***
	Joint moments contributi-on to body lift (%)		Hip joint 0	Knee joint 23.8	Ankle joint 72.6 ***

CM—center of mass; GRF—ground reaction forces; (a) normalized to body weight; (b) joint data normalized to the mass and the leg length of the sprinter; (c) ankle joint stiffness was calculated during the increase in ankle joint moment; (d) ankle joint stiffness was calculated during the decrease in ankle joint moment; (e) only elite female data.

References

1. Aerenhouts, D.; Delecluse, C.; Hagman, F.; Taeymans, J.; Debaere, S.; Van Gheluwe, B.; Clarys, P. Comparison of anthropometric characteristics and sprint start performance between elite adolescent and adult sprint athletes. *Eur. J. Sport Sci.* **2012**, *12*, 9–15. [CrossRef]
2. Bezodis, N.E.; Salo, A.I.; Trewartha, G. Choice of sprint start performance measure affects the performance-based ranking within a group of sprinters: Which is the most appropriate measure? *Sports Biomech.* **2010**, *9*, 258–269. [CrossRef]
3. Bezodis, N.E.; Salo, A.I.; Trewartha, G. Relationships between lower-limb kinematics and block phase performance in a cross section of sprinters. *Eur. J. Sport Sci.* **2015**, *15*, 118–124. [CrossRef]
4. Brazil, A.; Exell, T.; Wilson, C.; Willwacher, S.; Bezodis, I.; Irwin, G. Lower limb joint kinetics in the starting blocks and first stance in athletic sprinting. *J. Sports Sci.* **2017**, *35*, 1629–1635. [CrossRef] [PubMed]
5. Coh, M.; Peharec, S.; Bacic, P.; Mackala, K. Biomechanical Differences in the Sprint Start Between Faster and Slower High-Level Sprinters. *J. Hum. Kinet.* **2017**, *56*, 29–38. [CrossRef] [PubMed]
6. Debaere, S.; Delecluse, C.; Aerenhouts, D.; Hagman, F.; Jonkers, I. From block clearance to sprint running: Characteristics underlying an effective transition. *J. Sports Sci.* **2013**, *31*, 137–149. [CrossRef] [PubMed]
7. Slawinski, J.; Bonnefoy, A.; Leveque, J.M.; Ontanon, G.; Riquet, A.; Dumas, R.; Chèze, L. Kinematic and kinetic comparisons of elite and well-trained sprinters during sprint start. *J. Strength Cond. Res.* **2010**, *24*, 896–905. [CrossRef]
8. Slawinski, J.; Dumas, R.; Cheze, L.; Ontanon, G.; Miller, C.; Mazure-Bonnefoy, A. 3D kinematic of bunched, medium and elongated sprint start. *Int. J. Sports Med.* **2012**, *33*, 555–560. [CrossRef] [PubMed]
9. Aeles, J.; Jonkers, I.; Debaere, S.; Delecluse, C.; Vanwanseele, B. Muscle-tendon unit length changes differ between young and adult sprinters in the first stance phase of sprint running. *R. Soc. Open Sci.* **2018**, *5*, 180332. [CrossRef] [PubMed]
10. Bezodis, N.; Walton, S.; Nagahara, R. Understanding the track and field sprint start through a functional analysis of the external force features which contribute to higher levels of block phase performance. *J. Sports Sci.* **2019**, *37*, 560–567. [CrossRef]
11. Brazil, A.; Exell, T.; Wilson, C.; Willwacher, S.; Bezodis, I.N.; Irwin, G. Joint kinetic determinants of starting block performance in athletic sprinting. *J. Sports Sci.* **2018**, *36*, 1656–1662. [CrossRef] [PubMed]
12. Cavedon, V.; Sandri, M.; Pirlo, M.; Petrone, N.; Zancanaro, C.; Milanese, C. Anthropometry-driven block setting improves starting block performance in sprinters. *PLoS ONE* **2019**, *14*, e0213979. [CrossRef] [PubMed]
13. Čoh, M.; Jost, B.; Škof, B.; Tomazin, K.; Dolenec, A. Kinematic and kinetic parameters of the sprint start and start acceleration model of top sprinters. *Gymnica* **1998**, *28*, 33–42.
14. Debaere, S.; Delecluse, C.; Aerenhouts, D.; Hagman, F.; Jonkers, I. Control of propulsion and body lift during the first two stances of sprint running: A simulation study. *J. Sports Sci.* **2015**, *33*, 2016–2024. [CrossRef] [PubMed]
15. Debaere, S.; Vanwanseele, B.; Delecluse, C.; Aerenhouts, D.; Hagman, F.; Jonkers, I. Joint power generation differentiates young and adult sprinters during the transition from block start into acceleration: A cross-sectional study. *Sports Biomech.* **2017**, *16*, 452–462. [CrossRef] [PubMed]
16. Fortier, S.; Basset, F.A.; Mbourou, G.A.; Faverial, J.; Teasdale, N. Starting Block Performance in Sprinters: A Statistical Method for Identifying Discriminative Parameters of the Performance and an Analysis of the Effect of Providing Feedback over a 6-Week Period. *J. Sports Sci. Med.* **2005**, *4*, 134–143. [PubMed]
17. Gutierrez-Davilla, M.; Dapena, J.; Campos, J. The effect of muscular pre-tensing on the sprint start. *J. Appl. Biomech.* **2006**, *22*, 194–201. [CrossRef] [PubMed]
18. Mero, A. Force-Time Characteristics and Running Velocity of Male Sprinters during the Acceleration Phase of Sprinting. *Res. Q. Exerc. Sport* **1988**, *59*, 94–98. [CrossRef]
19. Mero, A.; Kuitunen, S.; Harland, M.; Kyrolainen, H.; Komi, P.V. Effects of muscle-tendon length on joint moment and power during sprint starts. *J. Sports Sci.* **2006**, *24*, 165–173. [CrossRef] [PubMed]
20. Nagahara, R.; Ohshima, Y. The Location of the Center of Pressure on the Starting Block Is Related to Sprint Start Performance. *Front. Sports Act Living* **2019**, *1*, 21. [CrossRef] [PubMed]
21. Otsuka, M.; Kurihara, T.; Isaka, T. Effect of a Wide Stance on Block Start Performance in Sprint Running. *PLoS ONE* **2015**, *10*, e0142230. [CrossRef] [PubMed]
22. Rabita, G.; Dorel, S.; Slawinski, J.; Saez-de-Villarreal, E.; Couturier, A.; Samozino, P.; Morin, J.-B. Sprint mechanics in world-class athletes: A new insight into the limits of human locomotion. *Scand. J. Med. Sci. Sports* **2015**, *25*, 583–594. [CrossRef] [PubMed]
23. Sado, N.; Yoshioka, S.; Fukashiro, S. Three-dimensional kinetic function of the lumbo-pelvic-hip complex during block start. *PLoS ONE* **2020**, *15*, e0230145. [CrossRef] [PubMed]
24. Sandamas, P.; Gutierrez-Farewik, E.M.; Arndt, A. The effect of a reduced first step width on starting block and first stance power and impulses during an athletic sprint start. *J. Sports Sci.* **2019**, *37*, 1046–1054. [CrossRef]
25. Schrodter, E.; Bruggemann, G.P.; Willwacher, S. Is Soleus Muscle-Tendon-Unit Behavior Related to Ground-Force Application During the Sprint Start? *Int. J. Sports Physiol. Perform.* **2017**, *12*, 448–454. [CrossRef]
26. Bezodis, N.E.; Willwacher, S.; Salo, A.I.T. The Biomechanics of the Track and Field Sprint Start: A Narrative Review. *Sports Med.* **2019**, *49*, 1345–1364. [CrossRef]
27. Harland, M.J.; Steele, J.R. Biomechanics of the sprint start. *Sports Med.* **1997**, *23*, 11–20. [CrossRef]

28. Page, M.J.; McKenzie, J.E.; Bossuyt, P.M.; Boutron, I.; Hoffmann, T.C.; Mulrow, C.D.; Shamseer, L.; Tetzlaff, J.M.; Akl, E.A.; Brennan, S.E.; et al. The PRISMA 2020 statement: An updated guideline for reporting systematic reviews. *BMJ* **2021**, *372*, 105906. [CrossRef]
29. Von Elm, E.; Altman, D.G.; Egger, M.; Pocock, S.J.; Gotzsche, P.C.; Vandenbroucke, J.P.; STROBE Initiative. The Strengthening the Reporting of Observational Studies in Epidemiology (STROBE) Statement: Guidelines for reporting observational studies. *Int. J. Surg.* **2014**, *12*, 1495–1499. [CrossRef]
30. Da Costa, B.R.; Cevallos, M.; Altman, D.G.; Rutjes, A.W.; Egger, M. Uses and misuses of the STROBE statement: Bibliographic study. *BMJ Open* **2011**, *1*, e000048. [CrossRef]
31. Sim, J.; Wright, C.C. The kappa statistic in reliability studies: Use, interpretation, and sample size requirements. *Phys. Ther.* **2005**, *85*, 257–268. [CrossRef] [PubMed]
32. Landis, J.R.; Koch, G.G. The measurement of observer agreement for categorical data. *Biometrics* **1977**, *33*, 159–174. [CrossRef]
33. Colyer, S.; Graham-Smith, P.; Salo, A. Associations between ground reaction force waveforms and sprint start performance. *Int. J. Sports Sci. Coach.* **2019**, *14*, 658–666. [CrossRef]
34. Guissard, N.; Duchateau, J.; Hainaut, K. EMG and mechanical changes during sprint starts at different front block obliquities. *Med. Sci. Sports Exerc.* **1992**, *24*, 1257–1263. [CrossRef] [PubMed]
35. Nagahara, R.; Gleadhill, S.; Ohshima, Y. Improvement in sprint start performance by modulating an initial loading location on the starting blocks. *J. Sports Sci.* **2020**, *38*, 2437–2445. [CrossRef] [PubMed]
36. Slawinski, J.; Bonnefoy, A.; Ontanon, G.; Leveque, J.M.; Miller, C.; Riquet, A.; Chèze, L.; Dumas, R. Segment-interaction in sprint start: Analysis of 3D angular velocity and kinetic energy in elite sprinters. *J. Biomech.* **2010**, *43*, 1494–1502. [CrossRef] [PubMed]
37. Chen, Y.; Wu, K.Y.; Tsai, Y.J.; Yang, W.T.; Chang, J.H. The kinematic differences of three types of crouched positions during a sprint start. *J. Mech. Med. Biol.* **2016**, *16*, 1650099. [CrossRef]
38. Ciacci, S.; Merni, F.; Bartolomei, S.; Di Michele, R. Sprint start kinematics during competition in elite and world-class male and female sprinters. *J. Sports Sci.* **2017**, *35*, 1270–1278. [CrossRef] [PubMed]
39. Graham-Smith, P.; Colyer, S.L.; Salo, A. Differences in ground reaction waveforms between elite senior and junior academy sprinters during the block phase and first two steps. *Int. J. Sports Sci. Coach.* **2020**, *15*, 418–427. [CrossRef]
40. Maulder, P.S.; Bradshaw, E.J.; Keogh, J.W. Kinematic alterations due to different loading schemes in early acceleration sprint performance from starting blocks. *J. Strength Cond. Res.* **2008**, *22*, 1992–2002. [CrossRef]
41. Milanese, C.; Bertucco, M.; Zancanaro, C. The effects of three different rear knee angles on kinematics in the sprint start. *Biol. Sport* **2014**, *31*, 209–215. [CrossRef] [PubMed]
42. Otsuka, M.; Shim, J.K.; Kurihara, T.; Yoshioka, S.; Nokata, M.; Isaka, T. Effect of expertise on 3D force application during the starting block phase and subsequent steps in sprint running. *J. Appl. Biomech.* **2014**, *30*, 390–400. [CrossRef]
43. Werkhausen, A.; Willwacher, S.; Albracht, K. Medial gastrocnemius muscle fascicles shorten throughout stance during sprint acceleration. *Scand. J. Med. Sci. Sports* **2021**, *31*, 1471–1480. [CrossRef] [PubMed]
44. Nagahara, R.; Kanehisa, H.; Fukunaga, T. Ground reaction force across the transition during sprint acceleration. *Scand. J. Med. Sci. Sports* **2020**, *30*, 450–461. [CrossRef] [PubMed]
45. Mero, A.; Komi, P.V.; Gregor, R.J. Biomechanics of sprint running. A review. *Sports Med.* **1992**, *13*, 376–392. [CrossRef] [PubMed]
46. Milloz, M.; Hayes, K.; Harrison, A.J. Sprint Start Regulation in Athletics: A Critical Review. *Sports Med.* **2021**, *51*, 21–31. [CrossRef] [PubMed]
47. Bezodis, N.E.; Trewartha, G.; Salo, A.I. Understanding the effect of touchdown distance and ankle joint kinematics on sprint acceleration performance through computer simulation. *Sports Biomech.* **2015**, *14*, 232–245. [CrossRef] [PubMed]
48. Colyer, S.L.; Nagahara, R.; Salo, A.I.T. Kinetic demands of sprinting shift across the acceleration phase: Novel analysis of entire force waveforms. *Scand. J. Med. Sci. Sports* **2018**, *28*, 1784–1792. [CrossRef]
49. Charalambous, L.; Irwin, G.; Bezodis, I.N.; Kerwin, D. Lower limb joint kinetics and ankle joint stiffness in the sprint start push-off. *J. Sports Sci.* **2012**, *30*, 1–9. [CrossRef]
50. McKay, A.K.A.; Stellingwerff, T.; Smith, E.S.; Martin, D.T.; Mujika, I.; Goosey-Tolfrey, V.L.; Sheppard, J.; Burke, L.M. Defining Training and Performance Caliber: A Participant Classification Framework. *Int. J. Sports Physiol. Perform.* **2022**, *17*, 317–331. [CrossRef] [PubMed]
51. Bissas, A.; Walker, J.; Paradisis, G.P.; Hanley, B.; Tucker, C.B.; Jongerius, N.; Thomas, A.; Merlino, S.; Vazel, P.; Girard, O. Asymmetry in sprinting: An insight into sub-10 and sub-11 s men and women sprinters. *Scand. J. Med. Sci. Sports* **2022**, *32*, 69–82. [CrossRef] [PubMed]
52. Nagahara, R.; Gleadhill, S. Catapult start likely improves sprint start performance. *Int. J. Sports Sci. Coach.* **2022**, *17*, 114–122. [CrossRef]
53. Robertson, D.G.E.; Caldwell, G.E.; Hamill, J.; Kamen, G.; Whittlesey, S.N. *Research Methods in Biomechanics*; Human Kinetics: Champaign, IL, USA, 2014.

International Journal of
Environmental Research and Public Health

MDPI

Article

Effect of Physical Training on Body Composition in Brazilian Military

Luis Alberto Gobbo [1], Raquel David Langer [2], Elisabetta Marini [3,*], Roberto Buffa [3], Juliano Henrique Borges [2], Mauro A. Pascoa [2], Vagner X. Cirolini [2], Gil Guerra-Júnior [2] and Ezequiel Moreira Gonçalves [2]

1 Skeletal Muscle Assessment Laboratory (LABSIM), School of Technology and Science, São Paulo State University (UNESP), Presidente Prudente 19060-900, SP, Brazil; luis.gobbo@unesp.br
2 Growth and Development Laboratory, Center for Investigation in Pediatrics (CIPED), School of Medical Sciences, University of Campinas (UNICAMP), Campinas 13083-887, SP, Brazil; raqueldlanger@gmail.com (R.D.L.); borgesedfisica@gmail.com (J.H.B.); pascoawaf@bol.com.br (M.A.P.); vxcirolini@hotmail.com (V.X.C.); gilguer@fcm.unicamp.br (G.G.-J.); emaildozeique@gmail.com (E.M.G.)
3 Department of Life and Environmental Sciences, University of Cagliari, 09042 Monserrato, Italy; rbuffa@unica.it
* Correspondence: emarini@unica.it; Tel.: +39-070-675-6607

Abstract: The military are selected on the basis of physical standards and are regularly involved in strong physical activities, also related to particular sports training. The aims of the study were to analyze the effect of a 7-month military training program on body composition variables and the suitability of specific 'bioelectrical impedance vector analysis' (spBIVA), compared to DXA, to detect the changes in body composition. A sample of 270 male Brazilian cadets (19.1 ± 1.1 years), composed of a group practicing military physical training routine only (MT = 155) and a group involved in a specific sport training (SMT = 115), were measured by body composition assessments (evaluated by means of DXA and spBIVA) at the beginning and the end of the military routine year. The effect of training on body composition was similar in SMT and MT groups, with an increase in LST. DXA and spBIVA were correlated, with specific resistance (Rsp) and reactance (Xcsp) positively related to fat mass (FM), FM%, LST, and lean soft tissue index (LSTI), and phase angle positively related to LST and LSTI. Body composition variations due to physical training were recognized by spBIVA: the increase in muscle mass was indicated by the phase angle and Xcsp increase, and the stability of FM% was consistent with the unchanged values of Rsp. Military training produced an increase in muscle mass, but no change in FM%, independently of the sample characteristics at baseline and the practice of additional sports. SpBIVA is a suitable technique for the assessment of body composition in military people.

Keywords: bioelectrical impedance; vector analysis; lean soft tissue; fat mass; muscle mass; phase angle

Citation: Gobbo, L.A.; Langer, R.D.; Marini, E.; Buffa, R.; Borges, J.H.; Pascoa, M.A.; Cirolini, V.X.; Guerra-Júnior, G.; Gonçalves, E.M. Effect of Physical Training on Body Composition in Brazilian Military. *Int. J. Environ. Res. Public Health* **2022**, *19*, 1732. https://doi.org/10.3390/ijerph19031732

Academic Editor: José Carmelo Adsuar Sala

Received: 15 December 2021
Accepted: 29 January 2022
Published: 2 February 2022

1. Introduction

The military paradigm is associated with healthy appearance, athletic bearing, and high-level physical performance. Indeed, the military are selected based on physical standards and are regularly involved in strong physical activities, also related to sports training, which requires monitoring for variations in body composition variants [1].

There are various methods usable to evaluate body composition, including anthropometry; bioimpedance; and more accurate techniques, such as potassium 40 counting, water isotope dilution, underwater weighing, imaging techniques, and dual energy X-ray absorptiometry (DXA) [2].

Due to the high suitability and low cost, the anthropometric techniques are the most used in many fields of application, including the routine military practice [3–5]. These methods, however, are not very accurate in detecting the main body compartments. For example, body mass index does not distinguish lean mass from fat mass [6–8] and so is incapable of evaluating muscle mass gain concomitant to fat weight loss (as generally occurs with intense military training) [9]. Accordingly, Pierce et al. [10] have recently demonstrated that BMI is not associated with performance on military relevant tasks in U.S. Army soldiers. Further, waist circumferences, largely used among the military [3–5] due to the associations with intra-abdominal fat and the related morbidity outcomes [11], are subject to intra- and inter-observer errors of measurement [12] and the need for a strong standardization because of the different possible measurement sites. Research results in military members are discordant, showing both a good and a poor agreement between the circumference measurement body composition method and dual-energy X-ray absorptiometry (DXA) [5,13].

Bioelectrical impedance analysis (BIA) is a non-invasive, low cost, and easy to operate technique, which needs a very short time compared to the more sophisticated body composition methods [14]. BIA has been rarely applied to research in the military showing a good agreement with DXA results [13,15,16]. The traditional two-component approach of BIA uses predictive equations, including bioelectrical values (generally resistance), and considers other variables (age, sex, and height) for the evaluation of fat mass and fat-free mass [17]. However, the application of predictive equations in samples differing from those where they have been calibrated can introduce a source of error. Otherwise, the use of population/group-specific equations reduces the comparability of results.

Alternative approaches, that have been proposed to avoid the use of equations and possible related errors, are based on the analysis of raw bioelectrical data of resistance (R, ohm) and reactance (Xc, ohm). The phase angle (arctan Xc/R $180/\pi$, degrees) is an indicator of nutritional status related to body cell mass and cell membrane integrity, that is largely used in clinical practice [18,19]. Phase angle has also been analyzed in relation to resistance training, and an increasing trend of its values has been registered [20]. However, as shown by Mereu et al. [21], the analysis of body composition based on the phase angle only can be inaccurate and is significantly improved if the information given by the vector length $(R^2 + Xc^2)^{0.5}$ is also considered.

Such a vectorial approach has been proposed by Piccoli et al. [22], who conceived the bioelectrical impedance vector analysis (BIVA). The classic BIVA procedure analyzes the bioelectrical values of resistance and reactance, standardized for body height (a proxy of conductor length). A BIVA variant defined as 'specific bioelectrical impedance vector analysis' (spBIVA) implies the standardization of resistance and reactance by length and by cross-sections of the body as well [22–24]. SpBIVA has been shown to be effective in the evaluation of fat mass percentage [23–25] and skeletal muscle mass [9,23,26,27]. Specific reference values have been proposed for 50 different populations, such as Italian-Spanish, U.S. young adults, and Italian healthy elderly [22,27,28]. The classic BIVA approach has been sporadically used in relation to sport and exercise [29], and specific BIVA even less [29–32]. Neither classic nor specific BIVA has been applied to the military samples.

The aims of the present study were two-fold: (1) to analyze the effect of a 7-month military training program on body composition variables, and (2) to analyze the correlation between the changes in body composition measured by spBIVA and DXA in a Brazilian military sample.

2. Materials and Methods

In accordance with the Helsinki Declaration, a written informed consent was obtained from all participants. The research was approved by the Ethics Committee of the School of Medical Sciences, University of Campinas. All procedures followed Resolution No. 466 of 2012 of the National Health Council of the Ministry of Health of Brazil.

2.1. The Sample

A sample of 270 young men (19.1 ± 1.1 years) from all the regions of Brazil (South, Southeast, Midwest, North, and Northeast) enrolled in the Preparatory School of Army

Cadets (EsPCEx) of the city of Campinas, SP, Brazil, was selected. Data were collected over two years (2013 and 2014), in two periods: at the beginning (March/April) and the end (October/November) of the military routine year.

The sample was divided into two groups: (1) the cadets who were involved in the military physical training routine only (MT, n = 155); (2) the cadets who were involved, by their own choice, in the military physical training routine plus a specific sport training for military competition (SMT, n = 115): track and field (n = 25), basketball (n = 16), fencing (n = 3), soccer (n = 18), judo (n = 3), swimming (n = 13), trekking (n = 4), shooting (n = 2), triathlon (n = 11), volleyball (n = 15), or chess (n = 5).

All cadets were included in the sample, except those who did not sign the consent form, who did not attend the day of the evaluations (even if only the second ones), who had a history of musculoskeletal injury at the time of the assessments, or were disconnected from the school.

2.2. *Military Physical Training*

Military physical training was performed 5 days/week during 90 min/day for 34 weeks, according to the academy military physical training manual, where the cadets were supposed to undergo a physical training that consisted of: (a) 2–3 sessions/week of continuous or interval running, with a weekly increased load; (b) 1 session/week of calisthenics exercises (7–15 repetitions of push-up, push-up/stand-up, sit-up, squat with hands on hip, lunge with hands on hip, and jumping jacks); (c) 1–2 sessions/week of circuit resistance training (2 sets of bench press, sit-up and its variations, half squat, barbell curl, pull-up, stair jumps, jump rope, and wrist roller, with 30 s of each exercise and 30 s of rest interval); (d) 1 session/week of swimming; and (e) 2 sessions/week of sports training. Before each session, all participants went through ~8 stretching exercises, ~7 neuromuscular warm-up, and ~7 general warm-up exercises. For sports training, each participant performed specific training for each modality [32,33].

2.3. *Measurements*

All subjects underwent anthropometric, BIA and DXA assessments, in the same sequence, in the morning.

Anthropometric measurements were performed following standard procedures [34], by an accredited International Society for the Advancement of Kinathropometry (ISAK) technician. Body weight (kg) and height (cm) were measured using a digital scale with precision of 0.1 kg (Filizola, São Paulo, Brazil) and a Harpenden stadiometer with precision of 1 mm (Holtain Limited, Crosswell, UK), respectively. Relaxed upper arm, waist and calf girths were measured using an anthropometric tape (precision of 1 mm). Body mass index in $kg{\cdot}m^{-2}$ was calculated by the ratio between body weight, in kilograms, and height squared, in meters (BMI).

A fan beam equipment model iDXA (GE Healthcare Lunar, Madison, WI, USA), enCore^{tm} 2011 software (version 13.6), was used to determine body composition. Total body composition was measured with the subject lying in the supine position, with the scanning time for the full length of approximately seven minutes. Total fat mass (FM, kg), fat percent (%FAT), lean soft tissue (LST, kg), and bone mineral content (BMC, kg) were measured. LSTI ($kg{\cdot}m^{-2}$) was calculated as LST/height squared in meters. To determine the reproducibility of the variables estimated by the equipment, coefficient of variation (CV%) and the technical error of measurement (TEM) were determined, based on the testing and retesting conducted with 23 subjects, and retested within 24 h. The values of CV% were 0.74%, 0.28%, and 0.26% for FM, BMC, and LST, respectively, and TEM were 0.25 kg (FM), 0.02 kg (BMC), and 0.25 kg (LST).

Bioelectrical measurements (resistance (R), ohm; reactance (Xc) ohm; at 50 kHz and 425 μA) were taken following the standard procedure [14]. With a Bioelectrical Body Composition Analyzer, tetrapolar device, single frequency (50 kHz), and model Quantum II (RJL Systems, Detroit, MI, USA). Specific bioelectrical impedance vector analysis was applied [24]. Specific bioelectrical values (resistivity (Rsp) ohm cm; reactivity (Xcsp) ohm cm) were obtained by multiplying resistance and reactance by a correction factor (A/L), where area (A, cm^2) and length (L, cm) were estimated as follows: A = (0.45 upper arm area + 0.10 waist area + 0.45 calf area) and L = 1.1 stature (in cm). The segment areas were calculated as $C2/4\pi$, where C (cm) is the girth of the upper arm, waist, or calf. The phase angle (degrees) was calculated as arctan ($Xc/R\ 180/\pi$) and the impedivity vector (Zsp, ohm cm) as $(Rsp^2 + Xcsp^2)^{0.5}$.

All participants should have fasted for at least 4 h, not ingest caffeinated foods or alcoholic beverages 24 h prior to the test, not perform strenuous physical activity less than 12 h before the test, not use any diuretics for at least 7 days before the test, urinate about 30 min before the test, and remove all metals (bracelets, watch, chains, etc.). During the assessment, the volunteers remained in the supine position, on a stretcher isolated from electrical conductors, in the supine position, with the legs abducted at an angle of approximately 45 degrees [35]. The values of CV% were 0.35% and 0.33% for R and Xc, respectively, and TEM were 3.54 Ω and 0.49 Ω, respectively, for R and Xc, for the same 23 subjects retested for DXA. Gonzalez et al. [36] validated the same equipment used in this study in a Brazilian sample, also using DXA as a reference method.

The within-sample variability was investigated by considering the distribution of bioelectrical values in the tolerance ellipses, representing the bivariate percentiles of the reference population. At this purpose, the tolerance ellipses for the Italo-Spanish adults (18–30 years) [28] have been used. The major axis of the ellipses refers to variations of FM% (higher values towards the upper pole) and the minor axis to variations of skeletal muscle mass and ECW/ICW (lower values on the left side).

2.4. Statistical Analyses

Bioelectrical values of MT and SMT groups were compared by mean of tolerance and confidence ellipses, using a two-sample Hotelling's T^2 test.

The consistency of the results obtained with specific BIVA and DXA was evaluated by means of Pearson's correlation between bioelectrical and DXA variables at baseline and comparing the trend of longitudinal body composition variations described by the two techniques. The effect of training (pre- vs. post-training) in the two sub-samples of SMT and MT was analyzed using two-way ANOVA (anthropometric, DXA output and bioelectrical values) and paired one-sample Hotelling's T^2 (confidence ellipses).

Statistical analyses were performed using IBM SPSS Statistics 19 (IBM SPSS Statistics for Windows, Version 19.0. Armonk, NY: IBM Corp) and the specific BIVA software (freely available at the website: http://specificbiva.unica.it/ (accessed on 17 August 2018).

3. Results

On average, at baseline, the sample of military people was in the normal weight BMI category and had a percent fat mass within the normal range for men (Table 1).

Table 1. Descriptive statistics for body composition of Brazilian Military (N = 270) at the beginning of the routine year.

Variables	Mean	SD	95% IC (Lower–Upper Limits)
Weight, kg	69.9	8.9	68.9–71.0
Height, cm	175.7	6.4	174.9–176.4
BMI, $kg \cdot m^{-2}$	22.6	2.4	22.4–22.9
Waist crf, cm	76.2	4.8	75.7–76.8
LSTI, $kg \cdot m^{-2}$	17.9	1.6	17.7–18.0
FM, kg	12.2	3.7	11.7–12.6
LST, kg	55.2	6.4	54.4–55.9
BMC, kg	3.0	0.4	2.9–3.0
FM%	17.1	3.8	16.6–17.5

Table 1. *Cont.*

Variables	Mean	SD	95% IC (Lower–Upper Limits)
Rsp, ohm·cm	313.0	28.8	309.8–316.8
Xcsp, ohm·cm	40.9	5.7	40.2–41.6
Phase Angle, degrees	7.4	0.8	7.3–7.5

Legend: SD = standard deviation; BMI = body mass index; LSTI = lean soft tissue index; FM = fat mass; LST = lean soft tissue; BMC = bone mineral content; FM% = fat mass percent; Rsp = specific resistance; Xcsp = specific reactance.

Bioelectrical values were quite totally within the reference tolerance ellipses, but slightly shifted toward the lower pole, indicative of low FM% (Figure 1).

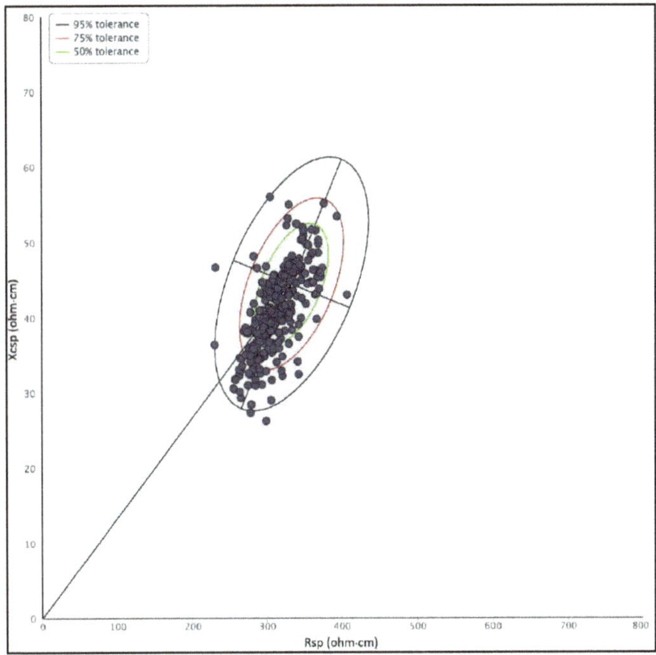

Figure 1. Distribution of bioelectrical values of Brazilian Military onto tolerance ellipses representing Italian-Spanish young adults, at the beginning of the routine year.

DXA and specific BIVA were correlated (Table 2). In fact, Rsp and Xcsp were positively related to FM, FM%, LST, and LSTI, while phase angle was positively related to LST and LSTI.

Table 2. Matrix of correlation between bioelectric and DXA variables (N = 270) at baseline.

	Rsp		Xcsp		PA	
	r	*p*	r	*p*	r	*p*
FM, kg	0.582	0.000	0.406	0.000	0.030	0.627
FM%	0.556	0.000	0.326	0.000	−0.049	0.418
LST, kg	0.229	0.000	0.300	0.000	0.189	0.002
LSTI, kg·m^{-2}	0.292	0.000	0.497	0.000	0.400	0.000

Legend: r = Pearson correlation coefficient; *p* = *p* value; FM = fat mass; FM% = fat mass percent; LST = lean soft tissue; LSTI = lean soft tissue index; Rsp = specific resistance; Xcsp = specific reactance; PA = phase angle.

Military people practicing sport activities (SMT) showed body composition differences with respect to those practicing military training only (MT). In fact, SMT group had higher values of LST and BMC, and lower values of FM and FM% (Table 3). The bioelectrical values of specific reactance and phase angle were significantly higher in SMT than in MT, indicating higher muscle mass (Table 3, Figure 2).

Table 3. Descriptive and comparative statistics.

	SMT (N = 115)				MT (N = 155)						
	Pre		Post		Pre		Post				
	Mean	Sd	Mean	Sd	Mean	Sd	Mean	Sd	Fg	Ft	Fgxt
Weight, kg	71.0	8.7	73.0	8.7	69.1	9.0	70.8	8.2	0.006	0.018	0.826
Height, cm	176.4	6.3	176.7	6.2	175.1	6.5	175.3	6.5	0.023	0.521	0.979
BMI, kg·m^{-2}	22.8	2.1	23.3	2.1	22.5	2.5	23.0	2.2	0.147	0.014	0.868
Waist crf, cm	77.2	4.9	78.5	5.1	75.5	4.6	77.2	4.3	0.000	0.000	0.581
FM, kg	11.5	3.2	12.2	3.1	12.6	4.0	12.7	3.3	0.011	0.253	0.305
LST, kg	56.9	6.6	58.1	6.6	53.9	5.9	55.5	5.8	0.000	0.009	0.671
LSTI, kg·m^{-2}	18.3	1.6	18.6	1.5	17.6	1.5	18.0	1.4	0.000	0.004	0.541
BMC, kg	3.1	0.4	3.1	0.4	2.9	0.4	3.0	0.4	0.000	0.118	0.991
FM%, %	16.0	3.3	16.5	3.1	17.9	3.9	17.6	3.3	0.000	0.720	0.199
Rsp, ohm	314.8	27.6	312.7	26.5	311.7	29.6	310.8	28.3	0.316	0.544	0.816
Xcsp, ohm	41.9	5.5	45.7	6.5	40.1	5.7	44.0	6.1	0.001	0.000	0.911
PA, degree	7.6	0.8	8.3	0.9	7.3	0.7	8.1	0.8	0.001	0.000	0.966

Legend: SMT: Sports and Military Training; MT: Military Training only; F, F test of two-way ANOVA for group (Fg), training (Ft), and group-training interaction (Fgxt); BMI = body mass index; FM = fat mass; FM% = fat mass percent; LST = lean soft tissue; LSTI = lean soft tissue index; Rsp = specific resistance; Xcsp = specific reactance; PA = phase angle.

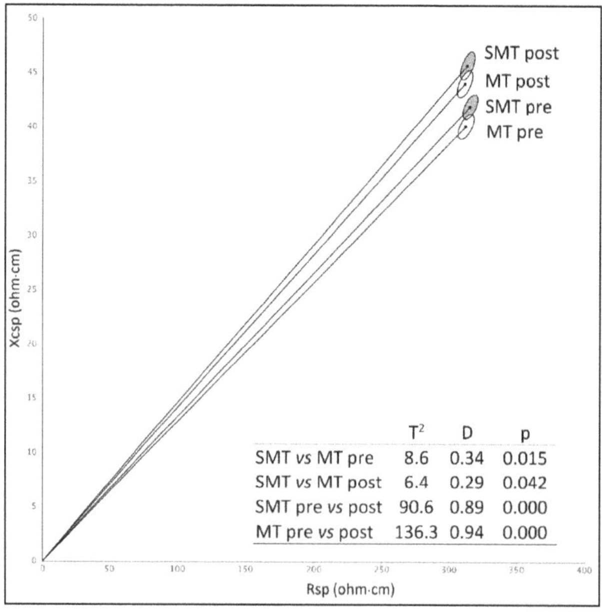

Figure 2. Confidence ellipses with T^2 Hotelling's test in the two groups before and after training. Legend: SMT: Sports and Military Training; MT: Military Training only. Comparisons between SMT and MT were performed using two-sample Hotelling's T2 tests, while those between pre- and post-training groups were performed with paired one-sample Hotelling's T^2 tests.

4. Discussion

In this sample of military personnel, DXA and specific BIVA showed a consistent scenario of body composition variations related to physical training. In fact, specific bioelectrical variables were correlated with DXA (FM, FM% and LST, and LSTI), and both the techniques showed: (a) different body composition in the military practicing physical training routine only (MT) or a specific sport training as well (SMT); (b) an increase in fat-free mass and a steady percentage of fat mass in relation to training, in both SMT and MT groups.

The sample, especially the SMT group, showed body composition characteristics adequate to the military standard, as suggested by the BMI indicative of normal weight [37], and the percentage of fat mass, which was lower than the body fat limits of approximately 20%, desirable for the U.S. army men [4]. The values of fat-free mass were higher in SMT than MT. The period of over ~7 months of military training induced, in both MT and SMT groups, a gain of lean soft tissue that contributed to the higher value of weight and BMI. However, the absolute and relative quantity of body fat did not change.

The observed differences of body composition are consistent with the effects of physical training described in the general population, and in the military [38,39]. Aerobic, stretching and resistance training are among the main interventions that can affect fat mass, fat free mass, and skeletal muscle mass, especially in young adults. These effects can be achieved in adults in a period of 3 to 12 months, depending on the characteristics of the sample and the volume of training, as well as other influent factors, such as daily habits and, particularly, alimentary style [40].

Research focused on military training has shown in general an increase in fat-free mass [15,16,34,41–43], but not in Margolis et al. [44], while the results on fat mass are less consistent among the studies. Mikkola et al. [16], in Finnish military performing regular, rather high-intensity, physical activity, over a period from 6 to 12 months, observed an increase in fat mass (in normal weight individuals), but a decrease in visceral fat. Indeed, intense physical activity promotes a greater reduction of visceral than subcutaneous adipose tissue, even if weight increases [45].

As previously presented, despite the inability of indicators such as BMI and waist circumference to validly identify lean and fat mass in physically active soldiers [6–12], there is still a gap in studies with DXA, for example, to test the agreement with the bioimpedance technique [5,13].

From a methodological point of view, similarly to the present research, previous studies realized in U.S. adults [9,23] and elderly Italians [24] detected a high correlation between DXA and specific BIVA variables. In particular, Rsp and Xcsp showed a positive correlation with FM% (especially Rsp; [9,24]) and with FFMI (especially Xcsp; [9]), while phase angle was positively related to FFMI only [9]. It is noteworthy that such convergent results have been obtained in samples characterized by different geographical provenience (Brazil, present study; US and Italy) [9,23,24], age class (Young adults, adults, and elders), and lifestyle (military, general population, and retirees). Indeed, the observed relationships are expected. In fact, resistance is negatively related to total body water and electrolytes, and hence, in normal-hydrated individuals, increases with the proportion of low conductive tissues, such as fat mass [9,24]. On the other side, the capacitive component (reactance) and phase angle are associated with the polarization produced by cell membranes and tissue interfaces and are positively related with body cell mass [22]. In this study, the correlation between DXA and spBIVA is shown by the trajectory of vector migration in relation to training, that, in both MT and SMT groups, is associated with increased values of reactance and phase angle (increasing muscle mass), but quite unchanged Rsp values (stable FM%). Such results can be comparable to those of Campa et al. [46], who analyzed three different sports modalities (volleyball, soccer, and rugby) and observed higher PhA values in athletes with a high mesomorphic component, which means, higher skeletal muscle component.

However, inconsistent results between DXA/spBIVA and waist circumference have been observed. In fact, SMT group showed higher values of waist circumferences, increasing in time, with respect to MT, but lower FM% levels, which remained stable after the military training. A similar disagreement between DXA and the circumference methods has already been described in the military [5]. In our research, we have also observed that spBIVA results, similarly to DXA, were not consistent with the pattern of waist circumference differences. However, compared to DXA, spBIVA did not recognize a lower percentage of fat mass in SMT with respect to MT. The observed gaps between abdominal circumference and DXA or spBIVA are noteworthy, considering the particular emphasis given to circumference measurement to calculate body fat percentage among the military [42]. The inconsistencies can be likely related to the different distribution of body components in the central and peripheral regions of the body and maybe to the greater effect of training on visceral than on subcutaneous fat, discussed above.

Despite the fact that the present study analyzed cadets who practiced 11 different sports, approximately 85% of the total SMT group practiced either teams' sports or individual sports, such as cyclical sports (swimming, athletics, or triathlon), therefore, the physiological characteristics were not so different when comparing practitioners of sports modalities, such as basketball, with practitioners of judo or fencing. In this way, the variations in the subjects' body composition, especially in the SMT group, are partly explained by training in some sports that total almost 90% of all the modalities practiced.

Considering, for example, the practice of team sports, such as soccer and volleyball, as was verified in different studies [47,48] that the longer the training time, the greater the phase angle and the lower the resistance values, indicating higher lean body mass values and, consequently, higher musculoskeletal mass. Micheli et al. [47] demonstrated that elite Italian professional football players (Series A and B) trained 9 weeks more throughout the year, with three more training sessions per week and one more game per week than amateur players, and consequently presented 7% higher and 8% lower phase angle and resistance values, respectively, indicating better body compositing status. In contrast, in our study, the differences between the phase angle (MT $\times$ SMT) were due to the greater values of reactance for those who practice sports activities. Such a situation can be explained by the amount of military training of both groups, which naturally provides good physical fitness, with lower resistance values. The SMT group, with specific sports training, presented higher values of phase angle, explained by higher reactance values, influenced mainly by greater amount of cell mass.

The main strengths of the present research are related to the application of a standardized protocol with cross-sectional and longitudinal measures, the use of reference techniques for body composition assessment in association with specific BIVA, and a well-controlled sample. However, some limitations are also present and are related to the poor representation of cases in the different disciplines, characterized by different training protocols, which made it impossible to recognize possible differences in body composition changes and in the underlying physiological mechanisms. Further, there was some disagreement between methods (anthropometry, DXA, and specific BIVA), likely related to regional differences of body components, which should be better analyzed by means of localized body composition analysis.

5. Conclusions

In conclusion, this research showed that spBIVA is a suitable technique for the assessment of body composition in the population studied. The effect of training on body composition was independent of sample characteristics or type of physical exercise: muscle mass increased, while the percentage of fat mass remained unchanged.

Author Contributions: Conceptualization, L.A.G., E.M. and R.B.; methodology, R.D.L. and E.M.G.; software, E.M.; formal analysis, E.M. and L.A.G.; investigation, R.D.L., E.M.G., J.H.B., M.A.P., V.X.C. and G.G.-J.; resources, V.X.C.; data curation, R.D.L., J.H.B. and E.M.G.; writing—original draft preparation, L.A.G., E.M. and R.B.; writing—review and editing, L.A.G., E.M. and R.B.; visualization, E.M.; project administration, G.G.-J. and E.M.G. All authors have read and agreed to the published version of the manuscript.

Funding: This research received no external funding.

Institutional Review Board Statement: The study was conducted according to the guidelines of the Declaration of Helsinki and approved by the Ethics Committee of the School of Medical Sciences, University of Campinas. All procedures followed Resolution No. 466 of 2012 of the National Health Council of the Ministry of Health of Brazil.

Informed Consent Statement: Informed consent was obtained from all subjects involved in the study.

Acknowledgments: The authors are grateful to the officers and the cadets of the "Preparatory School Cadets Army" (EsPCEx) of Campinas-SP for their authorization and collaboration in this study.

Conflicts of Interest: The authors declare no conflict of interest.

References

1. Jamro, D.; Zurek, G.; Lachowicz, M.; Lenart, D. Influence of Physical Fitness and Attention Level on Academic Achievements of Female and Male Military Academy Cadets in Poland. *Healthcare* **2021**, *9*, 1261. [CrossRef] [PubMed]
2. Heymsfield, S.B.; Ebbeling, C.B.; Zheng, J.; Pietrobelli, A.; Strauss, B.J.; Silva, A.; Ludwig, D. Multi-component molecular-level body composition reference methods: Evolving concepts and future directions. *Obes. Rev.* **2015**, *16*, 282–294. [CrossRef] [PubMed]
3. Leu, J.R.; Friedl, K.E. Body Fat Standards and Individual Physical Readiness in a Randomized Army Sample: Screening Weights, Methods of Fat Assessment, and Linkage to Physical Fitness. *Mil. Med.* **2002**, *167*, 994–1000. [CrossRef]
4. Friedl, K.E. Body Composition and Military Performance—Many Things to Many People. *J. Strength Cond. Res.* **2012**, *26*, S87–S100. [CrossRef]
5. Mitchell, K.M.; Pritchett, R.C.; Gee, D.L.; Pritchett, K.L. Comparison of Circumference Measures and Height–Weight Tables with Dual-Energy X-Ray Absorptiometry Assessment of Body Composition in R.O.T.C. Cadets. *J. Strength Cond. Res.* **2017**, *31*, 2552–2556. [CrossRef]
6. De Lorenzo, A.; Deurenberg, P.; Pietrantuono, M.; Di Daniele, N.; Cervelli, V.; Andreoli, A. How fat is obese? *Acta Diabetol.* **2003**, *40*, 254–257. [CrossRef]
7. Ode, J.J.; Pivarnik, J.M.; Reeves, M.J.; Knous, J.L. Body Mass Index as a Predictor of Percent Fat in College Athletes and Nonathletes. *Med. Sci. Sports Exerc.* **2007**, *39*, 403–409. [CrossRef]
8. Buffa, R.; Mereu, E.; Succa, V.; Latini, V.; Marini, E. Specific BIVA recognizes variation of body mass and body composition: Two related but different facets of nutritional status. *Nutrition* **2017**, *35*, 1–5. [CrossRef]
9. Malavolti, M.; Battistini, N.C.; Dugoni, M.; Bagni, B.; Bagni, I.; Pietrobelli, A. Effect of Intense Military Training on Body Composition. *J. Strength Cond. Res.* **2008**, *22*, 503–508. [CrossRef]
10. Pierce, J.R.; DeGroot, D.W.; Grier, T.L.; Hauret, K.G.; Nindl, B.C.; East, W.B.; McGurk, M.S.; Jones, B.H. Body mass index predicts selected physical fitness attributes but is not associated with performance on military relevant tasks in U.S. Army Soldiers. *J. Sci. Med. Sport* **2017**, *20*, S79–S84. [CrossRef]
11. Lee, W.-S. Body fatness charts based on BMI and waist circumference. *Obesity* **2015**, *24*, 245–249. [CrossRef] [PubMed]
12. Ulijaszek, S.J.; Kerr, D.A. Anthropometric measurement error and the assessment of nutritional status. *Br. J. Nutr.* **1999**, *82*, 165–177. [CrossRef] [PubMed]
13. Combest, T.M.; Howard, R.S.; Andrews, L.A.M. Comparison of Circumference Body Composition Measurements and Eight-Point Bioelectrical Impedance Analysis to Dual Energy X-Ray Absorptiometry to Measure Body Fat Percentage. *Mil. Med.* **2017**, *182*, e1908–e1912. [CrossRef]
14. NIH. *Bioelectrical Impedance Analysis in Body Composition Measurement: National Institutes of Health Technology Assessment Conference Statement*; National Institutes of Health: Washington, DC, USA, 1996.
15. Williams, A.G. Effects of Basic Training in the British Army on Regular and Reserve Army Personnel. *J. Strength Cond. Res.* **2005**, *19*, 254–259. [CrossRef]
16. Mikkola, I.; Jokelainen, J.J.; Timonen, M.J.; Härkönen, P.K.; Saastamoinen, E.; Laakso, M.A.; Peitso, A.J.; Juuti, A.-K.; Keinänen-Kiukaanniemi, S.M.; Mäkinen, T.M. Physical Activity and Body Composition Changes during Military Service. *Med. Sci. Sports Exerc.* **2009**, *41*, 1735–1742. [CrossRef] [PubMed]
17. Campa, F.; Toselli, S.; Mazzilli, M.; Gobbo, L.A.; Coratella, G. Assessment of Body Composition in Athletes: A Narrative Review of Available Methods with Special Reference to Quantitative and Qualitative Bioimpedance Analysis. *Nutrients* **2021**, *13*, 1620. [CrossRef]
18. Norman, K.; Stobäus, N.; Pirlich, M.; Bosy-Westphal, A. Bioelectrical phase angle and impedance vector analysis—Clinical relevance and applicability of impedance parameters. *Clin. Nutr.* **2012**, *31*, 854–861. [CrossRef]
19. Gonzalez, M.C.; Barbosa-Silva, T.G.; Bielemann, R.M.; Gallagher, D.; Heymsfield, S.B. Phase angle and its determinants in healthy subjects: Influence of body composition. *Am. J. Clin. Nutr.* **2016**, *103*, 712–716. [CrossRef]
20. Ribeiro, A.S.; Avelar, A.; Dos Santos, L.; Silva, A.M.; Gobbo, L.A.; Schoenfeld, B.J.; Sardinha, L.B.; Cyrino, E. Hypertrophy-type Resistance Training Improves Phase Angle in Young Adult Men and Women. *Int. J. Sports Med.* **2016**, *38*, 35–40. [CrossRef]
21. Mereu, E.; Buffa, R.; Lussu, P.; Marini, E. Phase angle, vector length, and body composition. *Am. J. Clin. Nutr.* **2016**, *104*, 845–847. [CrossRef]

22. Piccoli, A.; Rossi, B.; Pillon, L.; Bucciante, G. A new method for monitoring body fluid variation by bioimpedance analysis: The RXc graph. *Kidney Int.* **1994**, *46*, 534–539. [CrossRef] [PubMed]
23. Buffa, R.; Saragat, B.; Cabras, S.; Rinaldi, A.C.; Marini, E. Accuracy of Specific BIVA for the Assessment of Body Composition in the United States Population. *PLoS ONE* **2013**, *8*, e58533. [CrossRef] [PubMed]
24. Marini, E.; Sergi, G.; Succa, V.; Saragat, B.; Sarti, S.; Coin, A.; Manzato, E.; Buffa, R. Efficacy of specific bioelectrical impedance vector analysis (BIVA) for assessing body composition in the elderly. *J. Nutr. Health Aging* **2012**, *17*, 515–521. [CrossRef]
25. Buffa, R.; Mereu, E.; Comandini, O.; Ibanez, E.M.; Marini, E. Bioelectrical impedance vector analysis (BIVA) for the assessment of two-compartment body composition. *Eur. J. Clin. Nutr.* **2014**, *68*, 1234–1240. [CrossRef] [PubMed]
26. Marini, E.; Buffa, R.; Saragat, B.; Coin, A.; Berton, L.; Manzato, E.; Sergi, G.; Toffanello, E.D. The potential of classic and specific bioelectrical impedance vector analysis for the assessment of sarcopenia and sarcopenic obesity. *Clin. Interv. Aging* **2012**, *7*, 585–591. [CrossRef] [PubMed]
27. Saragat, B.; Buffa, R.; Mereu, E.; De Rui, M.; Coin, A.; Sergi, G.; Marini, E. Specific bioelectrical impedance vector reference values for assessing body composition in the Italian elderly. *Exp. Gerontol.* **2014**, *50*, 52–56. [CrossRef] [PubMed]
28. Ibáñez, M.E.; Mereu, E.; Buffa, R.; Gualdi-Russo, E.; Zaccagni, L.; Cossu, S.; Rebato, E.; Marini, E. Newspecificbioelectrical impedance vector reference values for assessing body composition in the Italian-Spanish young adult population. *Am. J. Hum. Biol.* **2015**, *27*, 871–876. [CrossRef] [PubMed]
29. Castizo-Olier, J.; Irurtia, A.; Jemni, M.; Carrasco-Marginet, M.; Fernández-García, R.; Rodríguez, F.A. Bioelectrical impedance vector analysis (BIVA) in sport and exercise: Systematic review and future perspectives. *PLoS ONE* **2018**, *13*, e0197957. [CrossRef]
30. Antoni, G.; Marini, E.; Curreli, N.; Tuveri, V.; Comandini, O.; Cabras, S.; Gabba, S.; Madeddu, C.; Crisafulli, A.; Rinaldi, A.C. Energy expenditure in caving. *PLoS ONE* **2017**, *12*, e0170853. [CrossRef]
31. Fukuda, D.H.; Stout, J.R.; Moon, J.R.; Smith-Ryan, A.E.; Kendall, K.L.; Hoffman, J.R. Effects of resistance training on classic and specific bioelectrical impedance vector analysis in elderly women. *Exp. Gerontol.* **2016**, *74*, 9–12. [CrossRef]
32. Borges, J.H.; Hunter, G.R.; Silva, A.M.; Cirolini, V.X.; Langer, R.D.; Páscoa, M.A.; Guerra-Júnior, G.; Gonçalves, E.M. Adaptive thermogenesis and changes in body composition and physical fitness in army cadets. *J. Sports Med. Phys. Fit.* **2019**, *59*, 94–101. [CrossRef] [PubMed]
33. Langer, R.D.; Silva, A.M.; Borges, J.; Cirolini, V.X.; Páscoa, M.A.; Guerra-Júnior, G.; Gonçalves, E.M. Physical training over 6 months is associated with improved changes in phase angle, body composition, and blood glucose in healthy young males. *Am. J. Hum. Biol.* **2019**, *31*, e23275. [CrossRef] [PubMed]
34. Lohman, T.J.; Roache, A.F.; Martorell, R. Anthropometric Standardization Reference Manual. *Med. Sci. Sports Exerc.* **1992**, *24*, 952. [CrossRef]
35. Kyle, U.G.; Bosaeus, I.; De Lorenzo, A.D.; Deurenberg, P.; Elia, M.; Gómez, J.M.; Heitmann, B.L.; Kent-Smith, L.; Melchior, J.-C.; Pirlich, M.; et al. Bioelectrical impedance analysis—Part II: Utilization in clinical practice. *Clin. Nutr.* **2004**, *23*, 1430–1453. [CrossRef]
36. Gonzalez, M.C.; Orlandi, S.P.; Santos, L.P.; Barros, A.J. Body composition using bioelectrical impedance: Development and validation of a predictive equation for fat-free mass in a middle-income country. *Clin. Nutr.* **2018**, *38*, 2175–2179. [CrossRef] [PubMed]
37. WHO Consultation Report. Obesity: Preventing and Managing the Global Epidemic. *World Health Organ. Tech. Rep. Ser.* **2000**, *894*, 1–253.
38. Westerterp, K.R. Exercise, energy balance and body composition. *Eur. J. Clin. Nutr.* **2018**, *72*, 1246–1250. [CrossRef]
39. Harty, P.S.; Friedl, K.E.; Nindl, B.C.; Harry, J.R.; Vellers, H.L.; Tinsley, G.M. Military Body Composition Standards and Physical Performance: Historical Perspectives and Future Directions. *J. Strength Cond. Res.* 2021; *In print*. [CrossRef]
40. Donnelly, J.E.; Blair, S.N.; Jakicic, J.M.; Manore, M.M.; Rankin, J.W.; Smith, B.K.; American College of Sports Medicine. American College of Sports Medicine Position Stand. Appropriate Physical Activity Intervention Strategies for Weight Loss and Prevention of Weight Regain for Adults. *Med. Sci. Sports Exerc.* **2009**, *41*, 459–471. [CrossRef]
41. Campos, L.C.B.; Campos, F.A.D.; Bezerra, T.A.R.; Pellegrinotti, L. Effects of 12 Weeks of Physical Training on Body Composition and Physical Fitness in Military Recruits. *Int. J. Exerc. Sci.* **2017**, *10*, 560–567.
42. Crombie, A.P.; Liu, P.-Y.; Ormsbee, M.J.; Ilich, J.Z. Weight and Body-Composition Change during the College Freshman Year in Male General-Population Students and Army Reserve Officer Training Corps (ROTC) Cadets. *Int. J. Sport Nutr. Exerc. Metab.* **2012**, *22*, 412–421. [CrossRef] [PubMed]
43. Wood, P.S.; Krüger, P.E.; Grant, C.C. DEXA-assessed regional body composition changes in young female military soldiers following 12-weeks of periodised training. *Ergonomics* **2010**, *53*, 537–547. [CrossRef] [PubMed]
44. Margolis, L.M.; Rood, J.; Champagne, C.; Young, A.J.; Castellani, J. Energy balance and body composition during US Army special forces training. *Appl. Physiol. Nutr. Metab.* **2013**, *38*, 396–400. [CrossRef] [PubMed]
45. Mourier, A.; Gautier, J.-F.; De Kerviler, E.; Bigard, A.X.; Villette, J.-M.; Garnier, J.-P.; Duvallet, A.; Guezennec, C.Y.; Cathelineau, G. Mobilization of Visceral Adipose Tissue Related to the Improvement in Insulin Sensitivity in Response to Physical Training in NIDDM: Effects of branched-chain amino acid supplements. *Diabetes Care* **1997**, *20*, 385–391. [CrossRef]
46. Campa, F.; Silva, A.M.; Matias, C.N.; Monteiro, C.P.; Paoli, A.; Nunes, J.P.; Talluri, J.; Lukaski, H.; Toselli, S. Body Water Content and Morphological Characteristics Modify Bioimpedance Vector Patterns in Volleyball, Soccer, and Rugby Players. *Int. J. Environ. Res. Public Health* **2020**, *17*, 6604. [CrossRef]

47. Micheli, M.L.; Pagani, L.; Marella, M.; Gulisano, M.; Piccoli, A.; Angelini, F.; Burtscher, M.; Gatterer, H. Bioimpedance and Impedance Vector Patterns as Predictors of League Level in Male Soccer Players. *Int. J. Sports Physiol. Perform.* **2014**, *9*, 532–539. [CrossRef]
48. Campa, F.; Toselli, S. Bioimpedance Vector Analysis of Elite, Subelite, and Low-Level Male Volleyball Players. *Int. J. Sports Physiol. Perform.* **2018**, *13*, 1250–1253. [CrossRef]

International Journal of
Environmental Research and Public Health

MDPI

Article

The Relationship between VO_2 and Muscle Deoxygenation Kinetics and Upper Body Repeated Sprint Performance in Trained Judokas and Healthy Individuals

André Antunes [1], Christophe Domingos [2], Luísa Diniz [1], Cristina P. Monteiro [1,3], Mário C. Espada [2,4], Francisco B. Alves [1,3] and Joana F. Reis [1,3,*]

1 Laboratory of Physiology and Biochemistry of Exercise, Faculdade de Motricidade Humana, Universidade de Lisboa, Cruz Quebrada-Dafundo, 1495-761 Lisboa, Portugal; andre.gs.antunes.97@gmail.com (A.A.); luuisamd@hotmail.com (L.D.); cmonteiro@fmh.ulisboa.pt (C.P.M.); falves@fmh.ulisboa.pt (F.B.A.)
2 Life Quality Research Centre, 2040-413 Rio Maior, Portugal; christophedomingos@esdrm.ipsantarem.pt (C.D.); mario.espada@ese.ips.pt (M.C.E.)
3 Interdisciplinary Centre for Human Performance Research (CIPER), Faculdade de Motricidade Humana, Universidade de Lisboa, Cruz Quebrada-Dafundo, 1495-761 Lisboa, Portugal
4 Polytechnic Institute of Setúbal, School of Education, 2914-514 Setúbal, Portugal
* Correspondence: joanareis@fmh.ulisboa.pt

Abstract: The present study sought to investigate if faster upper body oxygen uptake (VO_2) and hemoglobin/myoglobin deoxygenation ([HHb]) kinetics during heavy intensity exercise were associated with a greater upper body repeated-sprint ability (RSA) performance in a group of judokas and in a group of individuals of heterogenous fitness level. Eight judokas (JT) and seven untrained healthy participants (UT) completed an incremental step test, two heavy intensity square-wave transitions and an upper body RSA test consisting of four 15 s sprints, with 45 s rest, from which the experimental data were obtained. In the JT group, VO_2 kinetics, [HHb] kinetics and the parameters determined in the incremental test were not associated with RSA. However, when the two groups were combined, the amplitude of the primary phase VO_2 and [HHb] were positively associated with the accumulated work in the four sprints (ΣWork). Additionally, maximal aerobic power (MAP), peak VO_2 and the first ventilatory threshold (VT_1) showed a positive correlation with ΣWork and an inverse correlation with the decrease in peak power output (Dec-PPO) between the first and fourth sprints. Faster VO_2 and [HHb] kinetics do not seem to be associated with an increased upper body RSA in JT. However, other variables of aerobic fitness seem to be associated with an increased upper body RSA performance in a group of individuals with heterogeneous fitness level.

Keywords: VO_2 kinetics; muscle oxygenation; judo; upper body; arm crank; near-infrared spectroscopy; repeated sprint ability

Citation: Antunes, A.; Domingos, C.; Diniz, L.; Monteiro, C.P.; Espada, M.C.; Alves, F.B.; Reis, J.F. The Relationship between VO_2 and Muscle Deoxygenation Kinetics and Upper Body Repeated Sprint Performance in Trained Judokas and Healthy Individuals. *Int. J. Environ. Res. Public Health* **2022**, *19*, 861. https://doi.org/10.3390/ijerph19020861

Academic Editor: Paul B. Tchounwou

Received: 5 November 2021
Accepted: 11 January 2022
Published: 13 January 2022

1. Introduction

Judo is a technically and tactically demanding sport, involving several intermittent efforts of high-intensity activity, interceded by short rest periods [1]. An official, senior-level match may last up to 4 min, and judokas may perform up to seven matches during a tournament, including preliminary rounds, main rounds and finals, all in the same day. This sport is reported to rely heavily on upper body strength and power [2], and it has also been suggested that the high-intensity efforts that occur throughout a match are mainly supported by anaerobic energy systems, while the oxidative energy system may be crucial to the recovery process in between high-intensity efforts and between matches [2].

It has been shown that at the elite level, judo contest winners have a higher activity profile over the course of a match, performing more offensive actions per match (56 offensive actions/match in gold medalists vs. 49 offensive actions/match in silver medalists) [3].

These results highlight that for high-level judo athletes, the ability to perform multiple high-intensity actions over time may be a crucial aspect in determining a contest winner. Therefore, the study of the factors underlying the ability to perform more high-intensity actions over the course of a match in this group of athletes seems to be of relevance.

Many sport activities rely on the ability to repeat several efforts over time. The ability to perform these activities seems to be dependent on the ability to quickly restore phosphocreatine (PCr) stores, ensuring that high rates of muscular work can be sustained over the course of several high-intensity bouts [4]. The ability to restore PCr stores back to near-resting level seems to be dependent on muscular oxidative capacity [5]. Moreover, as these short-duration efforts are repeated over time, the contribution of the oxidative energy system seems to increase [6]. The ability to quickly attain a high rate of ATP resynthesis from oxidative phosphorylation, as expressed by the rate at which oxygen uptake (VO_2) rises, seems to be an important aspect in delaying the onset of fatigue [7]. As this quick rise is associated with a reduction in oxygen deficit [8], it can contribute to an increased high-intensity exercise tolerance.

The speed at which VO_2 rises to attain a given value necessary to support the exercise workload can be characterized by the time constant of the primary component of VO_2 kinetics ($\tau_{phase\ II}$), which represents the time necessary for 63% of the final oxygen uptake response to be complete [9–11]. In response to moderate-intensity workloads, the VO_2 response is characterized by three phases [12,13]: Phase I is the cardiodynamic phase, and corresponds to the phase during which VO_2 rises as a consequence of increased pulmonary blood flow. Phase II corresponds to the primary component, where VO_2 rises in an exponential manner, until a steady-state VO_2 is achieved, corresponding to phase III. The profile of VO_2 response during phase II seems to closely reflect muscle VO_2 profile [13–15]. During heavy-intensity constant load exercise, the attainment of a steady-state VO_2 is delayed due to the rise in VO_2, which exceeds the expected values of VO_2 based on the VO_2–exercise intensity relationship established during submaximal (moderate intensity domain) workloads, coinciding with the emergence of a VO_2 slow component (VO_{2SC}).

Faster VO_2 kinetics, characterized by shorter values for $\tau_{phase\ II}$, have been observed in trained individuals [16–20] and have been associated with a smaller decrease in speed over a repeated-sprint ability test (RSA) in a group of soccer players [21]. Moreover, a shorter τ has been associated with longer high-speed running distances in a group of young high-level soccer athletes [22]. Trained individuals have also been shown to have a smaller VO_{2SC} at a given workload compared to untrained individuals [23,24], which has been associated with an increase in the ability to sustain high-intensity exercise workloads over time at a given absolute workload [25]. Some studies have also shown that upper body trained individuals have faster VO_2 kinetics compared to untrained subjects [18,19]. To our knowledge, no study to date has sought to understand the characteristics of the response of VO_2 kinetics during heavy-intensity upper body exercise, nor attempted to establish a relationship between VO_2 kinetics variables and upper body high-intensity exercise performance in a group of judo athletes.

Near infrared spectroscopy (NIRS) has been used to examine the relative matching of O_2 delivery with tissue oxygen utilization during constant-workload exercise transitions [26]. The hemoglobin/myoglobin deoxygenation ([HHb]) signal derived from NIRS measurements is reported to reflect the balance between O_2 delivery and O_2 utilization and has been used as a non-invasive index of O_2 extraction from muscle capillaries during exercise [27,28].

Given that several variables of aerobic fitness, such as maximal aerobic speed [29] and $\tau_{phase\ II}$ [21], have been associated with increased RSA, we hypothesized that participants with faster upper body VO_2 and [HHb] kinetics would achieve a higher upper body RSA performance. Therefore, the present study sought to understand if faster VO_2 and [HHb] kinetics were associated with a higher performance, expressed as a lower decrease in peak power output (PPO) and mean power output (MPO), as well as a higher accumulated work

(ΣW), over the course of an upper body RSA test in a group of judo athletes and in a group of healthy individuals of heterogenous fitness level.

2. Materials and Methods

Eight male judo athletes (JT) (age 21.1 ± 3.0 years, height 172.3 ± 4.5 cm, body mass 71.5 ± 7.1 kg, triceps skinfold thickness 4.5 ± 0.7 mm) and seven male untrained healthy participants (UT) (age 22.6 ± 1.0 years, height 172.7 ± 4.5 cm, body mass 64.3 ± 5.8 kg, triceps skinfold thickness 6.6 ± 1.6 mm) volunteered to participate in the study. The JT were all black belts, of national (placed in 1st–7th place in the national championships) and international (placed 3rd–9th place in European and World cups) level and had been training (13.1 ± 2.8 year) and competing regularly (6.0 ± 1.3 competitions in the previous year) for at least three years; the UT were not involved in any upper body exercise modalities, although they were all healthy, active (at least 150 min. of physical activity/wk) individuals [30].

None of the participants were suffering from any upper body injuries at the time of testing or recovering from any major upper body injury that had occurred in the past 12 months, nor taking any medicine. None of the individuals of JT were cutting weight nor preparing for a major competition at the time that the testing sessions were undertaken.

In order to determine the sample size for the present study, a priori statistical power analysis was performed with G-Power [31] based on the studies of Dupont [21] and McNarry [32], aiming for a power of 85% (alpha = 0.05, two-tailed). The sample size suggested was of 10 individuals for correlations and 7 for each group for the comparisons. Given the strenuous nature of the tests that were undertaken, and that participants could drop out of the study at any time, additional participants were recruited for a total sample of 15 individuals.

All the participants were fully informed of any risks before giving their written informed consent to participate in the study, in accordance with the requirements outlined by the Ethics Committee of the Faculty of Human Kinetics of the University of Lisbon (approval code 42/2021) and in accordance with the Declaration of Helsinki [33].

The participants were required to report to the laboratory on three occasions. To avoid circadian rhythm effects, testing occurred at the same time of day, with each session separated by at least 48 h and all testing sessions were completed within 2 weeks. All subjects were required to present themselves in the laboratory with comfortable clothes, in a rested and hydrated state, to refrain from drinking any sort of alcoholic beverages at least 24 h prior to each testing session, and from eating or taking caffeine 3 h prior to each test.

All tests were performed on an electronically braked arm crank ergometer (Lode Angio, Groningen, Netherlands). In their first session, participants performed an incremental step test to determine maximal aerobic power (MAP), peak oxygen uptake (peak VO_2), first ventilatory threshold (VT_1) and its respective workload (W_VT_1). In the second session, participants performed two heavy-intensity square-wave exercise transitions to determine VO_2 kinetics and [HHb] kinetics of triceps brachii. Each transition began with 3 min of baseline cranking at 0 W, following which the transition workload was imposed. Each square-wave transition was separated by 1 h of passive recovery. In the third session, participants completed a standardized warm-up, followed by an RSA test, consisting of four 15 s upper body all-out sprints, each interceded by 45 s of passive recovery in between.

2.1. Gas Exchange and Muscle Deoxygenation Measurements

Gas exchange variables were collected breath-by-breath with a gas analyzer (MetaMax 3B, Cortex Biophysik, Leipzig, Germany), after calibration according to the manufacturer's instructions.

The changes in the [HHb] signal in the local circulation of the long head of the triceps brachii were monitored using a continuous-wave tissue oximeter (NIMO, Nirox, Brescia, Italy) using the NIRS technique. In order to reliably collect the NIRS signal, the local skin of each participant's upper arm was initially shaved and cleaned. A probe consisting of a photon emitter and a photon receptor, emitting and detecting near-IR beams with three different wavelengths (685 nm, 850 nm and 980 nm), was attached to the skin surface, and secured with tape and then covered with an optically dense elastic bandage in order to minimize movement, prevent loss of near-IR signal and stray light interference, and also to constrain the signal emission-reception site. The signal was sampled at a frequency of 40 Hz. To account for the effects of adipose tissue thickness on the NIRS signal, the skinfold thickness at the site where NIRS probes were placed was measured with a caliper (Slim Guide Caliper, Creative Health, Ann Arbor, MI, USA) and a correction factor was used in the analysis software (Nimo Data Analysis Peak). All NIRS measurements were conducted on the right limb, and [HHb] was monitored during the second and third testing sessions (square-wave transitions and RSA test, respectively). The validity and limitations associated with the measurements obtained via this oximeter have been reviewed by Rovati and associates [34].

2.2. Incremental Step Test

Participants performed an incremental exercise test for determination of MAP, peak VO_2, VT_1 and W_VT_1. Participants performed a 3 min step of baseline cranking at 0 W, following which the power was increased 15 W each min (step) until participants reached voluntary exhaustion. The participants were instructed to crank the wheel at the rate of 70 rotations per minute (rpm), grabbing the handles of the ergometer in a standard position, in which they stood upright with their feet shoulder width apart, flat on the floor, and with their shoulder joint levelled with the pedal crank axle. Handle height and ergometer configuration were recorded and reproduced in subsequent tests. The present incremental test protocol's characteristics were based on the protocols of Koppo and associates [20] and Schneider and associates [35], which also studied upper body VO_2 kinetics of a group of heterogenous fitness level. Breath-by-breath pulmonary gas-exchange data were collected continuously during the incremental step test. The peak VO_2 was taken as the highest 30 s average value attained before the participants reached volitional exhaustion. The MAP was defined as the minimal workload which elicited peak VO_2.

The VT_1 was estimated by monitoring the ventilatory equivalents for oxygen (VE/VO_2) and carbon dioxide (VE/VCO_2), determined by inspection to define the point at which an increase in VE/VO_2 was observed, with no concomitant increase in VE/VCO_2 [36]. The workload over which these responses were observed was defined as the W_VT_1. Throughout the test, heart rate (HR) was monitored continuously (ONRHYTHM 500, Kalenji, France) and the highest HR value observed in the last stage of exercise was registered as peak HR. The workload associated with VT_1 was used to determine the intensity for the square-wave transitions, which was set at 20%Δ, calculated as W_VT_1 plus 20% of the difference between the W_VT_1 and the MAP.

2.3. Square-Wave Transitions

The participants performed two square-wave constant workload transitions for the determination of VO_2 and [HHb] kinetics, with a workload of 20%Δ, corresponding to a heavy-intensity workload. After a 3 min period of baseline cranking at 0 W, the target workload was imposed. Each square-wave transition lasted 6 min and the transitions were separated by 1 h of passive rest. Given the lower exercise tolerance associated with upper body exercise, a 20%Δ workload was chosen, to ensure that the subjects were working in the heavy-intensity exercise domain without incurring excessive fatigue, which would compromise performance in the subsequent square-wave transition, for both groups, and therefore confound the underlying physiological response.

The VO_2 data were collected breath-by-breath from each transition and were examined to exclude errant breaths and values lying more than 4 standard deviations from the local mean (based on 5 breaths), and subsequently linearly interpolated to provide 1 s values. The data from the two transitions were then time aligned to the start of exercise and averaged to reduce signal noise and enhance the underlying physiological response characteristics [37].

VO_2 kinetics parameters were calculated by an iterative procedure, minimizing the sum of the residuals, according to the following bi-exponential model:

$$VO_2\ (t) = VO_{2baseline} + A\ [1 - e^{-(t-TDp)/\tau p}] + Asc\ [1 - e^{-(t-TDsc)/\tau sc}]$$

where VO_2 (t) represents the absolute VO_2 at a given time t, $VO_{2baseline}$ represents the mean VO_2 under unloaded conditions 30 s prior to the work transition; A, TDp, and τ represent the amplitude, time delay, and time constant, of the phase II of the increase in VO_2 after the onset of exercise, and Asc, TDsc, and τsc represent the amplitude of the slow component, time delay before the onset of, and time constant of the slow component phase of VO_2 kinetics, respectively [38].

The end-exercise VO_2 was defined as the mean VO_2 value obtained in the last 30 s of the 6 min constant workload transitions. The first 20 s of VO_2 data were excluded from the analysis to remove the influence of the cardiodynamic phase on the subsequent response [39]. Because the asymptotic value of the second function is not necessarily reached at the end of the exercise, the amplitude of the slow component was defined as

$$A'_{SC} = Asc\ [1 - e^{-(te-TDsc/\tau sc)}]$$

where te was the time at the end of the exercise bout [20].

Throughout each square-wave transition, the [HHb] signal was monitored in order to provide a non-invasive surrogate of the changes in O_2 saturation of the hemoglobin/myoglobin in the local circulation of the long head of the triceps brachii.

The [HHb] data were normalized to resting values, considering the average of the 3 min rest before the unloaded pedaling, and the [HHb] response was characterized according to a monoexponential model, with a timed-delay (TD) at the onset of exercise, followed by an exponential increase [27] until the end of the exercise period:

$$[HHb]\ (t) = [HHb]_{baseline} + AHHb\ [1 - e^{-(t-TDHHb)/\tau HHb}]$$

where [HHb] (t) represents the [HHb] at a given time t, $[HHb]_{baseline}$ represents the 60 s average [HHb] prior to the participant gripping the handles, and A [HHb] and τ [HHb] correspond to the amplitude and time constant of the exponential phase of [HHb] kinetics, respectively. The TD was defined as the time between the onset of exercise and the time at which a first increase in the [HHb] signal was observed [40], which was determined by visual inspection. [HHb] data were fit from the time of initial increase in [HHb] to 180 s. The exponential-like phase of the [HHb] kinetics was also characterized by an "effective" time constant (τ'), which corresponded to the sum of TD and τ [40].

2.4. Repeated Sprint Exercise

The upper body RSA test consisted of four 15 s all-out sprints, each separated by 45 s of passive rest. Participants performed a 6 min warm-up at 30 W with a cadence of 70 rpm, with three brief sprints (<5 s duration) during the last 3 min of the warm-up. Participants were then given 2 min of rest before commencing the upper RSA test. Thirty seconds before the start of the test, the participants were asked to grip the ergometer handles. Throughout the whole test, the participants were verbally encouraged to give their maximum effort.

The exercise workload was set at 5% of the body mass of each individual [41]. The peak power output (PPO) and mean power output (MPO) attained during each sprint were monitored, and the total work performed (Work) during each sprint was derived as the integral of power output over the 15 s period.

The 15 s work period was chosen by considering the data reported by Soriano and associates [42] based on the sum of average time it took male judokas to come to grips (8.4 ± 3.1 s), establish a grip and control their opponent (6.1 ± 3.5 s) and execute a throw (1.3 ± 0.5 s). The 45 s rest period duration was chosen as a compromise between what is observed in a typical judo match (2:1 work-to-rest ratio) [1,2], and the typical work density that has been reported in several RSA studies (1:6-8 work-to-rest ratio) [4,6], to ensure that the work period matched what typically occurs throughout a match, as exercise intensity and duration are the main determinants of energy system specificity [43], while allowing participants to sustain the power output over the course of several high-intensity bouts.

A set of variables were computed, in order to characterize overall RSA performance:

$$\text{Dec-PPO (Decrease in PPO)} = \text{PPO 1st sprint} - \text{PPO 4th sprint}$$

$$\text{Dec-MPO (Decrease in MPO)} = \text{MPO 1st sprint} - \text{MPO 4th sprint}$$

$$\Sigma\text{Work (Accumulated work)} = \sum_{1st\ sprint}^{4th\ sprint} \text{Work performed}$$

Throughout each sprint, the [HHb] signal was monitored, and the data collected were used to compute the maximal [HHb] attained in each sprint (Max. A [HHb]).

2.5. Statistical Analysis

The results are presented as means ± SD. The Shapiro–Wilk test was used to verify the normal distribution of the data for each variable [44]. Unpaired t-tests were used to compare the differences between groups regarding each variable. Pearson product correlations were used to determine the correlation between variables. In order to determine if a linear relationship could be established between a variable of interest and other independent variables, a stepwise regression analysis was performed, using Peak VO_2, MAP, $VT_1\_VO_2$, $A_{phase\ II}$, $\tau_{phase\ II}$, the effective slow component amplitude (A'SC), A [HHb], τ' [HHb], Max. A [HHb] 1, Max. A [HHb] 2, Max. A [HHb] 3 and Max. A [HHb] 4 as independent variables, and Σ Work as the dependent variable. The collected data regarding each individual variable were analyzed as a whole, considering all participants as a single heterogeneous group, and were analyzed separately by groups. The effect size for the differences between groups was calculated based on the ratio between the difference in the mean values and the weighted pooled SD. The threshold values for Hedges' effect size (ES, g) statistics were characterized according to the following scale [45]: <0.20 = negligible effect, 0.20–0.49 = small effect, 0.50–0.79 = moderate effect, ≥0.80 = large effect.

3. Results

3.1. Incremental Step Test

Mean and standard deviation of the variables obtained by each group of participants in the incremental test are depicted in Table 1.

Table 1. Physiological responses attained by UT and JT participants in the incremental step test.

Variables	UT	JT	*p*
Peak VO_2 ($mL \cdot kg^{-1} \cdot min^{-1}$)	32.3 ± 5.1	40.4 ± 3.7	0.004 *
MAP (W)	101.4 ± 13.8	149.3 ± 17.4	<0.001 *
$VT_1\_VO_2$ ($mL \cdot kg^{-1} \cdot min^{-1}$)	11.9 ± 1.1	16.9 ± 1.3	0.289
W_VT_1 (W)	42.9 ± 20.2	69.4 ± 11.2	0.007 *
20%Δ (W)	57.1 ± 18.5	86.5 ± 11.7	0.003 *
Peak HR (beats/min)	172.4 ± 11.4	177.0 ± 6.4	0.349

UT, untrained participants; JT, judo athletes; Peak VO_2, peak oxygen consumption; MAP, maximal aerobic power; $VT_1\_VO_2$, oxygen consumption rate at the onset of the first ventilatory threshold; VT_1_W, workload at the onset of the first ventilatory threshold; 20% ΔW, Workload corresponding to the sum of VT_1_W plus 20% of the difference between the MAP and VT_1_W; Peak HR, Peak heart rate achieved during the incremental test; * Significant differences between groups for $p < 0.05$.

The JT group displayed higher Peak VO_2, MAP, W_VT_1 and 20%Δ than the UT group. A large effect size was observed for Peak VO_2 (g = 1.8), MAP (g = 3.0), W_VT1 (g = 1.7) and 20%Δ (g = 1.9). $VT_1\_VO_2$ and peak HR were not different between groups.

3.2. Square-Wave Transitions

Table 2 shows the VO_2 kinetics variables obtained by the two groups of participants in the heavy-intensity square-wave transitions.

Table 2. VO_2 kinetics parameters in the heavy-intensity square-wave transitions for each group.

Variables	UT	JT	p
$VO_{2baseline}$ (mL·kg^{-1}·min^{-1})	9.4 ± 1.2	9.1 ± 1.1	0.627
$A_{phase\ II}$ (mL·kg^{-1}·min^{-1})	11.5 ± 5.5	15.1 ± 2.8	0.128
$TD_{phase\ II}$ (s)	11.4 ± 9.2	10.5 ± 8.0	0.850
$\tau_{phase\ II}$ (s)	61.6 ± 8.2	47.5 ± 13.4	0.032 *
A'_{SC} (mL·kg^{-1}·min^{-1})	2.6 ± 0.7	4.9 ± 3.4	0.097
TD_{SC} (s)	204.4 ± 74.9	175.6 ± 49.0	0.388
τ_{SC} (s)	53.6 ± 26.5	102.2 ± 52.1	0.045 *
EE VO_2 (mL·kg^{-1}·min^{-1})	23.9 ± 6.1	28.4 ± 3.4	0.094
A'_{SC}/EE VO_2	0.1 ± 0.1	0.2 ± 0.1	0.191
Sum of residuals	533.2 ± 222.9	369.0 ± 185.6	0.141

UT, untrained participants; JT, judo athletes; $VO_{2baseline}$, baseline oxygen consumption rate; $A_{phase\ II}$, Amplitude of the primary phase; $\tau_{phase\ II}$, Time constant of the primary phase; $TD_{phase\ II}$, Time delay of the primary phase; Asc, Amplitude of the slow component phase; A'sc, Effective amplitude of the slow component; TD_{SC}, Time delay of the slow component phase; τ_{sc}, Time constant of the slow component phase; EE VO_2, oxygen uptake rate observed at the end of the square-wave transitions; A'SC/EE VO_2, Effective amplitude of the slow component relative to the oxygen consumption rate observed at the end of the square-wave transitions; Sum of residuals, Discrepancy in a dataset that is not explained by the model. * Significant differences between groups for $p < 0.05$.

The JT group presented significantly lower $\tau_{phase\ II}$ and higher τ_{SC} than UT. None of the other VO_2 kinetics parameters were different between groups. Large effect sizes were observed for $\tau_{phase\ II}$ (g = 1.2) and τ_{SC} (g = 1.2).

The normalized parameters of the response of [HHb] in the heavy-intensity exercise transitions for the two groups are presented in Table 3.

Table 3. Observed [HHb] kinetics during the heavy-intensity exercise square-wave transitions.

Variables	UT	JT	p
τ' [HHb] (s)	36.6 ± 17.1	36.2 ± 10.7	0.587
A [HHb] (A.U.)	18.5 ± 13.8	35.4 ± 15.9	0.049 *

τ' [HHb], Effective time constant of [HHb] kinetics; A [HHb], Amplitude of response of hemoglobin/myoglobin deoxygenation. * Significant differences between groups for $p < 0.05$.

The JT group presented significantly higher A [HHb] than the untrained participants, whereas τ' was not significantly different between groups. A large effect size was observed for A [HHb] (g = 1.1)

3.3. Upper Body RSA Protocol

The Dec-PPO, Dec-MPO and ΣWork of the RSA test obtained for each group are shown in Table 4.

Table 4. Variables of RSA for the two groups of participants.

Variables	UT	JT	p
Dec-PPO (W)	42.6 ± 8.8	22.1 ± 14.4	0.006 *
Dec-MPO (W)	14.2 ± 12.0	9.4 ± 9.6	0.406
ΣWork (KJ)	163.0 ± 5.5	175.8 ± 4.8	<0.001 *

Dec-PPO, decrease in the peak power output between the first and fourth sprint, Dec-MPO, decrease in mean power output between the first and fourth sprint; ΣWork, accumulated work in the RSA test. * Significantly different from UT values for $p < 0.05$.

There were significant differences between groups in the Dec-PPO, with the JT group displaying a lower Dec-PPO over the course of the upper body RSA protocol. Significant differences were also observed between groups in the ΣWork, with a larger mean value of ΣWork being observed in the JT group. A large effect size was observed for Dec-PPO (g= 1.7) and ΣWork (g = 2.5).

3.4. Correlations between RSA and the Physiological Variables Obtained in the Square-Wave Transitions and Incremental Step Test

Analysis considering the separate groups did not show any correlation between the performance parameters in the upper body RSA protocol and the parameters determined in the incremental test or square-wave transitions, neither in the JT group nor in the UT group. However, considering the heterogeneous group consisting of the whole sample of participants, significant correlations were found between MAP, peak VO_2 and $VT_1\_VO_2$ and Dec-PPO or ΣWork (Figure 1).

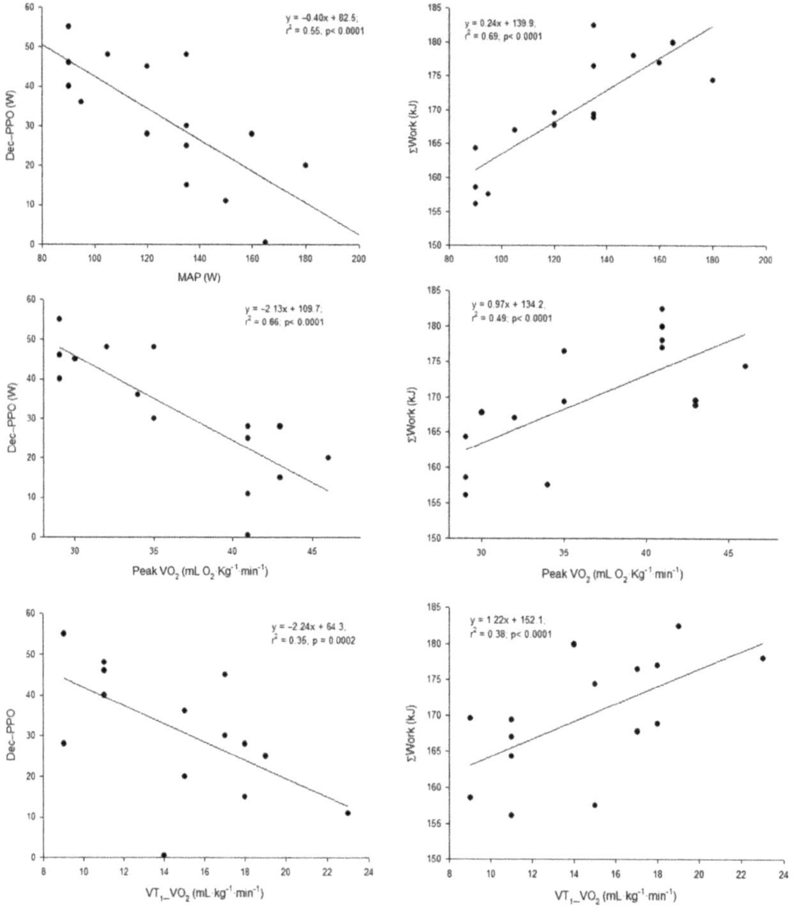

Figure 1. Relationships (correlations) between maximal aerobic power (MAP), peak oxygen consumption (peak VO_2) and oxygen consumption at the first ventilatory threshold (VT1_VO_2) achieved in the incremental test and the decrement in peak power output (Dec-PPO) and accumulated work (ΣWork) over the course of the upper body repeated sprint (RSA) test, observed in the group of heterogeneous fitness level (whole sample).

Additionally, for the UT, a significant correlation was found between $A_{phase\ II}$ and Dec-PPO or ΣWork and between A [HHb] and ΣWork, as highlighted in Figure 2.

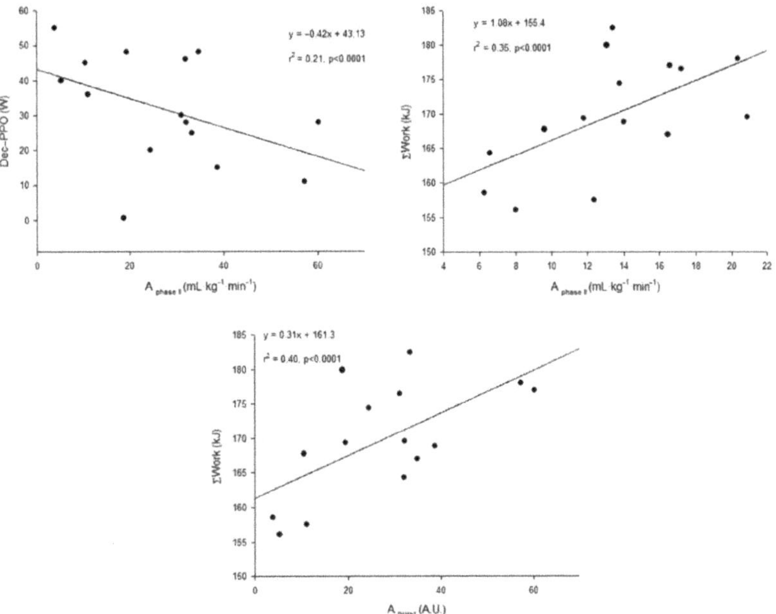

Figure 2. Relationships (correlations) between A of the phase II VO_2 kinetics, A [HHb] observed during the square-wave transitions and the decrement in peak power output (Dec-PPO) or accumulated work (ΣWork) over the course of the upper body RSA test, observed in the group of heterogeneous fitness level (whole sample).

No other significant correlations were observed between any of the performance variables and the VO_2 kinetics and [HHb] kinetics variables, whether when we consider each group separately or when we analyze the whole sample as a single heterogeneous group.

3.5. Predictive Model for ΣWork over the Course of the RSA Protocol

Considering the whole sample as a single heterogeneous group, a significant regression equation was found (F (2.12) = 12.737; $p < 0.001$) with an r^2 of 0.68, presented in Table 5, for which the main predictors were Peak VO_2 and Max. A [HHb] 4. The model had a y-intercept at 132.9 kJ, with the ΣWork increasing 0.8 kJ per each unit of increase in Peak VO_2 and increasing 0.16 kJ per each unit of increase in Max. A [HHb]. These two variables were found to explain 68% of the ΣWork during the RSA protocol.

Table 5. Predictors of accumulated work over the course of the upper body repeated sprint test for the whole group of participants.

Accumulated Work over the Course of the Upper Body RSA Protocol (kJ)	R	R^2	Adj. R^2	SEE	*p*
ΣWork (kJ) = 132.9 + 0.8 Peak VO_2 + 0.16 Max. A [HHb] 4	0.82	0.68	0.63	5.1	0.001

Peak VO_2, the highest 30 s average VO_2 attained over the course of the incremental test, Max. A [HHb] 4, the maximal [HHb] achieved in the fourth repetition of the upper body RSA test. All other variables were excluded from the model.

When each group was analyzed separately, no significant regression equation to predict ΣWork was found.

4. Discussion

To our knowledge, this was the first study to date which analyzed the relationship between parameters of aerobic fitness and upper body RSA performance, both in a heterogeneous sample consisting of trained and untrained participants, and more specifically, in a group of trained judo athletes. The present study revealed that a shorter $\tau_{\text{phase II}}$ and τ' [HHb] were not correlated to a lower decrease in PO over the course of an upper body RSA protocol, nor with a higher ΣWork. However, other variables of aerobic fitness, namely Peak VO_2, MAP, VT_1_ VO_2, and A of the phase II of VO_2 kinetics, were inversely correlated with the decrease in PO and directly correlated with the ΣWork over the course of the upper body RSA protocol, in the sample comprising both UT and JT groups. A [HHb] was also directly correlated with a higher ΣWork over the course of the upper body RSA protocol, in the sample comprising both UT and JT groups.

It has been proposed that VO_2 kinetics influences high-intensity exercise performance [46–49]. Several authors proposed that faster VO_2 kinetics, as expressed by a shorter $\tau_{\text{phase II}}$, are associated with the ability to support a given workload without tapping into O_2 deficit-related metabolic processes [7] and that faster VO_2 kinetics are related with faster [PCr] recovery kinetics following exercise [48], two potential aspects that may determine exercise tolerance during repeated high-intensity exercise.

The results observed in the current study indicate that there is no significant correlation between pulmonary $\tau_{\text{phase II}}$ and upper body RSA performance, namely between $\tau_{\text{phase II}}$ and the Dec-PPO, Dec-MPO or ΣWork over the course of the four sprints, either when we consider the sample of participants as a single heterogeneous group, or when we analyze the JT separately. These results contradict our main hypothesis, in which we proposed that a shorter $\tau_{\text{phase II}}$, would be associated with improved RSA performance variables, namely a higher ΣWork and smaller Dec-PPO and Dec-MPO.

Dupont and associates [21] have previously reported a significant direct correlation between $\tau_{\text{phase II}}$ and relative decrease in speed and total work performed over the course of an RSA protocol in a group of soccer players. Rampini and associates [49] also found a direct and significant ($r = 0.62$; $p < 0.05$) association between $\tau_{\text{phase II}}$ and the relative decrease in sprint speed over the course of six 40 m (20 m run-and-back) shuttle sprints separated by 20 s of passive recovery.

However, in line with our results, Buchheit [50] found no correlations between RSA performance and $\tau_{\text{phase II}}$, reporting that stepwise multiple regression analysis showed that mean repeated-sprint time, best sprint time and maximal aerobic speed were the only significant predictors of RSA performance. Accordingly, Christensen and associates [51] also found that $\tau_{\text{phase II}}$ was not associated with better RSA performance in a group of soccer players, although the changes in $\tau_{\text{phase II}}$ after a speed-endurance training program were associated with changes in RSA performance.

The studies mentioned above [21,50,51] involved running activities, utilizing a different set of muscle groups, in a different set of participants, exposed to very different training regimens compared to the individuals involved in the present study. However, they allow us to make some assertions regarding the observed results. The protocol used in the study by Dupont and associates [21] involved significantly more volume, and involved active recovery periods between sprints (15 × 40 m sprints, interceded with 25 s of active recovery), which may have biased the contribution of the aerobic energy system to the total work performed, and therefore the degree of association between RSA performance and $\tau_{\text{phase II}}$, while Buchheit [50] using a set of lower volume RSA protocols (10 × 30 m; 6 × 2 × 15 m; 6 × 16 m; 6 × 16 m; 20 × 15 m; 6 × 25 m) did not find such correlations. It is possible that an association may only be found between $\tau_{\text{phase II}}$ and RSA performance involving a high volume of RSA activity, given the increased contribution of the aerobic system as the number of sprints increases [6]. It is possible that an association between

$\tau_{phase\ II}$ and RSA performance would be found if a protocol involving a higher volume of sprints had been used. Moreover, the possibility that upper body VO_2 kinetics variables may play a more important role in judo contests that drag over a longer period of time should also be considered.

The results of the present study indicate that there is an inverse correlation between the MAP attained in the incremental step test and the Dec-PPO for the whole group of participants. Given that MAP is associated with training status [52], these results reveal that participants who are "aerobically" trained to a greater extent display an increased ability to resist decreases in PO over the course of repeated high-intensity exercise. Interestingly, this association was not observed for the JT group. It may be that the MAP of the individuals of the JT group was too similar (low range or spread of values) for any significant correlation to be established. It seems that there is a certain fitness threshold for which this association is valid, and that above this fitness threshold, other variables are more important in determining upper body RSA performance.

An inverse correlation was also found between the peak VO_2 and the Dec-PPO and a direct correlation between peak VO_2 and ΣWork. Similar findings have been reported by other authors [4,6]. The importance of peak VO_2 to RSA performance seems to be two-fold: (1) Across multiple sprints, aerobic ATP provision progressively increases such that aerobic metabolism may contribute as much as 40% of the total energy supply during the final repetitions of an RSA protocol [53]; (2) Enhanced oxygen delivery to muscles post-exercise potentially accelerates the rate of PCr resynthesis, an oxygen-dependent process [53,54], facilitating a faster recovery from high-intensity exercise.

Bishop and associates [55] observed a significant negative correlation ($r = -0.50$; $p < 0.05$) between VO_{2max} and % decrease in work over the course of 5 x 6 s sprints in a group of female basketball athletes. The authors proposed that athletes with greater VO_{2max} would be able to achieve a higher VO_2 rate throughout each sprint, reducing the contribution of substrate-level phosphorylation to ATP resynthesis, and therefore allowing more work to be done over the course of the RSA protocol [56]. Aguiar and associates [57] also reported a significant negative correlation ($r = -0.58$, $p < 0.05$) between VO_{2max} and the decrease in performance over the course of an RSA protocol (10 × 35 m sprints, 20 s recovery between sprints) in a heterogeneous group composed of endurance runners, sprinters and healthy individuals. Collectively, these results seem to emphasize the relationship between VO_{2max} and the ability to maintain a high power output over the course of several RSA efforts. This observed association between VO_{2max} and the decrease in work capacity over the course of an RSA task may also be associated with a higher cardiac output (Q) and subsequent increase in muscle blood flow, which may aid post-exercise recovery [58].

Furthermore, since a positive correlation was found between A [HHb] and ΣWork, it seems that repeated sprint performance is enhanced in individuals with higher oxygen extraction during heavy intensity exercise. The NIRS-derived [HHb] signal has been considered to reflect the ratio between muscle O_2 delivery and demand, and therefore has been considered an index of muscle O_2 extraction [27,59]. A higher A [HHb] has been associated with a greater muscle oxygen extraction following the onset of exercise [60,61] and has been shown to increase following training [62]. This indicates that a greater oxygen extraction at the onset of exercise and in repeated sprints may have an important role in the ability to maintain a constant performance over the course of several upper body high-intensity efforts.

Moreover, peak VO_2 and maximal [HHb] achieved in the fourth sprint were found to be significant predictors of ΣWork over the course of the four sprints, which seems to indicate that these aerobic fitness variables contribute to performance in repeated sprints. Specifically, both central and peripheral determinants of oxygen uptake contribute to performance during high intensity efforts where the anaerobic component is predominant. These associations were not observed when we consider the JT group separately, probably due to the more homogeneous response in these parameters in this restricted sample. In light of the observed results, it seems that aerobic fitness variables are associated with and increased upper body RSA performance in a group of individuals of heterogenous fitness. However, it is possible that once a certain level of upper body aerobic fitness is attained, other physiological and fitness variables may play a more important role in determining upper body RSA performance.

Complementary to this, the present study seems to indicate that there are significant differences between JT and UT participants in regard to upper body VO_2 kinetics parameters. No previous study has reported upper body VO_2 kinetics parameters in a group of JT. The $\tau_{phase\ II}$ values observed in this group of athletes are similar to those found by Koppo and associates [20] in a group of physically active males ($\tau_{phase\ II}$ = 48 $\pm$ 12 s). The values of VO_2 kinetics parameters for the group of UT participants observed in this study are similar to those reported by Schneider and associates [35] in a group of untrained participants ($\tau_{phase\ II}$ = 66 $\pm$ 3 s). Both mentioned studies analyzed the VO_2 kinetics response across the same exercise intensity range that was used in the present study. By comparison, the $\tau_{phase\ II}$ values observed by Invernizzi and associates [18] in a group of specifically upper body trained participants (elite competitive swimmers; $\tau_{phase\ II}$ = 34.3 $\pm$ 8.5 s), determined in an arm crank ergometer test, were much shorter than those which were observed in the JT group.

It is possible that judo-specific training may have induced sufficient training adaptations that resulted in a faster VO_2 kinetics response to exercise relative to untrained participants. However, given that judo-specific drills involve a different skeletal muscle function regimen, where isometric muscular actions of the upper body are emphasized, and muscle actions are performed in an intermittent way, the physiological adaptations that occur may involve very different mechanisms than those which are associated with the performance of high-volume, continuous exercise training of moderate–heavy exercise intensity, typical of swimming. This may also explain the similar results observed in the upper body VO_2 kinetics in the JT group compared to a group of physically active males [20].

Several studies have observed that different training programs, performed at different training intensities have the potential to induce adaptations compatible with shorter $\tau_{phase\ II}$ of VO_2 kinetics. Studies have observed improvements in VO_2 kinetics with training protocols ranging from low-intensity work at 60% VO_2 max. [62] to sprint-interval training performed at supramaximal intensities [60]. Nevertheless, these observations have been reported for studies involving dynamic, running or cycling exercise, which involve a different set of muscle groups and muscle action regimen compared to judo-specific training. As it has already been noted, judo-specific modalities seem to rely more on upper body musculature [2]. Given that upper body exercise has been associated with different hemodynamic [63] and metabolic responses [64], physiological adaptations may vary considerably compared to other forms of exercise.

McNeil and associates [65] have observed that performing isometric dorsiflexions at 100% of the maximal voluntary contraction (MVC) resulted in significant decreases in NIRS-derived tissue oxygenation compared to performing isometric dorsiflexions at 30% of MVC. Even though these authors reported no significant changes in tibial artery mean blood flow during the course of 60 s of sustained contraction, they suggested that the capillary mean blood flow might have been severely compromised over the course of the sustained exercise, and that this may have compromised tissue oxygenation dynamics throughout the 100% MVC exercise periods [66]. Given that, over the course of judo training and competition drills, athletes are likely to be exposed to similar conditions, muscle oxygen uptake dynamics might be compromised, and in turn, this may influence the type of physiological adaptations that take place.

The current study presents some limitations, which may limit the degree to which we can generalize the observations that were made. The sample size was relatively small, which affects the statistical meaningfulness of the observed results, as well as the degree to which we can extrapolate our conclusions. Nonetheless, it satisfies the requirements of statistical power determined a priori. Moreover, the athletes that participated in the present study were mostly national-level athletes, although it included some international-level athletes, and was therefore quite heterogenous. It is possible that different associations may have been found if the study had included judo athletes with a higher performance level, and therefore, further conclusions might be drawn regarding the physiological and fitness parameters which may be associated with upper body RSA performance in this group of athletes.

Furthermore, accessing body composition variables (% fat and fat-free mass), and other physical fitness variables (maximal upper body strength, anaerobic power) could have provided further insight regarding the determinants of upper body RSA performance and help explain the differences observed between groups. Although we designed the RSA test according to the observations reported by Soriano and associates [42], where the sum of time to kumi kata (8.4 s), kumi kata (6.1 s) and throwing time (1.3 s) corresponded to approximately 15 s, the duration chosen for the working periods, we recognize that the number of sprints may have been insufficient to match the number of sequences of attacks observed in more prolonged judo matches. It should also be noted that these tests were undertaken in an arm-crank ergometer, which is a general form of upper body dynamic exercise, and therefore the observed performance achieved by each individual in this task bears little resemblance to what actually happens over the course of a judo match. However, given that judo, as a grappling sport, relies heavily on the upper body musculature, a greater ability to preserve the work capacity of these muscle groups throughout a match seems to be a relevant aspect for potentially achieving a greater performance in the context of a judo match.

No studies to date had attempted to study the fitness variables underlying upper body RSA performance, particularly involving judo athletes. Therefore, the present study provides valuable information for further research regarding the variables that determine upper body RSA performance, particularly in this group of athletes, which may shed light regarding the factors that contribute to maintaining a high activity/attack profile over the course of a match.

Future research should consider using different repeated-sprint protocols with more repetitions and possibly a different work:rest period in order to reveal which factors may be associated with improved repeated-sprint ability under different match conditions. Moreover, future studies should include a larger sample of individuals, whether in a group of judokas of heterogenous fitness level/competitive status or in specific groups (international/elite vs. national level athletes), in order to better understand the factors that may determine RSA performance capacity in different groups. Future studies should also seek to include female individuals, in order to understand if different fitness variables influence upper body RSA performance for both sexes.

Practical Applications for Coaches

In lower-level athletes, developing a higher upper body aerobic fitness through higher volume general or specific exercises/drills would benefit their ability to maintain a higher performance over the course of the several high-intensity sequences of activity that take place during a match. However, for higher level athletes who already have reasonable upper body aerobic fitness, it may be more pertinent to devote time to developing other fitness variables in order to improve the ability to sustain a higher performance over the course of a judo match. The variables that ought to be developed in order to improve upper body RSA ability in higher-level judo athletes warrant further investigations.

5. Conclusions

The main conclusions of the present study are the following: (1) There seems to be no significant correlation between $\tau_{\text{phase II}}$ or τ' [HHb] and upper body RSA performance in a group of judokas or in a group of subjects with a heterogeneous fitness level; (2) No significant correlations seem to exist between peak VO_2, MAP or $VT_1\_VO_2$ and upper body RSA variables in a group of trained judokas; (3) Judokas displayed significantly faster VO_2

kinetics and higher muscle oxygen extraction in heavy-intensity exercise than untrained participants and; (4) There seems to be a positive association between MAP, peak VO_2, $VT_1\_VO_2$, $A_{phase\ II}$ and A [HHb] and ΣWork over the course of an upper body RSA task in a group of participants of heterogeneous fitness level. Therefore, aerobic fitness variables seem to play an important role in upper body RSA performance in individuals with a heterogenous fitness level.

Author Contributions: Conceptualization: A.A., C.P.M., F.B.A. and J.F.R.; methodology: A.A., L.D., C.P.M., M.C.E., F.B.A. and J.F.R.; formal analysis, A.A, L.D., C.D. and M.C.E.; investigation, A.A., L.D. and C.D.; supervision: F.B.A. and J.F.R.; data curation: A.A., L.D. and J.F.R.; writing—original draft preparation: A.A., C.D., C.P.M. and J.F.R.; writing—review and editing: A.A., C.D., C.P.M., M.C.E., F.B.A. and J.F.R. All authors have read and agreed to the published version of the manuscript.

Funding: This project was supported by the Portuguese Foundation for Science and Technology, I.P., Grant/Award Number UID/CED/04748/2020. This study was also partially supported by Fundação para a Ciência e Tecnologia, under grant UIDB/00447/2020 to CIPER–Centro Interdisciplinar para o estudo da performance humana (Unit 447).

Institutional Review Board Statement: The study was conducted according to the guidelines of the Declaration of Helsinki and approved by the Ethics Committee of the Faculty of Human Kinetics of the University of Lisbon (approval code 42/2021).

Informed Consent Statement: Informed consent was obtained from all subjects involved in the study.

Data Availability Statement: The data that support the findings of this study are available from the corresponding author upon reasonable request.

Acknowledgments: The authors gratefully acknowledge the participants for their time and effort and national judo coaches for helping recruit the participants.

Conflicts of Interest: The authors declare no conflict of interest that are directly relevant to the content of this manuscript.

References

1. Franchini, E.; Artioli, G.G.; Brito, C.J. Judo combat: Time-motion analysis and physiology. *Int. J. Perform. Anal. Sport* **2013**, *13*, 624–641. [CrossRef]
2. Franchini, E.; Del Vecchio, F.B.; Matsushigue, K.A.; Artioli, G.G. Physiological Profiles of Elite Judo Athletes. *Sports Med.* **2011**, *41*, 147–166. [CrossRef] [PubMed]
3. Boguszewski, D. Analysis of the final fights of the judo tournament at Rio 2016 Olympic Games. *J. Combat. Sports Martial Arts* **2016**, *7*, 63–68. [CrossRef]
4. Bogdanis, G.C.; Nevill, M.E.; Boobis, L.H.; Lakomy, H.K. Contribution of phosphocreatine and aerobic metabolism to energy supply during repeated sprint exercise. *J. Appl. Physiol.* **1996**, *80*, 876–884. [CrossRef] [PubMed]
5. Sahlin, K. Control of energetic processes in contracting human skeletal muscle. *Biochem. Soc. Trans.* **1991**, *19*, 353–358. [CrossRef]
6. Gaitanos, G.C.; Williams, C.; Boobis, L.H.; Brooks, S. Human muscle metabolism during intermittent maximal exercise. *J. Appl. Physiol.* **1993**, *75*, 712–719. [CrossRef] [PubMed]
7. Jones, A.M.; Burnley, M. Oxygen Uptake Kinetics: An Underappreciated Determinant of Exercise Performance. *Int. J. Sports Physiol. Perform.* **2009**, *4*, 524–532. [CrossRef]
8. Demarle, A.P.; Slawinski, J.J.; Laffite, L.P.; Bocquet, V.G.; Koralsztein, J.P.; Billat, V.L. Decrease of O_2 deficit is a potential factor in increased time to exhaustion after specific endurance training. *J. Appl. Physiol.* **2001**, *90*, 947–953. [CrossRef] [PubMed]
9. Glancy, B.; Barstow, T.; Willis, W.T. Linear relation between time constant of oxygen uptake kinetics, total creatine, and mitochondrial content in vitro. *Am. J. Physiol. Cell Physiol.* **2008**, *294*, C79–C87. [CrossRef] [PubMed]
10. Grassi, B. Oxygen uptake kinetics: Why are they so slow? And what do they tell us? *J. Physiol. Pharmacol.* **2006**, *57*, 53–65.
11. Burnley, M.; Jones, A.M. Oxygen uptake kinetics as a determinant of sports performance. *Eur. J. Sport Sci.* **2007**, *7*, 63–79. [CrossRef]
12. Poole, D.C.; Jones, A.M. Oxygen Uptake Kinetics. *Compr. Physiol.* **2012**, *2*, 933–996. [CrossRef]
13. Grassi, B.; Pogliaghi, S.; Rampichini, S.; Quaresima, V.; Ferrari, M.; Marconi, C.; Cerretelli, P. Muscle oxygenation and pulmonary gas exchange kinetics during cycling exercise on-transitions in humans. *J. Appl. Physiol.* **2003**, *95*, 149–158. [CrossRef]
14. Krustrup, P.; Jones, A.M.; Wilkerson, D.P.; Calbet, J.A.; Bangsbo, J. Muscular and pulmonary O_2 uptake kinetics during moderate- and high-intensity sub-maximal knee-extensor exercise in humans. *J. Physiol.* **2009**, *587*, 1843–1856. [CrossRef] [PubMed]
15. Wagner, P.D. A theoretical analysis of factors determining VO_{2MAX} at sea level and altitude. *Respir. Physiol.* **1996**, *106*, 329–343. [CrossRef]

16. Carter, H.; Jones, A.M.; Barstow, T.J.; Burnley, M.; Williams, C.; Doust, J.H. Effect of endurance training on oxygen uptake kinetics during treadmill running. *J. Appl. Physiol.* **2000**, *89*, 1744–1752. [CrossRef] [PubMed]
17. Hickson, R.C.; Bomze, H.A.; Hollozy, J.O. Faster adjustment of O_2 uptake to the energy requirement of exercise in the trained state. *J. Appl. Physiol.* **1978**, *44*, 877–881. [CrossRef] [PubMed]
18. Invernizzi, P.L.; Caporaso, G.; Longo, S.; Scurati, R.; Alberti, G. Correlations between upper limb oxygen kinetics and performance in elite swimmers. *Sport Sci. Health* **2008**, *3*, 19–25. [CrossRef]
19. Cerretelli, P.; Pendergast, D.; Paganelli, W.C.; Rennie, D.W. Effects of specific muscle training on VO_2 on-response and early blood lactate. *J. Appl. Physiol.* **1979**, *47*, 761–769. [CrossRef]
20. Koppo, K.; Bouckaert, J.; Jones, A.M. Oxygen uptake kinetics during high-intensity arm and leg exercise. *Respir. Physiol. Neurobiol.* **2002**, *133*, 241–250. [CrossRef]
21. Dupont, G.; Millet, G.P.; Guinhouya, C.; Berthoin, S. Relationship between oxygen uptake kinetics and performance in repeated running sprints. *Eur. J. Appl. Physiol.* **2005**, *95*, 27–34. [CrossRef]
22. Doncaster, G.; Marwood, S.; Iga, J.; Unnithan, V. Influence of oxygen uptake kinetics on physical performance in youth soccer. *Eur. J. Appl. Physiol.* **2016**, *116*, 1781–1794. [CrossRef] [PubMed]
23. Hoppeler, H.; Howald, H.; Conley, K.; Lindstedt, S.L.; Claassen, H.; Vock, P.; Weibel, E.R. Endurance training in humans: Aerobic capacity and structure of skeletal muscle. *J. Appl. Physiol.* **1985**, *59*, 320–327. [CrossRef] [PubMed]
24. Womack, C.J.; Davis, S.E.; Blumer, J.L.; Barrett, E.; Weltman, A.L.; Gaesser, G.A. Slow component of O_2 uptake during heavy exercise: Adaptation to endurance training. *J. Appl. Physiol.* **1995**, *79*, 838–845. [CrossRef] [PubMed]
25. Billat, V.L.; Richard, R.; Binsse, V.M.; Koralsztein, J.P.; Haouzi, P. The VO_2 slow component for severe exercise depends on type of exercise and is not correlated with time to fatigue. *J. Appl. Physiol.* **1998**, *85*, 2118–2124. [CrossRef]
26. Ferreira, L.F.; Townsend, D.K.; Lutjemeier, B.J.; Barstow, T.J. Muscle capillary blood flow kinetics estimated from pulmonary O_2 uptake and near-infrared spectroscopy. *J. Appl. Physiol.* **2005**, *98*, 1820–1828. [CrossRef]
27. DeLorey, D.S.; Kowalchuk, J.M.; Paterson, D.H. Relationship between pulmonary O_2 uptake kinetics and muscle deoxygenation during moderate-intensity exercise. *J. Appl. Physiol.* **2003**, *95*, 113–120. [CrossRef]
28. Ferreira, L.F.; Koga, S.; Barstow, T.J. Dynamics of noninvasively estimated microvascular O_2 extraction during ramp exercise. *J. Appl. Physiol.* **2007**, *103*, 1999–2004. [CrossRef]
29. Da Silva, J.F.; Guglielmo, L.G.A.; Bishop, D. Relationship between Different Measures of Aerobic Fitness and Repeated-Sprint Ability in Elite Soccer Players. *J. Strength Cond. Res.* **2010**, *24*, 2115–2121. [CrossRef]
30. World Health Organization. Global Recommendations on Physical Activity for Health. 2010. Available online: http://whqlibdoc.who.int/publications/2010/9789241599979_eng.pdf (accessed on 13 February 2020).
31. Erdfelder, E.; Faul, F.; Buchner, A. GPOWER: A general power analysis program. *Behav. Res. Methods Instrum. Comput.* **1996**, *28*, 1–11. [CrossRef]
32. McNarry, M.A.; Welsman, J.R.; Jones, A.M. Influence of training status and exercise modality on pulmonary O_2 uptake kinetics in pubertal girls. *Eur. J. Appl. Physiol.* **2011**, *111*, 621–631. [CrossRef]
33. World Medical Association. World Medical Association Declaration of Helsinki: Ethical Principles for Medical Research Involving Human Subjects. *JAMA* **2013**, *310*, 2191–2194. [CrossRef] [PubMed]
34. Rovati, L.; Fonda, S.; Bulf, L.; Ferrari, R.; Biral, G.; Salvatori, G.; Bandera, A.; Corradini, M. Functional cerebral activation detected by an integrated system combining CW-NIR spectroscopy and EEG. In *Optical Biopsy V, Proceedings of the Biomedical Optics 2004, San Jose, CA, USA, 24–29 January 2004*; SPIE: Bellingham, WA, USA, 2004; Volume 5326, pp. 118–126. [CrossRef]
35. Schneider, D.A.; Wing, A.N.; Morris, N.R. Oxygen uptake and heart rate kinetics during heavy exercise: A comparison between arm cranking and leg cycling. *Eur. J. Appl. Physiol.* **2002**, *88*, 100–106. [CrossRef] [PubMed]
36. Reinhard, U.; Müller, P.H.; Schmülling, R.-M. Determination of Anaerobic Threshold by the Ventilation Equivalent in Normal Individuals. *Respiration* **1979**, *38*, 36–42. [CrossRef]
37. Barstow, T.J.; Lamarra, N.; Whipp, B.J. Modulation of muscle and pulmonary O_2 uptakes by circulatory dynamics during exercise. *J. Appl. Physiol.* **1990**, *68*, 979–989. [CrossRef]
38. Whipp, B.J.; Davis, J.A.; Torres, F.; Wasserman, K. A test to determine parameters of aerobic function during exercise. *J. Appl. Physiol.* **1981**, *50*, 217–221. [CrossRef]
39. Whipp, B.J.; Ward, S.A.; Lamarra, N.; Davis, J.A.; Wasserman, K. Parameters of ventilatory and gas exchange dynamics during exercise. *J. Appl. Physiol.* **1982**, *52*, 1506–1513. [CrossRef] [PubMed]
40. McKay, B.R.; Paterson, D.H.; Kowalchuk, J.M. Effect of short-term high-intensity interval training vs. continuous training on O_2 uptake kinetics, muscle deoxygenation, and exercise performance. *J. Appl. Physiol.* **2009**, *107*, 128–138. [CrossRef]
41. Tobias, G.; Benatti, F.B.; de Salles Painelli, V.D.; Roschel, H.; Gualano, B.; Sale, C.; Harris, R.C.; Lancha, A.H., Jr.; Artioli, G.G. Additive effects of beta-alanine and sodium bicarbonate on upper-body intermittent performance. *Amino Acids* **2013**, *45*, 309–317. [CrossRef]
42. Soriano, D.; Irurtia, A.; Tarragó, R.; Tayot, P.; Milà-Villaroel, R.; Iglesias, X. Time-motion analysis during elite judo combats (defragmenting the gripping time). *Arch. Budo* **2019**, *15*, 33–43.
43. Baker, J.S.; McCormick, M.C.; Robergs, R.A. Interaction among Skeletal Muscle Metabolic Energy Systems during Intense Exercise. *J. Nutr. Metab.* **2010**, *2010*, 905612. [CrossRef] [PubMed]
44. Navidi, W. *Statistics for Engineers and Scientists*, 3rd ed.; Mcgraw-Hill Education: New York, NY, USA, 2008.

45. Cohen, J. A power primer. *Psychol. Bull.* **1992**, *112*, 155–159. [CrossRef]
46. Jones, A.M.; Vanhatalo, A.; Burnley, M.; Morton, R.H.; Poole, D.C. Critical Power: Implications for Determination of VO_{2max} and Exercise Tolerance. *Med. Sci. Sports Exerc.* **2010**, *42*, 1876–1890. [CrossRef]
47. Murgatroyd, S.R.; Ferguson, C.; Ward, S.A.; Whipp, B.J.; Rossiter, H.B. Pulmonary O_2 uptake kinetics as a determinant of high-intensity exercise tolerance in humans. *J. Appl. Physiol.* **2011**, *110*, 1598–1606. [CrossRef] [PubMed]
48. Chilibeck, P.D.; Paterson, D.H.; McCreary, C.R.; Marsh, G.D.; Cunningham, D.A.; Thompson, R.T. The effects of age on kinetics of oxygen uptake and phosphocreatine in humans during exercise. *Exp. Physiol.* **1998**, *83*, 107–117. [CrossRef] [PubMed]
49. Rampinini, E.; Sassi, A.; Morelli, A.; Mazzoni, S.; Fanchini, M.; Coutts, A.J. Repeated-sprint ability in professional and amateur soccer players. *Appl. Physiol. Nutr. Metab.* **2009**, *34*, 1048–1054. [CrossRef]
50. Buchheit, M. Repeated-Sprint Performance in Team Sport Players: Associations with Measures of Aerobic Fitness, Metabolic Control and Locomotor Function. *Int. J. Sports Med.* **2012**, *33*, 230–239. [CrossRef]
51. Christensen, P.M.; Krustrup, P.; Gunnarsson, T.P.; Kiilerich, K.; Nybo, L.; Bangsbo, J. VO_2 Kinetics and Performance in Soccer Players after Intense Training and Inactivity. *Med. Sci. Sports Exerc.* **2011**, *43*, 1716–1724. [CrossRef]
52. Poole, D.C.; Ward, S.A.; Whipp, B.J. The effects of training on the metabolic and respiratory profile of high-intensity cycle ergometer exercise. *Eur. J. Appl. Physiol. Occup. Physiol.* **1990**, *59*, 421–429. [CrossRef]
53. McGawley, K.; Bishop, D.J. Oxygen uptake during repeated-sprint exercise. *J. Sci. Med. Sport* **2015**, *18*, 214–218. [CrossRef]
54. Colliander, E.B.; Dudley, G.A.; Tesch, P.A. Skeletal muscle fiber type composition and performance during repeated bouts of maximal, concentric contractions. *Eur. J. Appl. Physiol. Occup. Physiol.* **1988**, *58*, 81–86. [CrossRef] [PubMed]
55. Harris, R.C.; Edwards, R.H.T.; Hultman, E.; Nordesjö, L.O.; Nylind, B.; Sahlin, K. The time course of phosphorylcreatine resynthesis during recovery of the quadriceps muscle in man. *Pflügers Arch.* **1976**, *367*, 137–142. [CrossRef] [PubMed]
56. Bishop, D.; Edge, J. Determinants of repeated-sprint ability in females matched for single-sprint performance. *Eur. J. Appl. Physiol.* **2006**, *97*, 373–379. [CrossRef] [PubMed]
57. De Aguiar, R.A.; Raimundo, J.A.G.; Lisbôa, F.D.; Salvador, A.F.; Pereira, K.L.; de Oliveira Cruz, R.S.; Turnes, T.; Caputo, F. Influence of aerobic and anaerobic variables on repeated sprint tests. *Rev. Bras. Educ. Fís. Esporte* **2016**, *30*, 553–563. [CrossRef]
58. Bangsbo, J.; Hellsten, Y. Muscle blood flow and oxygen uptake in recovery from exercise. *Acta Physiol. Scand.* **1998**, *162*, 305–312. [CrossRef]
59. Grassi, B. Delayed Metabolic Activation of Oxidative Phosphorylation in Skeletal Muscle at Exercise Onset. *Med. Sci. Sports Exerc.* **2005**, *37*, 1567–1573. [CrossRef]
60. Bailey, S.J.; Wilkerson, D.P.; DiMenna, F.J.; Jones, A.M. Influence of repeated sprint training on pulmonary O_2 uptake and muscle deoxygenation kinetics in humans. *J. Appl. Physiol.* **2009**, *106*, 1875–1887. [CrossRef]
61. Krustrup, P.; Hellsten, Y.; Bangsbo, J. Intense interval training enhances human skeletal muscle oxygen uptake in the initial phase of dynamic exercise at high but not at low intensities. *J. Physiol.* **2004**, *559*, 335–345. [CrossRef]
62. Berger, N.J.A.; Tolfrey, K.; Williams, A.G.; Jones, A.M. Influence of Continuous and Interval Training on Oxygen Uptake On-Kinetics. *Med. Sci. Sports Exerc.* **2006**, *38*, 504–512. [CrossRef]
63. Calbet, J.A.L.; González-Alonso, J.; Helge, J.W.; Søndergaard, H.; Munch-Andersen, T.; Saltin, B.; Boushel, R. Central and peripheral hemodynamics in exercising humans: Leg vs. arm exercise. *Scand. J. Med. Sci. Sports* **2015**, *25*, 144–157. [CrossRef]
64. Pendergast, D.R. Cardiovascular, respiratory, and metabolic responses to upper body exercise. *Med. Sci. Sports Exerc.* **1989**, *21*, S126–S131. [CrossRef]
65. McNeil, C.J.; Allen, M.D.; Olympico, E.; Shoemaker, J.K.; Rice, C.L. Blood flow and muscle oxygenation during low, moderate, and maximal sustained isometric contractions. *Am. J. Physiol. Regul. Integr. Comp. Physiol.* **2015**, *309*, R475–R481. [CrossRef] [PubMed]
66. Sjøgaard, G.; Savard, G.; Juel, C. Muscle blood flow during isometric activity and its relation to muscle fatigue. *Eur. J. Appl. Physiol. Occup. Physiol.* **1988**, *57*, 327–335. [CrossRef] [PubMed]

International Journal of
Environmental Research and Public Health

MDPI

Article

Postactivation Performance Enhancement (PAPE) Increases Vertical Jump in Elite Female Volleyball Players

Lamberto Villalon-Gasch , Alfonso Penichet-Tomas , Sergio Sebastia-Amat , Basilio Pueo * and Jose M. Jimenez-Olmedo

Physical Education and Sports, Faculty of Education, University of Alicante, 03690 Alicante, Spain; lamberto.villalon@ua.es (L.V.-G.); alfonso.penichet@ua.es (A.P.-T.); sergio.sebastia@ua.es (S.S.-A.); j.olmedo@ua.es (J.M.J.-O.)
* Correspondence: basilio@ua.es

Abstract: The purpose of this study was to verify if a conditioning activity was effective to elicit postactivation performance enhancement (PAPE) and to increase the performance in vertical jump (VJ) in elite female volleyball players. Eleven national Superliga-2 volleyball players (22.6 ± 3.5 years) were randomly assigned to an experimental and control group. Countermovement jumps (CMJ) were performed on eight occasions: before (Pre-PAPE) and after activation (Post-PAPE), after the match (Pre-Match), and after each of the five-match sets (Set 1 to 5). ANOVA showed significantly increased jump performance for the experiment between baseline (Pre-PAPE) and all the following tests: +1.3 cm (Post-PAPE), +3.0 cm (Pre-Match), +4.8 cm (Set 1), +7.3 cm (Set 2), +5.1 cm (Set 3), +3.6 cm (Set 4), and +4.0 cm (Set 5), all showing medium to large effect size (0.7 < ES < 2.4). The performance of the control group did not show significant increases until Set 3 (+3.2 cm) and Set 5 (+2.9 cm), although jump heights were always lower for the control group than the experimental. The use of conditioning activity generates increased VJ performance in Post-PAPE tests and elicited larger PAPE effects that remain until the second set of a volleyball match.

Keywords: back squat; countermovement jump; sports performance; PAP; RM; training

Citation: Villalon-Gasch, L.; Penichet-Tomas, A.; Sebastia-Amat, S.; Pueo, B.; Jimenez-Olmedo, J.M. Postactivation Performance Enhancement (PAPE) Increases Vertical Jump in Elite Female Volleyball Players. *Int. J. Environ. Res. Public Health* **2022**, *19*, 462. https://doi.org/10.3390/ijerph19010462

Academic Editors: Catarina Nunes Matias, Stefania Toselli, Cristina Monteiro and Francesco Campa

Received: 26 November 2021
Accepted: 29 December 2021
Published: 1 January 2022

Publisher's Note: MDPI stays neutral with regard to jurisdictional claims in published maps and institutional affiliations.

1. Introduction

Vertical jump (VJ) is a good prognosticator of performance in numerous sports that involve explosive actions, including volleyball [1]. The jump height reached by players can be considered a key factor in volleyball. An improvement in height in VJ allows obtaining enhancements in technical actions such as sets, hits, services, or blocks [2] which are decisive to achieve success in a volleyball game [3]. Service, attack, and block effectiveness are the skills more correlated with winning games in volleyball [4–6].

In addition, jumping capacity is correlated to muscular strength [7] since greater muscular strength can lead to modifications in force–time profile resulting in better VJ performance. Numerous strength training methods have been used to improve VJ performance in volleyball, being most of them strength-based methods such as plyometrics, combined training methods as contrast and complex training [8], or routines based on weightlifting and powerlifting [9].

While these VJ improvement methods are long-term effect procedures, other practices are aiming at achieving acute effects on performance, on certain occasions during the competition, (e.g., warm-up). One of these short-term methods to enhance VJ performance is the Postactivation Performance Enhancement (PAPE) [10–12]. This concept has recently been proposed to be used when high-intensity voluntary conditioning contractions lead to enhancement in voluntary muscular performance, and therefore activation is produced in different ways as with postactivation potentiation (PAP) [10,11]. Although PAP and PAPE are related, they can be considered as a different phenomenon, since the mechanisms that produce PAP are different from those for PAPE. PAP implies an enhancement in the effectiveness of contraction due to a better pairing of actin and myosin, and is generated by electrostimulation. On the other hand, PAPE is related to phenomena such as muscle temperature, the proportion of water in the muscle fibers, and the number of activated motor units among other causes [12]. Therefore, their effects may appear at different times and intensities [13]. The presence of PAP does not have to imply that PAPE is generated [14], even so, PAPE could be evoked by PAP, or occur simultaneously [12], and there have also been cases where PAPE is produced without PAP, which confirms that the mechanisms that generate these phenomena are different [13].

From an ecological point of view, it seems more precise to use PAPE than PAP to refer to the performance improvement in volleyball, since the dependent variables used to verify its existence are directly related to performance, such as force, speed, or jump [12]. Furthermore, its possible effects last longer and are more applicable to real volleyball match conditions, where, due to the game's rules [15], it is not possible to generate 8-min strength-related pre-activity before the start of the match, but activations before starting the warm-up that could elicit greater PAPE during the match are plausible. On the other hand, the effects of PAP can also increase sports performance, and in volleyball, this could be achieved by including PAP in resistance workouts that allow obtaining improvements in strength through complex training [16,17].

Since the magnitude of PAPE depends on the levels of fatigue and potentiation [18], the magnitude of the activation will depend on this relation, and therefore, the performance will be increased if the effect of the potentiation is larger than fatigue [19]. This relationship is influenced by other individual factors such as individual physiological characteristics of the subject, experience, age, type of muscular fibers' distribution (i.e., fast-twitch vs. slow-twitch fibers), maximum strength, strength to power ratio, level of training, among others [20].

The design of the activation protocols will greatly affect the result of the enhancement achieved. A resting period between activation and potentiation elicits better performance [21,22]. Similarly, other determining factors of PAPE are the intensity, volume, and protocols of the activation loads and the intensity of jumps or displacements after the potentiation [23].

It has also been suggested that the best increases in VJ are obtained with strength exercises such as the squat, with protocols of 1 to 3 sets of 1 to 5 repetitions and loads greater than 80% of 1RM, obtaining the best results in between 1 to 9 min after activation [24,25]. In the review carried out on vertical jump improvement, Suchomel et al. (2016) arrive at similar conclusions adding the cumulative fatigue of the athlete as individual factors to those already mentioned.

All of these studies have used both trained and untrained subjects as a sample [26]. In this meta-analysis, it was observed that the greatest effects of PAPE occur between 3 and 7 min in trained subjects, obtaining better results than studies for less than 3 min. Also, for studies carried out between 3 and 12 min or more, always for loads greater than 80% of 1RM and in trained subjects, the same authors noted that the longer times included in other meta-analysis are suitable for untrained subjects with smaller loads, where the effect of fatigue is greater.

However, contradictory results were found in the reviewed literature: the improvements in VJ found by Dobbs et al. (2018) were not statistically significant. In addition, some authors did not find effects on jumping performance after a PAP protocol [14,26–30]. Furthermore, the persistence of PAP is significant only for a limited period of time from 28 s to less than 3 min [31], obtaining the performance peak improvement (PAPE) at 6–20 min [25,32].

After reviewing the studies of PAPE protocols applied to volleyball players, it was found that the samples in all the studies are mostly composed of university or college players [22,33–36], with most of them being male players. The physiological difference between sexes [37] may elicit different responses to PAPE. In general, male players have greater type II fibers cross-sectional area and shorter twitch contraction times, whereas female players show more fatigue resistance [38]. Therefore, the outcomes of PAPE may be different depending on gender [38]. Thus, there is a lack of studies on female athletes, particularly in elite female players [39]. Furthermore, none of the studies in the literature has been conducted in real game conditions with volleyball players.

Therefore, the purpose of this study was to observe the effects of PAPE throughout a match in professional female volleyball players. The initial hypothesis was that squat-based pre-activation can trigger PAPE which is displayed as an improvement of VJ height 8 min after the application of the activation and that PAPE lasts for several minutes in a volleyball match of female national Spanish Superliga 2 players.

2. Materials and Methods

2.1. Subjects

Twelve Superliga 2 players of University of Alicante volleyball team volunteered to participate in this study (Table 1). Informed consent was obtained from all subjects involved in the study, who read and signed the document before any action in the study was taken. The study was conducted according to the guidelines of the Declaration of Helsinki [40], and approved by the Ethics Committee of University of Alicante (UA-17 November 2018).

Table 1. Characteristics of the subjects aggregated by group (mean ± SD).

	Experimental (n = 6)	Control (n = 5)	Total
Age (years)	21.33 ± 3.0	23.2 ± 3.8	22.2 ± 3.3
Height (cm)	171.3 ± 7.0	172.4 ± 8.7	171.8 ± 7.8
Body mass (kg)	64.0 ± 5.3	63.0 ± 3.8	63.5 ± 4.5
BMI (kg/m^2)	21.8 ± 5.3	21.3 ± 2.0	21.6 ± 1.6
Volleyball Experience (years)	8.8 ± 2.7	11.0 ± 2.6	9.8 ± 2.7
Strength Experience (years)	3.2 ± 1.8	3.2 ± 2.0	3.2 ± 1.9

BMI: body mass index, n: number of subjects, Volleyball experience: years the subjects have been playing volleyball; strength experience: time that subjects have been doing specific workouts.

The inclusion criteria were to have 4 years of experience minimum in the practice of volleyball in a national competition and to have previous knowledge in both strength training and half-squat exercise. The exclusion criteria were not to participate in all the tests involved in the study or to suffer injury or illness that prevents the performance of the tests. A control group participant suffered an injury during the experimental process, therefore, she was excluded from the experimental procedure and subsequent analysis.

2.2. Instruments

For the determination of the force–velocity profile and the vertical jump height, a linear encoder was used (Chronojump-Boscosystem, Barcelona, Spain). To estimate the vertical jump height, a jump mat was used (Chronojump-Boscosystem, Barcelona, Spain), from which to measure the flight time and, thus, estimate the jump height [41]. Both instruments worked at 1000 Hz.

2.3. Procedure

The experimental design shown in Figure 1 consisted of three phases: individualization, activation, and match, which are described in more detail as follows.

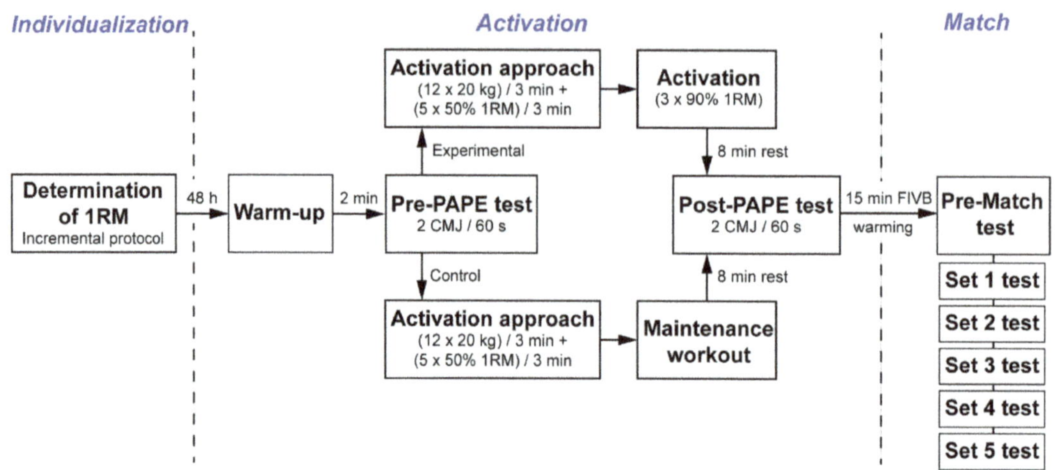

Figure 1. Experimental design of the study¸ RM: Repetition maximum; CMJ: Countermovement jump; PAPE: Post-activation performance enhancement FIVB: Fedération Internationalle de Volleyball.

2.3.1. Estimation of 1RM in the Half-Squat Exercise

In order to determine the load corresponding to the 1RM percentage in the half-squat exercise for the PAP protocol, the relationship between force and velocity was analyzed, since the speed of execution and the percentage of 1RM are proportional to each other [41]. An incremental loading test was carried out, in which the initial load was established at 30 kg and was gradually increased in 10 kg steps until mean barbell velocity was below 0.50 m/s (i.e., around 80% of 1RM). Afterward, the load was increased from 5 kg and at the end of the test, with speeds close to 0.30 m/s, increments of 1 kg were made to reach 1RM in the most precise way [42,43]. The value of 1RM was considered the load interpolated in the force–velocity profile with the average acceleration velocity value for the half-squat exercise of 0.30 m/s [44]. The players were refrained from performing physical activity 48 h previous to the test to ensure the absence of fatigue.

2.3.2. Vertical Jump

To determine the possible effect of the activation on PAPE in the lower train, countermovement jump (CMJ) heights were measured using a jump mat [45]. CMJ was performed starting from the standing position, with their feet in the center of the jump mat and hands positioned at the hips in akimbo position. After an auditory signal, subjects performed a knee flexion before jumping vertically to maximum height and were instructed to land in the center of the jump mat. A video camera monitoring players in sagittal plane was used to control that knee flexion reached the right joint angle. Three attempts were carried out with a 60-s resting time [44] and the highest value was considered for data analysis [45].

2.3.3. Activation Protocol

Considering that the sample were players with four years of minimum experience in volleyball training, they can be considered as trained subjects, so the guidelines set by Dobbs et al. (2018) in their meta-analysis were observed, as well as the corresponding 3-repetition activation protocol at 90% intensity of 1RM [35,46] with a resting time of 8 min. Such a protocol follows the margins indicated by these authors [26] and also the rest of the studies consulted [23,24,32,34,45].

Prior to activation, a standardized warm-up was performed for both groups, control and experimental, consisting of 4 min of soft running followed by 4 min of dynamic stretching; then 2 min of speed and changes of rhythm and direction inside the playground, and 5 consecutive CMJ jumps to finish [23]. After the warm-up, 2-min rest period was performed, followed by an initial evaluation of jump height before activation (Pre-PAPE test).

The experimental group performed the activation protocol, consisting of an approaching phase (12 repetitions with 20 kg, 3-min rest, 5 repetitions at 50% of 1RM, 3-min rest), followed by a conditioning phase (3 repetitions at 90% of 1RM). The control group executed the same approaching phase as the experimental group, but when the experimental group performed the conditioning phase control group executed a maintenance workout, consisting of smooth running interspersed with slight changes of direction and 3 vertical jumps. After 8 min, CMJ was measured (Post-PAPE test) to both groups with an identical methodology to that of the Pre-PAP data collection.

2.3.4. PAPE Monitoring during a Volleyball Match

After warm-up was finished, both groups performed a CMJ test before starting the match (Pre-match) and also just at the end of every set of the match (Set 1 to Set 5). This procedure allows describing the evolution in the height reached for volleyball players in a match, as well as to check whether the experimental group shows enhancement derived from PAPE and how much this condition lasts.

2.4. Statistical Analysis

Descriptive data are presented as mean and standard deviation. Due to the small sample size, the Shapiro–Wilk normality test was used, which resulted in a normal distribution. The differences in jump height between Pre-PAPE, Post-PAPE, Pre-match, Set 1, Set 2, Set 3, Set 4, and Set 5, in regards to experimental and control groups were evaluated using a repeated-measures ANOVA, including the different tests in time points as an intragroup variable and group as a between-subjects factor. Variance homogeneity and homogeneity of the error variances were verified via the Mauchly's test ($p = 0.304$) and the Levene's Test of Equality of Error (p range between 0.192 and 0.892 for all comparisons). In addition, a t-test for independent samples was conducted to compare differences on improvement percentage between experimental and control groups. The level of significance was set at $p < 0.05$. The d index was analyzed to determine the magnitude of an effect independent of sample size [47] and to classify the effect size, the criteria of Rhea for elite trained athletes were applied (d < 0.25 trivial; 0.25 ≤ d > 0.50 low; 0.50 ≤ d > 1.0 moderate; d ≥ 1.0 Large) [48]. In this quasi-experimental study, the sample is composed of volleyball elite players competing at a national level. Power analysis conducted with G*Power (v3.1.9.7, Heinrich-Heine-Universität Düsseldorf, Düsseldorf, Germany) indicated a minimum sample size of $n = 11$ subjects in order to detect an effect size of Cohen's d = 1.6 with 80% power ($\alpha = 0.05$, two-tailed) [49].

3. Results

Table 2 shows the compared results between control and experimental groups for CMJ. There were no significant differences between groups in Pre-PAPE tests in the height reached for both groups (p-value > 0.05), indicating that before the intervention the groups were homogeneous. Furthermore, significant intergroup differences and large ES can be observed in CMJ in Post-PAPE, Pre-Match, Set 1, Set 2, and Set 5 always being greater for the experimental group, therefore, the behavior is different for the groups until Set 2 and return to be different in Set 5 but with a reduced ES, Set 3 and Set 4 did not show significant differences. As well, there was a significant difference in the improvement percentage (Δ%) between the control and experimental groups from Post-PAPE until Set 2 test, but in the tests Set 3, Set 4, and Set 5 no differences were found (p-value < 0.05).

Table 2. Vertical jump height performance (mean ± SD).

	CMJ Experimental (cm) $n = 6$	CMJ Control (cm) $n = 5$	p	ES (d)
Pre-PAPE	34.08 ± 3.98	31.35 ± 4.28	0.302	0.66 [Moderate]
Post-PAPE	35.40 ± 3.69 *	29.61 ± 4.10	0.036	1.49 [Large]
Pre-Match	37.10 ± 4.09 *#	31.38 ± 3.99	0.045	1.41 [Large]
Set 1	38.84 ± 4.74 *#	31.22 ±2.61	0.011	1.94 [Large]
Set 2	41.37 ± 4.91 *#	32.75 ±4.47	0.015	1.83 [Large]
Set 3	39.15 ± 4.19 #	34.60 ± 4.43 #	0.115	1.05 [Large]
Set 4	37.66 ± 3.98 #	32.76 ± 2.44	0.073	1.23 [Large]
Set 5	38.11 ± 5.40 *#	34.32 ± 3.26 #	0.205	0.83 [Moderate]

* Significant difference between control and experimental groups at the same time point ($p < 0.05$); # Intragroup significant difference between Pre-PAPE and the other post-intervention tests.

As it can be observed in Table 3 the ES for the comparison between the pre-intervention and all post-intervention tests always are larger in the experimental group than in control, except in the Pre-PAPE test where both groups have moderate effect sizes, being slightly higher in the control group. However, in the control group, these values are negative, which occurred in a decrease in vertical jump performance.

Table 3. Effect size of intragroup differences in CMJ for Pre-PAPE and Pre-Match vs. the rest of the tests for control and experimental groups.

	Experimental		Control	
	p	ES (d)	p	ES (d)
Pre-PAPE vs. Post-PAPE	0.147	0.70 [Moderate]	0.127	0.87 [Moderate]
Pre-PAPE vs. Pre-Match	0.005	1.94 [Large] #	0.922	0.04 [Trivial]
Pre-PAPE vs. Set 1	0.002	2.31 [Large] #	0.903	0.05 [Trivial]
Pre-PAPE vs. Set 2	0.004	2.08 [Large] #	0.069	1.10 [Large]
Pre-PAPE vs. Set 3	0.002	2.40 [Large] #	0.009	2.14 [Large] #
Pre-PAPE vs. Set 4	0.012	1.60 [Large] #	0.313	0.51 [Moderate]
Pre-PAPE vs. Set 5	0.013	1.53 [Large] #	0.046	1.28 [Large] #
Pre-Match vs. Set 1	0.106	0.62 [Moderate]	0.834	0.09 [Trivial]
Pre-Match vs. Set 2	0.022	1.50 [Large] #	0.050	0.74 [Moderate]
Pre-Match vs. Set 3	0.057	0.79 [Moderate]	0.003	1.76 [Large] #
Pre-Match vs. Set 4	0.508	0.01 [Trivial]	0.268	0.76 [Moderate]
Pre-Match vs. Set 5	0.503	0.36 [Low]	0.015	1.61 [Large] #

Intragroup Significant difference between Pre-PAPE and Pre -Match with the rest of post-intervention tests.

In Table 3, intragroup differences in CMJ can be appreciated for experimental and control groups. The control group presents lower values for CMJ in the Post-PAPE, Pre-Match, and Sets one to five tests, whereas the experimental group increases the jump height in comparison with Pre-PAPE (baseline) values in all tests. On the other hand, another baseline at the beginning of the match (Pre-Match) allows for the analysis of jump performance in the five sets of the match, which showed different behaviors between both groups. The experimental group showed significant differences with the second set, while the control group did in the third and fifth sets.

Therefore, the improvement percentage values are lower in the control group than the experimental group as observed in Figure 2. There are significant differences between groups until Set 2. Also, it could be an increase in the values of improvement percentage in the experimental group up to Set 2, where there is a drop in performance until the end of the intervention, Set 5. On the other hand, the control group does not show significant improvement percentages until Set 3.

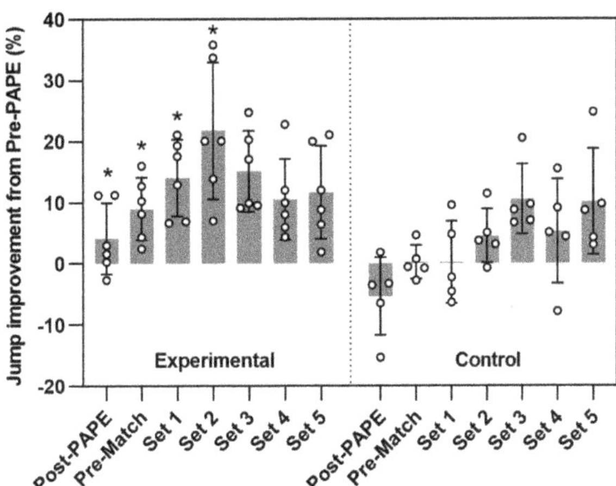

Figure 2. Comparison of the improvement values from Pre-PAPE performance expressed as percentage. * Significant difference between experimental and control groups at the time points: pre-PAPE, Pre-match and Set 1, Set 2, Set 3, Set 4, and Set 5. Bars, whiskers, and dots represent mean, standard deviation and individual values, respectively.

4. Discussion

The aim of this study was to analyze the effects of PAPE for professional female volleyball players during a match. In general, the results highlighted that squat-based pre-activation stimulates higher levels of PAPE shown as improvements in VJ height after activation, these improvements in VJ remained for several minutes during the match. To our knowledge, this is the first study of this kind with elite volleyball players.

PAP is an electrically evoked mechanism that produces an increase of muscle strength and twitch forces, as a result of its contractile history [12,47]. According to this, PAP produces improvements in the rate of force development (RFD) and maximal voluntary contractions (MVC) for a specified level of neural activation [12]. PAP is induced by the rise in the phosphorylation of regulatory light chains, which renders actin-myosin more sensitive to submaximal Ca^{2+} concentrations [48]. This activation occurs with more intensity in the fibers with the isoform II, which are involved in high intensity and short duration actions such as the VJ [19]. Other factors, such as the reduction in the pennation angle after a maximal voluntary contraction, are also suggested as possible mechanisms of PAP.

On the other hand, PAPE is associated whit an intensification in force production induced by previous muscle activity (i.e., voluntary contraction), and its presence is confirmed by performance outcomes [10,12]. Mechanisms proposed for PAPE are different from PAP, nevertheless, there are not well defined yet, but PAPE may be associated whit more lasted processes such as an increase in muscle temperature [49,50]. Also, the Muscle flow or/and water content and muscle activation (Partly through motivation) are mechanisms proposed for PAPE [12]. Finally, the increase in plasma catecholamines induced by exercise [51], and intensification in excitability of high order motor units [48,52,53] are proposed as mechanisms of PAPE, their effects may be observed until 20 min after Pre-activation at least. However, more investigation is needed in order to confirm those effects.

Most studies showed that PAPE protocol increased the performance in VJ in volleyball players [35,52–55]. Similar results can be found in our study with elite players as the experimental group showed improvements in VJ performance, while the control group has an opposite trend. Nevertheless, these differences between Pre-PAP and Post-PAPE tests are not statistically significant ($p > 0.05$). These results are in concordance with the study by [26,36] in which improvements in VJ were found, although not statistically significant. Due to the difficulty of access to elite athletes, the low sample size in our study may limit the statistical power to show differences between measures taken before and after activation. However, a moderate effect size of the VJ performance was observed, confirming the jump improvement tendency observed.

Volleyball players usually perform a typical explosive strength workout [56]. Those workouts include intensity loads ranging from 40 to 70% of 1RM, which are far from 90% of 1RM, and therefore, the classification as trained subjects [26,57] must be questioned. intensity loads of 90% of 1RM could produce an excess of fatigue in volleyball players and, as a result, the subjects could become non-responders, according to the criteria of [21,58]. Under these circumstances, the load may not fully adjust to the characteristics of the group, and therefore the response obtained is a smaller quantity than expected. Therefore, it is necessary to individualize the PAPE very carefully in order to adjust the activation intensity and volume loads to the individual characteristics of female volleyball players. Previous studies comparing routines based on peak strength and hypertrophy find that explosive-based workouts generate less fatigue [59].

On the other hand, if the improvement percentage values of both groups are compared, as shown in Figure 2, there are significant differences between control and experimental groups in the improvement reached in the post-PAPE tests. Positive improvement percentage values were observed for the experimental group (4.12%) while the control group showed an opposite trend (−5.37%). In addition, these statistically significant results are consistent with their moderate effect sizes, as depicted in Table 3, showing practical significance for the improvement percentage and the jump height reached in the CMJ in the Post-test. Hence, our study suggests that a conditioning activity would generate a positive effect on VJ performance, i.e., PAPE, as a result of an increase in muscle strength obtained 8 min after activation protocol [29,31,47,52,56,59].

The jumps distribution profile during the match was clearly different for control and experimental groups, which suggests that the effect of the activation could be one of the causes of this difference. After the peak of PAPE had occurred in the Post-PAPE test for the experimental group, the CMJ heights still progress, as shown in Table 2, peaking at Set 2 and reaching the end of the match with values similar to those at the start. These results agree with studies in which CMJ is used to evaluate fatigue after using loads and intensities higher than those used in our study (3 sets of 3 repetitions, 90% 1 RM) [59], in the analyzed study a decrease of 6% is observed immediately after the load, but an increase of 2% is observed in the CMJ 24 h after workout. Significant differences and large effect sizes were observed for all test occasions compared to the Pre-PAP test. However, the control group showed a different trend: all VJ heights were lower than the experimental, only Set 3 and Set 5 showed significant differences in regards to the Pre-PAPE test, and peaking at Set 3, later than experimental. This trend can also be analyzed through improvement percentage, shown in Figure 2. The experimental group achieved larger values and, again, showed a peak in Set 2, followed by a decrease in improvement percentage. The control group did not show improvement until Set 3.

The difference between groups could be explained by the presence of PAPE in the experimental group, which effect would extend beyond the time window of 7–12 min [25,26], increasing the jump performance, and also making the athlete more sensitive to future stimulus. Therefore, if PAP is combined with more explosive actions in warm-up routines, a summative effect may occur and therefore a performance improvement. As a result, the effect of PAPE combined with a standard volleyball warm-up, in which numerous jumps are executed [60], may be effective for elicited VJ enhancements.

The effects of PAPE last longer than those of PAP [12], but as in PAP, these effects will depend on the relationship with accumulated fatigue [11]. For the experimental group, the effects of PAPE could be largest than fatigue until Set 2 and consequently, a better improvement in jump performance than in the control group is observed The possible effects of PAPE were evaluated from activation at times ranging from 2 to 20 min maximum in other protocols [30,36]. In our study, CMJ tests were taken in longer time spans: after activation (8 min), at the beginning of the match (23 min), and in Sets 1 to 5 (46, 68, 95, 120, and 123 min, respectively). The experimental group peaked at Set 2 that occurred at 45 min from the beginning of the match and 68 min from activation, while the control group peaked at 90 min after activation. From Set 3 onwards, both groups appear to have similar conditions, and the values for improvement percentage are similar. The possible effect of accumulated fatigue, in addition to the dissipation of activation, causes the behavior of both groups to be more similar, which could be understood as the effect of PAPE is no longer present in the experimental group from Set 3 to the end of the match (i.e., from 90 to 130 min). Performance improvements are probably not mostly due to PAPE, but it is intuited that PAPE helps to generate a summation effect that produces an increase in the performance in VJ of volleyball players.

However, attributing the improvement percentage exclusively to the effect of PAPE generated by an initial conditioning activity and warm-up would not be entirely correct. Numerous physical and cognitive factors can affect the final performance, which is very difficult to control in a real game situation. For these reasons, the individualization of stimuli is very important. Despite this, the two groups in this study were in similar situations and only the group that performed a previous potentiation obtained a better improvement percentage in all sets with a greater effect size magnitude, and therefore, sports performance will be greater in this group.

The main limitation of this study is the sample size due to limited access to elite players in match conditions. The restricted statistical power because of the sample size in this study may have influenced the significance of some of the statistical comparisons conducted. A post hoc power analysis revealed that, for the lowest effect size of interest observed in the present study ($d = 0.7$), the number of players would have been approximately 25 for each group to obtain statistical power at the recommended 80% level. The results of this study serve as a basis that can be generalized to larger populations. Thus, more investigation with larger samples is needed to determine the effects of PAPE in volleyball female players and related sports.

5. Conclusions

The use of conditioning activity consisting of three repetitions of 90% of 1RM in the back half-squat exercise generates differences in the increase in CMJ heights between control and experimental group in Post-PAPE tests and elicited larger PAPE effects that remain until the second set of a volleyball match. The results of this study suggest that, if the activation is fitted individually on the correct form, and combined with an optimal warm-up, PAPE may be used to improve vertical jump performance, a key feature in volleyball and other related sports. Therefore, the inclusion of such protocols in volleyball warm-ups should be considered by coaches and physical trainers of volleyball teams. However, further investigation should be carried out, following different warm-up strategies with a wider sample in order to generalize the results achieved in the present study.

Author Contributions: Conceptualization, L.V.-G., B.P. and J.M.J.-O.; Data curation, A.P.-T.; Formal analysis, S.S.-A. and B.P.; Funding acquisition, A.P.-T.; Investigation, L.V.-G., A.P.-T., S.S.-A. and J.M.J.-O.; Methodology, A.P.-T., S.S.-A., B.P. and J.M.J.-O.; Supervision, J.M.J.-O.; Writing-original draft, L.V.-G. and B.P.; Writing-review & editing, L.V.-G., A.P.-T., S.S.-A., B.P. and J.M.J.-O. All authors have read and agreed to the published version of the manuscript.

Funding: This research was funded by Generalitat Valenciana, grant number GV/2021/098.

Institutional Review Board Statement: The study was conducted according to the guidelines of the Declaration of Helsinki, and approved by the Ethics Committee of University of Alicante (UA-17 November 2018, date of approval 20 December 2018).

Informed Consent Statement: Informed consent was obtained from all subjects involved in the study.

Data Availability Statement: Data can be obtained through the corresponding author on reasonable request.

Conflicts of Interest: The authors declare no conflict of interest.

References

1. Kraska, J.M.; Ramsey, M.W.; Haff, G.G.; Fethke, N.; Sands, W.A.; Stone, M.E.; Stone, M.H. Relationship between strength characteristics and unweighted and weighted vertical jump height. *Int. J. Sports Physiol. Perform.* **2009**, *4*, 461–473. [CrossRef]
2. Rodriguez-Ruiz, D.; Palmas, L.; Quiroga, M.E.; Palmas, L.; Miralles, J.A.; Palmas, L.; Canaria, D.G.; García-Manso, J.M.; Palmas, L.; Rodriguez-Ruiz, D.; et al. Study of the Technical and Tactical Variables Determining Set Win or Loss in Top-Level European Men's Volleyball. *J. Quant. Anal. Sports* **2011**, *7*, 7. [CrossRef]
3. Palao, J.M.; Manzanares, P.; Valadés, D. Way of scoring of Spanish first division volleyball teams in relation to winning/losing, home/away, final classification, and type of confrontation. *J. Hum. Sport Exerc.* **2015**, *10*, 36–46. [CrossRef]
4. Silva, M.; Marcelino, R.; Lacerda, D.; João, P.V. Match Analysis in Volleyball: A systematic review. *Monten. J. Sports Sci.* **2016**, *5*, 35–46.
5. Yu, Y.; García-de-alcaraz, A.; Wang, L.; Liu, T. Analysis of winning determinant performance indicators according to teams level in Chinese women' s volleyball. *Int. J. Perform. Anal. Sport* **2018**, *18*, 750–763. [CrossRef]
6. Toselli, S.; Campa, F. Anthropometry and Functional Movement Patterns in Elite Male Volleyball Players of Different Competitive Levels. *J. Strength Cond. Res.* **2018**, *32*, 2601–2611. [CrossRef] [PubMed]
7. Suchomel, T.J.; Nimphius, S.; Stone, M.H. The Importance of Muscular Strength in Athletic Performance. *Sports Med.* **2016**, *46*, 1419–1449. [CrossRef]
8. Forza, J. Complex Training for Volleyball: An Original Article. *Strength Cond.* **2019**, *27*, 71–77.
9. Holmberg, P.M. Weightlifting to Improve Volleyball Performance. *Strength Cond. J.* **2017**, *35*, 79–88. [CrossRef]
10. Boullosa, D.; Beato, M.; Dello Iacono, A.; Cuenca-Fernández, F.; Doma, K.; Schumann, M.; Zagatto, A.M.; Loturco, I.; Behm, D.G. A New Taxonomy for Postactivation Potentiation in Sport. *Int. J. Sports Physiol. Perform.* **2020**, *15*, 1–4. [CrossRef]
11. Cuenca-Fernández, F.; Smith, I.C.; Jordan, M.J.; MacIntosh, B.R.; López-Contreras, G.; Arellano, R.; Herzog, W. Nonlocalized postactivation performance enhancement (PAPE) effects in trained athletes: A pilot study. *Appl. Physiol. Nutr. Metab.* **2017**, *42*, 1122–1125. [CrossRef]
12. Blazevich, A.J.; Babault, N. Post-activation Potentiation Versus Post-activation Performance Enhancement in Humans: Historical Perspective, Underlying Mechanisms, and Current Issues. *Front. Physiol.* **2019**, *10*, 1359. [CrossRef] [PubMed]
13. Zimmermann, H.B.; Macintodh, B.R.; Pupo, J.D. Does postactivation potentiation (PAP) increase voluntary performance? *Appl. Physiol. Nutr. Metab.* **2020**, *45*, 349–356. [CrossRef] [PubMed]
14. Prieske, O.; Maffiuletti, N.A.; Granacher, U. Postactivation potentiation of the plantar flexors does not directly translate to jump performance in female elite young soccer players. *Front. Physiol.* **2018**, *9*, 276. [CrossRef] [PubMed]
15. FIVB. *Official Volleyball Rules 2017–2020*; de Volleyball, F.I., Ed.; FIVB: Lausanne, Switzerland, 2014.
16. Forza, J. Complex training for volleyball: A Critical Review. *Strength Cond.* **2018**, *26*, 57–63.
17. Pagaduan, J.; Pojskic, H. A Meta-Analysis on the Effect of Complex Training on Vertical Jump Performance. *J. Hum. Kinet.* **2020**, *71*, 255–265. [CrossRef]
18. Sale, D. Postactivation potentiation: Role in performance. *Br. J. Sports Med.* **2004**, *33*, 196–198. [CrossRef]
19. Tillin, N.A.; Bishop, D. Factors modulating post-activation potentiation and its effect on performance of subsequent explosive activities. *Sports Med.* **2009**, *39*, 147–166. [CrossRef]
20. Xenofondos, A.; Laparidis, K.; Kyranoudis, A.; Bassa, E.; Kotzamanidis, C.; Performance, S.; Science, S.; Science, S. Post-activation potentiation: Factors affeecting it and the effect on performance. *J. Phys. Educ. Sport* **2010**, *28*, 32–38.
21. Seitz, L.; Trajano, G.S.; Maso, F.D.; Haff, G.G.; Blazevich, A.J. Post-activation potentiation during voluntary contractions after continued knee extensor task-specific practice. *Appl. Physiol. Nutr. Metab.* **2015**, *40*, 230–237. [CrossRef] [PubMed]
22. Ah Sue, R.; Adams, K.; DeBeliso, M. Optimal Timing for Post-Activation Potentiation in Women Collegiate Volleyball Players. *Sports* **2016**, *4*, 27. [CrossRef] [PubMed]
23. Suchomel, T.J.; Lamont, H.S.; Moir, G.L. Understanding Vertical Jump Potentiation: A Deterministic Model. *Sports Med.* **2016**, *46*, 809–828. [CrossRef]
24. Picón-martínez, M.; Chulvi-medrano, I.; Cortell-tormo, J.M.; Cardozo, L.A. La potenciación post-activación en el salto vertical: Una revisión Post-activation potentiation in vertical jump: A review. *Retos: Nuevas Tendencias en Educación Física, Deporte y Recreación* **2019**, *2041*, 44–51. [CrossRef]
25. Wilson, J.M.; Duncan, N.M.; Marin, P.J.; Brown, L.E.; Loenneke, J.P.; Wilson, S.M.; Jo, E.; Lowery, R.P.; Ugrinowitsch, C. Met-Analysis of Postactivation Potentiation and Power: Effects of Conditioning Activity, Volume, gender, rest periods, and training status. *J. Strength Cond. Res.* **2013**, *27*, 854–859. [CrossRef] [PubMed]
26. Dobbs, W.; Tolusso, D.; Fedewa, M.; Esco, M. Effect of Postactivation Potentiation on Explosive Vertical Jump: A Systematic Review and Meta-Analysis. *J. Strength Cond. Res.* **2019**, *33*, 2009–2018. [CrossRef] [PubMed]
27. Healy, R.; Comyns, T.M. The application of postactivation potentiation methods to improve sprint speed. *Strength Cond. J.* **2019**, *33*, 2009–2018. [CrossRef]

28. Hilfiker, R.; Hubner, K.; Lorenz, T.; Marti, B. Effects of drop jumps added to the warm-up of elite sport athletes with a high capacity for explosive force development. *J. Strength Cond. Res.* **2007**, *21*, 550–555.
29. Sañudo, B.; de Hoyo, M.; Haff, G.G.; Muñoz-López, A. Article influence of strength level on the acute post-activation performance enhancement following flywheel and free weight resistance training. *Sensors* **2020**, *20*, 7156. [CrossRef] [PubMed]
30. Vargas-Molina, S.; Salgado-Ramírez, U.; Chulvi-Medrano, I.; Carbone, L.; Maroto-Izquierdo, S.; Benítez-Porres, J. Comparison of post-activation performance enhancement (PAPE) after isometric and isotonic exercise on vertical jump performance. *PLoS ONE* **2021**, *16*, e0260866. [CrossRef]
31. Maroto-Izquierdo, S.; Bautista, I.J.; Rivera, F.M. Post-activation performance enhancement (PAPE) after a single bout of high-intensity flywheel resistance training. *Biol. Sport* **2020**, *37*, 343–350. [CrossRef] [PubMed]
32. Oliveira, J.J.; Silva, A.S.; Baganha, R.J.; Barbosa, C.G.R.; Silva, J.A.O.; Dias, R.M.; Oliveira, L.H.S.; Pereira, A.A.; Ribeiro, A.G.S.V.; Pertille, A. Effect of Different Post-Activation Potentiation Intensities on Vertical Jump Performance in University Volleyball Players. *Off. Res. J. Am. Soc. Exerc. Physiol.* **2018**, *21*, 90–100.
33. Chen, Z.R.; Lo, S.L.; Wang, M.H.; Yu, C.F.; Peng, H. Te Can Different Complex Training Improve the Individual Phenomenon of Post-Activation Potentiation? *J. Hum. Kinet.* **2017**, *56*, 167–175. [CrossRef]
34. Maraboli, P.Q.; Garrido, A.B.; Hernández, C.A.; Guerra, S.C.; González, S.U. Jump height increase in university voleyball players. *Apunt. Educ. Física Y Deport.* **2016**, *4*, 64–71. [CrossRef]
35. Krzysztofik, M.; Kalinowski, R.; Trybulski, R.; Filip-Stachnik, A.; Stastny, P. Enhancement of Countermovement Jump Performance Using a Heavy Load with Velocity-Loss Repetition Control in Female Volleyball Players. *Int. J. Environ. Res. Public Health* **2021**, *18*, 11530. [CrossRef]
36. Arabatzi, F.; Patikas, D.; Zafeiridis, A.; Kotzamanidis, C.M. The Post-Activation Potentiation Effect on Squat Jump Performance: Age and Sex Effect. *Pediatr. Exerc. Sci.* **2014**, *26*, 187–194. [CrossRef]
37. Rixon, K.P.; Lamont, H.S.; Bemben, M.G. Influence of type of muscle contraction, gender, and lifting experience on postactivation potentiation performance. *J. Strength Cond. Res.* **2007**, *21*, 500–505.
38. López Villar, C.; Alvariñas Villaverde, M. Análisis muestrales desde una perspectiva de género en revistas de investigación de Ciencias de la Actividad Física y del Deporte españolas. *Apunt. Educ. Física Y Deport.* **2011**, 62–70. [CrossRef]
39. World Medical Association. World Medical Association Declaration of Helsinki. Ethical principles for medical research involving human subjects. *Bull. World Health Organ.* **2001**, *79*, 373. [CrossRef]
40. Pueo, B.; Penichet-Tomas, A.; Jimenez-Olmedo, J.M. Reliability and validity of the Chronojump open-sourcejump mat system. *Biol. Sport* **2020**, *37*, 255–259. [CrossRef] [PubMed]
41. Pérez-Castilla, A.; Jerez-Mayorga, D.; Martínez-García, D.; Rodríguez-Perea, Á.; Chirosa-Ríos, L.J.; García-Ramos, A. Comparison of the bench press one-repetition maximum obtained by different procedures: Direct assessment vs. lifts-to-failure equations vs. two-point method. *Int. J. Sport. Sci. Coach.* **2020**, *15*, 337–346. [CrossRef]
42. Martínez-Cava, A.; Morán-Navarro, R.; Sánchez-Medina, L.; González-Badillo, J.J.; Pallarés, J.G. Velocity- and power-load relationships in the half, parallel and full back squat. *J. Sports Sci.* **2019**, *37*, 1088–1096. [CrossRef]
43. Nibali, M.L.; Chapman, D.W.; Robergs, R.A.; Drinkwater, E.J. Validation of jump squats as a practical measure of post-activation potentiation. *Appl. Physiol. Nutr. Metab.* **2013**, *313*, 306–313. [CrossRef]
44. de Oliveira, J.J.; Verlengia, R.; Barbosa, C.G.R.; Sindorf, M.A.G.; da Rocha, G.L.; Lopes, C.R.; Crisp, A.H. Effects of post-activation potentiation and carbohydrate mouth rinse on repeated sprint ability. *J. Hum. Sport Exerc.* **2018**, *14*, 159–169. [CrossRef]
45. McCann, M.R.; Flanagan, S.P. The effects of exercise selection and rest interval on postactivation potentiation of vertical jump performance. *J. Strength Cond. Res.* **2010**, *24*, 1285–1291. [CrossRef]
46. Sanchez-Lopez, S.; Rodriguez-Perez, M.A. Effects of different protocols of Post-Activation Potentiation on Performance in the Vertical Jump, in relation to the F-V Profile in Female Elite Handball Players. *E-Balonmano.com Rev. Ciencias Deport.* **2018**, *14*, 16–17.
47. Robbins, D.W. Postactivation potentiation and its practical applicability: A brief review. *J. Strength Cond. Res.* **2005**, *19*, 453–458. [CrossRef] [PubMed]
48. Vandenboom, R. Modulation of skeletal muscle contraction by myosin phosphorylation. *Compr. Physiol.* **2017**, *7*, 171–212. [CrossRef]
49. Suchomel, T.; Sato, K.; DeWeese, B.; Ebben, W.; Stone, M. Potentiation effects of half-squats performed in a ballistic or nonballistic manner. *J. Strength Cond. Res.* **2016**, *30*, 1652–1660. [CrossRef]
50. McGowan, C.J.; Pyne, D.B.; Thompson, K.G.; Rattray, B. Warm-Up Strategies for Sport and Exercise: Mechanisms and Applications. *Sport. Med.* **2015**, *45*, 1523–1546. [CrossRef] [PubMed]
51. Cairns, S.P.; Borrani, F. β-Adrenergic modulation of skeletal muscle contraction: Key role of excitation-contraction coupling. *J. Physiol.* **2015**, *593*, 4713–4727. [CrossRef]
52. Evetovich, T.K.; Conley, D.S.; McCawley, P.F. Postactivation potentiation enhances upper- and lower-body atheltic performance in collegiate male and female athletes. *J. Strength Cond. Res.* **2015**, *29*, 336–342. [CrossRef]
53. Wallace, B.J.; Shapiro, R.; Wallace, K.L.; Mark, G.A.; Symons, T.B. Muscular and neural contributions to postactivation potentiation. *J. Strength Cond. Res.* **2019**, *33*, 615–625. [CrossRef] [PubMed]
54. Choon, Y.N.; Chen, S.E.; Lum, D. Inducing Postactivation Potentiation With Different Modes of Exercise. *Strength Cond. J.* **2019**, *42*, 63–81. [CrossRef]

55. Saez Saez de Villarreal, E.; González-Badillo, J.J.; Izquierdo, M. Optimal warm-up stimuli of muscle activation to enhance short and long-term acute jumping performance. *Eur. J. Appl. Physiol.* **2007**, *100*, 393–401. [CrossRef]
56. de Freitas, M.C.; Rossi, F.E.; Colognesi, L.A.; de Oliveira, J.V.N.S.; Zanchi, N.E.; Lira, F.S.; Cholewa, J.M.; Gobbo, L.A. Postactivation potentiation improves acute resistance exercise performance and muscular force in trained men. *J. Strength Cond. Res.* **2021**, *35*, 1357–1363. [CrossRef]
57. Seitz, L.; Villareal, E.; Haff, G. The temporal profile os postactivation potentiation is related to streght level. *J. Strength Cond. Res.* **2014**, *28*, 706–715. [CrossRef] [PubMed]
58. Hiscock, D.J.; Dawson, B.; Clarke, M.; Peeling, P. Can changes in resistance exercise workload influence internal load, countermovement jump performance and the endocrine response? *J. Sports Sci.* **2017**, *36*, 191–197. [CrossRef]
59. Boullosa, D.; Abreu, L.; Beltrame, L.G.; Behm, D.G. The acute effect of different half squad set configurations on jump potentiation. *J. Strength Cond. Res.* **2013**, *27*, 2059–2066. [CrossRef]
60. Pérez-López, A.; Valadés, D. Bases fisiológicas del calentamiento en voleibol: Propuesta práctica (Physiological Basis of Volleyball Warm-Up: Practical Proposal). *Cult. Cienc. Y Deporte* **2013**, *8*, 31–40. [CrossRef]

International Journal of
Environmental Research and Public Health

MDPI

Article

Effects of Acute Microcurrent Electrical Stimulation on Muscle Function and Subsequent Recovery Strategy

Alessandro Piras [1], Lorenzo Zini [2], Aurelio Trofè [2], Francesco Campa [2,*] and Milena Raffi [2]

[1] Department of Biomedical and Neuromotor Sciences, University of Bologna, 40127 Bologna, Italy; alessandro.piras3@unibo.it
[2] Department for Life Quality Studies, University of Bologna, 47921 Rimini, Italy; lorenzo.zini2@studio.unibo.it (L.Z.); aurelio.trofe2@unibo.it (A.T.); milena.raffi@unibo.it (M.R.)
* Correspondence: francesco.campa3@unibo.it

Abstract: Microcurrent electrical neuromuscular stimulation (MENS) is believed to alter blood flow, increasing cutaneous blood perfusion, with vasodilation and hyperemia. According to these physiological mechanisms, we investigated the short-term effects of MENS on constant-load exercise and the subsequent recovery process. Ten healthy subjects performed, on separate days, constant-load cycling, which was preceded and followed by active or inactive stimulation to the right quadricep. Blood lactate, pulmonary oxygen, and muscle deoxyhemoglobin on-transition kinetics were recorded. Hemodynamic parameters, heart rate variability, and baroreflex sensitivity were collected and used as a tool to investigate the recovery process. Microcurrent stimulation caused a faster deoxyhemoglobin (4.43 ± 0.5 vs. 5.80 ± 0.5 s) and a slower VO_2 (25.19 ± 2.1 vs. 21.94 ± 1.3 s) on-kinetics during cycling, with higher lactate levels immediately after treatments executed before exercise (1.55 ± 0.1 vs. 1.40 ± 0.1 mmol/L) and after exercise (2.15 ± 0.1 vs. 1.79 ± 0.1 mmol/L). In conclusion, MENS applied before exercise produced an increase in oxygen extraction at muscle microvasculature. In contrast, MENS applied after exercise improved recovery, with the sympathovagal balance shifted toward a state of parasympathetic predominance. MENS also caused higher lactate values, which may be due to the magnitude of the muscular stress by both manual treatment and electrical stimulation than control condition in which the muscle received only a manual treatment.

Keywords: MENS; oxygen consumption; deoxyhemoglobin kinetics; near-infrared spectroscopy; lactate; cycling

Citation: Piras, A.; Zini, L.; Trofè, A.; Campa, F.; Raffi, M. Effects of Acute Microcurrent Electrical Stimulation on Muscle Function and Subsequent Recovery Strategy. *Int. J. Environ. Res. Public Health* **2021**, *18*, 4597. https://doi.org/10.3390/ijerph18094597

Academic Editor: Olga Scudiero

Received: 10 March 2021
Accepted: 24 April 2021
Published: 26 April 2021

1. Introduction

Microcurrent electrical neuromuscular stimulation (MENS) involves series of stimuli delivered superficially, in the microampere range, through special transducer gloves that allow managing microcurrent signals through manipulation techniques. It is a key component for many medical and sport applications, and it is largely used for rehabilitation, training, and recovery purposes [1].

Nowadays, interest in the use of low-intensity current such as MENS is increasing, as its effects take place at the cell level (protein synthesizing activity; increased ATP generation), with sub-sensory application (i.e., painless), besides the absence of collateral effect, low cost, and easy utilization [2]. The utilization of electric field and currents comparable to different cells results in the stimulation of growth and tissue restoration [3] and diminution of edema [4]. Electric stimulations ranging from 10 to 1000 μA increase ATP levels and protein synthesis of rat skin, without having an impact on DNA metabolism [2]. The consequences on ATP production are described by proton actions [5], whereas the amino acids transport through the cell are facilitated by the alterations of the electrical gradients across the membranes [2]. Throughout stimulation of damaged muscles, MENS manages the modified membrane function by various processes, such as the preservation of intracellular Ca^{2+} homeostasis and with the augmented production of ATP levels [6].

Prior studies demonstrated that muscle damage treatment through microcurrent at low amperage (<500 μA) can decrease the severity of muscle symptoms [7] and a quicker regrowth of atrophied animals leg muscles [8]. In addition, microcurrents tone up the smooth muscles of blood vessels, improve skin turgor and tissue temperature, with an increase in blood flow through area treated [9]. All these characteristics are associated with vasodilation, then stimulating the metabolism of waste and toxins from the blood, therefore increasing healing and decreasing pain [2].

Based on these health-related cellular effects, a combination of stimulation plus exercise might improve exercise performance, and it might also be valuable to accelerate the subsequent recovery, thanks to the increased muscle blood flow that accelerates muscle metabolites removal [10]. One of the most physiological variables used to evaluate the recovery process is the heart rate variability (HRV). At the end of exercise, HRV returns exponentially to control value, and its increment is functionally related to athletes' training status and to the exercise intensity previously executed. HRV is the tool used to analyze the cardiac autonomic responses in combination with the baroreflex sensitivity (BRS), which is a reflex that adapts the heart period in response to variations in systolic blood pressure. These parameters have been used to evaluate the different adaptations to exercise and the recovery times after exercise [11–15]. Regardless of several research and medical applications, few studies have investigated the MENS effect before or after endurance exercise [7,16]. To date, only one study has investigated MENS effects in combination with aerobic exercise in reducing abdominal fat [16]. Authors found that microcurrent application with a frequency range of 25–50 Hz, combined with aerobic exercise, led to a significant decrease in subcutaneous abdominal fat thickness through the lipolysis stimulation [16]. Furthermore, the majority of studies performed with MENS reported a significant reduction of delayed onset muscle soreness after strength exercise [7,17,18]. In elderly people, Kwon et al. [19] found that MENS, after 40 min of short-term application, has an effect on muscle function, enhancing handgrip strength and single leg heel-rise.

Considering the influence of MENS on microcirculation, vascularization, and cellular energy production described above, and that endurance exercise stimulates the microvascular oxygenation following the onset of contractions [20], it could be interesting to investigate the effect of MENS stimulation on muscle tissue oxygenation and its influence on pulmonary oxygen kinetics during cycling. The rapid increase of the pulmonary oxygen kinetics at the transition between rest and exercise is a determinant of aerobic performance and an indicator of a well-done state of oxidative energetic system activity [21]. Additionally, the faster rise in VO_2 after the onset of exercise indicates a higher muscle oxygen utilization, which is a characteristic of elite athletes and trained subjects [22].

Until now, to our knowledge, no studies have investigated the effects of MENS at the human muscle tissue level, and more precisely, on factors of endurance capacity that are related to performance, such as faster oxygen kinetics, higher muscle oxygen release, or reduced blood lactate level at higher aerobic intensity. Thus, it is possible to hypothesize that the instantaneous and short-term effects of MENS might enhance the individual's capabilities on exercise at submaximal intensities and to accelerate the subsequent recovery process. Therefore, the aim of our study was to evaluate the acute effects of MENS on the muscle endurance capacity and subsequent recovery in sport. The results could be of great importance for elucidation of the O_2 release to acute, localized MENS exposure and the development of efficacious performance and recovery modalities.

2. Materials and Methods

2.1. Participants and Inclusion Criteria

Experiments were performed in 10 healthy subjects (2 females, 8 males; mean ± SD: age 27.2 ± 3.6 years; body mass index (BMI) 23.4 ± 2.5 Kg/m^2; VO_2peak 49.9 ± 7.9 mL/kg/min). The subjects were recreationally active but not highly trained. All subjects were volunteers, healthy, non-smokers, and none of them were taking medications or supplements. None of the subjects reported physical deficit or injuries during the study.

All participants received a verbal explanation of experimental procedures, and informed consent was obtained before the beginning of recordings. In agreement with the Declaration of Helsinki, the experimental protocol was approved by our University Institutional Ethic Committee.

2.2. Study Design and Test Protocol

The study design was a cross-sectional, single-blind, randomized controlled trial. For the realization of this study, the participants visited our laboratory five times, with at least three days between each visit, in which we performed different recordings. In the first visit, the participants performed an incremental test on a cycle-ergometer (H-300-R Lode), to determine ventilatory threshold (VT), respiratory compensation point (RCP), and peak oxygen consumption (VO_2peak) to identify their individual workload for the succeeding four recording sessions. Expirated gases were analyzed using a Quark b^2 breath-by-breath metabolic system (Cosmed, Rome, Italy). After the incremental session, the subjects came to our laboratory and performed two repetitions of each conditions of ON (MENS stimulation) and OFF (sham stimulation). The cycling exercise protocol consisted of 1 min of unloaded exercise followed by 5-min of heavy-intensity exercise. On the following days, the four recordings were performed in random order, and the participants were never informed about the status of stimulation, because it was not perceived by the subjects at the cutaneous level (single-blind).

The incremental test consisted of one minute of unloaded pedaling, followed by a warm-up of 5 min at 50 W. Then, at a constant cycling frequency of 75 rpm, the power output started at 80 W and was increased of 20 W/min until volitional exhaustion was reached or the required pedal rate could not be maintained [23]. Ventilatory and gas exchange variables were measured continuously breath by breath throughout the test. The highest VO_2 averaged over a 20 s interval was taken as VO_2peak. The LT and RCP were estimated from gas exchange measurements using the V-slope method, ventilatory equivalents, and end-tidal gas tensions [24]. Briefly, VT was determined from a different measurement, such as (i) the first unbalanced increase in CO_2 production (VCO_2) with respect to VO_2; (ii) an increase in expired ventilation (VE/VO_2) with no increase in VE/VCO_2; and (iii) an increase in end-tidal oxygen tension with no fall in end-tidal carbon dioxide tension. RCP was determined from a number of measurements including (i) an increase in VE/VCO_2; and (ii) an increase in end-tidal CO_2 tension.

Then, participants come to our laboratory for the exercise sessions. The procedure followed 4 stages: baseline; MENS-pre-exercise; exercise; and MENS-post-exercise, as illustrated below (Figure 1).

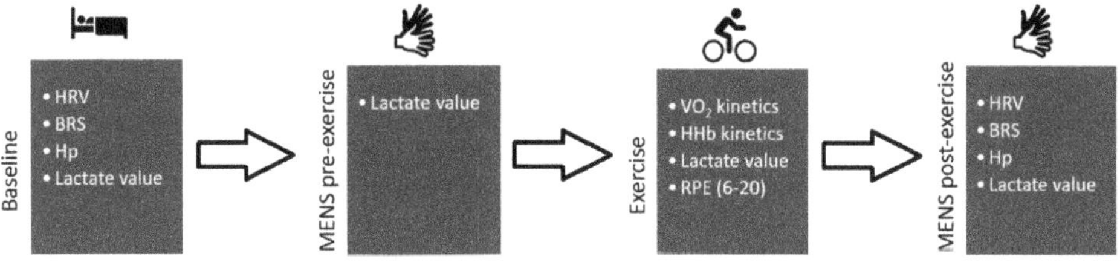

Figure 1. Graphical overview of the experimental protocol. HRV = heart rate variability, BRS = baroreflex sensitivity, Hp = hemodynamic parameters, MENS = microcurrent electrical neuromuscular stimulation, RPE = rate of perceived exertion, VO_2 = oxygen consumption, HHb = deoxyhemoglobin value. Gloves black and white represent active (ON) and inactive (OFF) stimulation, respectively.

Baseline. Participants stayed in a supine position in a quiet room, with a comfortable temperature (22–25 °C), for 10 min. They underwent noninvasive continuous blood pressure monitoring using servo-controlled infrared finger plethysmography (Portapres device; TNO/BMI, Amsterdam, the Netherlands) for analysis of heart rate variability

(HRV) and baroreflex sensitivity (BRS). HRV is the amount of heart rate fluctuations around the mean heart rate, and it reflects the cardiorespiratory control system. The BRS is an established tool for the assessment of the sympathetic and parasympathetic role of the autonomic nervous system [25]. Tests were performed under a standardized procedure at the same time of the day (9:00–12:00) to prevent circadian effects. Then, we took a blood sample at ear lobe for lactate measurement (Lactate Scout, SensLab, Leipzig, Germany). The reliability of the portable blood lactate analyzer was <0.5 mM for concentrations in the range of ≈1.0–10 mM [26].

MENS pre-exercise. Participants, in the same supine position, after the baseline procedure and before exercise, were manipulated with MENS (Electra Microlab, LED, Via Selciatella, Italy). The operator, one of the authors, applied electric pulses through the machine using special transducer gloves to manage microcurrent signals over the most common manipulation techniques, which is necessary to massage and stimulate the quadriceps of the right leg. During sham stimulation (MENS OFF), the operation followed the identical procedure used during the real stimulation, except for the fact that the instrument was turned off. It was possible because the electric pulse was not perceived by the subjects at the cutaneous level (single-blind). The entire massage and stimulation had a duration of 20 min with a Faradic current with rectangular waveform (1 s of impulse duration; frequency of 256 Hz; amplitude of 400 μA; Positive/Negative polarity with change direction), with the intention to stimulate hyperemic vasodilation. After stimulation, we took a second blood sample for lactate measurement; then, participants were ready for the exercise test.

Exercise. The data collected during the incremental test were used to calculate the work rates used during the subsequent constant-load exercise tests. Specifically, the individualized workload for each athlete (mean ± SD: 311.4 ± 70.1 watt) corresponded to ~ ≈50% of the difference between power (watt) reached at VT and at the RCP (≈50% Δ RCP-VT). Pedaling frequency was kept at about 70–80 revolutions/min. On-transitions were from unloaded pedaling to the imposed load, which was attained in about 3 s. Pulmonary ventilation (VE), oxygen consumption (VO_2), and carbon dioxide output (VCO_2) were determined with a Quark b^2 breath-by-breath metabolic system (Cosmed, Rome, Italy) previously calibrated according to the manufacturer's guidelines (included room air calibration, reference gas calibration, and turbine calibration with a 3-L syringe).

The changes in the vastus lateralis muscle oxygenation were evaluated by near-infrared spectroscopy (NIRS). A portable NIRS single-distance continuous-wave photometer (NIMO, Nirox Srl, Brescia, Italy) was utilized for the present study. In brief, the procedure is based on the changes of oxygen absorption with near infrared light, and it includes an emission probe that emits 3 wavelengths (685, 850, and 980 nm) and a photon detector. The transmitted light was recorded continuously at 40 Hz and utilized to quantity deoxygenated myoglobin and hemoglobin levels [27]. The deoxygenated value is less conditioned by the variations of the blood flow, and it is considered as a measure of fractional oxygen extraction inside the microvascular tissue [28]. After that, we had carefully shaven the skin, we attached the probe on it, covering, about 10–12 cm above the knee joint, the lower extremity of the right leg vastus lateralis muscle [29]. Then, the probe and the skin were wrapped with black cloth to prevent corruption from ambient light.

At the third minute of effort, we took a third blood sample for lactate measurement. At the end of the cycling, rate of perceived exertion was recorded with 6–20 Borg scale. Participants were asked how hard they felt the exercise [30]. The constant load exercise had a duration of about 10 min.

MENS-post-exercise. Immediately after exercise, participants were back positioned in a supine position, in the same room used before the exercise. To assess the effect of MENS on recovery, we stimulated the quadriceps of the right leg for 20 min with the same protocol described for MENS-pre-exercise. After that, in order to quantify the recovery level, we took a fourth blood lactate and a second continuous blood pressure monitoring using the same plethysmography used before exercise (Portapres device) for HRV and BRS analysis.

2.3. Data Analysis

VO_2 and HHb Kinetics. Breath by breath VO_2 and muscle oxygenation (from this point forward identified as HHb, expressed in μM) data obtained in the different repetitions of the exercise protocol (ON; OFF) were time aligned, interpolated on a second-by-second basis, and then superimposed for every athlete. Average values (every 1 s) were calculated and utilized for kinetics analysis. Data equivalent to the "cardiodynamic phase", recorded during the first 20 s of the on-transition were not included from the analysis [31]. To evaluate mathematically the VO_2 and HHb on-transition kinetics, data were fitted with two-exponential terms (primary and slow component of Equations (1) and (2)):

$$VO_2(t) = VO_2(b) + AP * (1 - e - (t\text{-}TDp/\tau p) \text{ (phase 2) (primary component)} + As * (1 - e - (t\text{-}TDs/\tau s) \text{ (phase 3) (slow component)} \quad (1)$$

and

$$HHb(t) = HHb(b) + AP * (1 - e - (t\text{-}TDp/\tau p) \text{ (phase 2) (primary component)} + As * (1 - e - (t\text{-}TDs/\tau s) \text{ (phase 3) (slow component).} \quad (2)$$

In Equations (1) and (2), Ap, As, TDp, TDs, and τp and τs denote the amplitude, time delay, and time constant, respectively, of the primary and slow component phases. Equations (1) and (2) were used on the basis of which equations yielded the lowest sum of squared residuals. We calculated also the percent contribution of the slow component with respect to the total amplitude of the response. Moreover, the gain of VO_2, as the increase in VO_2 above baseline to the reached steady state, and corrected for individualized workload (WL), was also calculated according to this equation (Equation (3)):

$$\text{Gain} = (VO_2\ [150\ s - 180\ s] - VO_2\ bas)/WL. \quad (3)$$

Heart rate variability. Time and frequency domain parameters were calculated regarding the HRV task force guidelines [25]. For the time domain, the square root of the mean squared differences of successive R-R intervals (RMSSD), and the standard deviation of successive R-R intervals (SDRR) were examined. Spectral analysis provides two main frequency parts: low frequency (LF) ranging between 0.04 and 0.15 Hz and high frequency (HF) positioned at the breathing frequency of 12 breath/minute. It has been revealed that HF is an index of the vagal tone, whereas LF reflects both sympathetic and vagal activities. Both indices (variables with skewed distributions) were log transformed (Ln). The LF/HF ratio provide quantitative markers of the cardiac sympathetic and the vagal modulation [25].

Baroreflex sensitivity. It was evaluated with Beatscope version 1.1 a (TNO/BMI, the Netherlands) with a BRS add-on module based on cross correlation analysis [13,14]. The slope of the regression line between SBP (systolic blood pressure) and R-R interval (all intervals between adjacent QRS complexes resulting from sinus node depolarizations) variations are considered as an index of BRS modulation of HR.

Hemodynamic parameters. The pulse contour method of Wesseling (the Modelflow method) was used to evaluate cardiac output (CO), stroke volume (SV), ejection time (EJT), and total peripheral vascular resistance (TPR) from the blood pressure waveform [32].

2.4. Statistical Analysis

All data are shown as means ± SD. All dependent parameters (VO_2; HHb; HRV; BRS; hemodynamic and lactate values) were compared between conditions (ON; OFF) with the paired sample t-test, in which means were considered significantly different at $p < 0.05$. To determine the magnitude of the stimulation effects, effect sizes (Cohen's d) were calculated as the mean difference standardized by the between-subject standard deviation and interpreted according to the thresholds: <0.20; small, >0.20–0.60; moderate,

>0.60–1.20; large, >1.20–2.00; very large, >2.00–4.00; extremely large, >4.0 [33]. Data were analyzed with SPSS v22.0 (IBM, New York, NY, USA). Regression analysis was done by the least squared residuals technique. Even though several powerful and dedicated software have been commercialized for regression analysis, we used the Solver add-in bundled of Microsoft Excel [34].

3. Results

All participants completed the protocol. Individualized oxygen uptake at maximal and submaximal level, with personalized workload obtained at exhaustion during the incremental exercise are shown in Table 1. VT occurred at 63% of the VO_2peak and at 56% of the maximum workload; consequently, during constant-load exercise, participants pedaled at a mean workload of ~65% (±5) of their maximum.

Table 1. Oxygen uptake and workload characteristics.

Athletes	VO_2peak (mL/kg/min)	RCP (mL/kg/min)	VT (mL/kg/min)	VO_2peak (Watt)	RCP (Watt)	VT (Watt)	Δ50% RCP + VT (Watt)
1	38.63	33.68	25.57	200.00	160.00	114.00	137.00
2	49.66	39.99	33.75	223.00	160.00	139.00	150.00
3	47.08	37.91	26.74	300.00	220.00	126.00	173.00
4	55.56	43.44	34.08	380.00	279.00	218.00	248.50
5	41.60	26.65	21.46	280.00	170.00	130.00	150.00
6	44.68	37.60	24.90	340.00	260.00	162.00	211.00
7	61.97	47.10	42.91	405.00	305.00	274.00	290.00
8	47.46	37.09	30.10	257.00	193.00	140.00	166.50
9	50.61	39.04	32.83	360.00	253.00	220.00	236.50
10	61.75	48.84	41.64	369.00	274.00	220.00	247.00
Mean	49.90	39.13	31.40	311.40	227.40	174.30	200.95
SD	7.80	6.40	7.10	70.10	53.90	54.30	52.60
SEM	2.50	2.00	2.20	22.20	17.10	17.20	16.60

Abbreviations: VO_2peak, peak oxygen consumption; RCP, respiratory compensation point; VT; ventilatory threshold.

Oxygen uptake kinetics. Figure 2 shows VO_2 on-kinetic analysis, in both conditions, for a typical subject. A slow component was observed in both experimental conditions, with non-significant slightly higher value during OFF than the ON condition (4.3% vs. 3.7%). Mean oxygen kinetic parameters for the exponential curve fitting are shown in Table 2. Analysis showed a slower primary component ($t(9) = -3.38$; $p = 0.004$; mean diff. = 3.25; $d = 0.60$), with a slower mean response time ($t(9) = -2.57$; $p = 0.015$; mean diff. = 4.88; $d = 0.66$) and a shorter time delay of the slow component ($t(9) = 1.90$; $p = 0.045$; mean diff. = −29.83; $d = -0.70$) during ON in comparison to the OFF condition. Moreover, VO_2 at the steady-state level was higher during ON than it was in the OFF condition ($t(9) = 2.90$; $p = 0.010$; mean diff. = 0.14; $d = 0.26$). We did not find any significant differences for the increase in VO_2 per unit increase in work rate (the gain of the primary phase), with values corresponding to 9.14 ± 0.5 and 9.05 ± 0.5 mL/min/W during the OFF and ON condition, respectively.

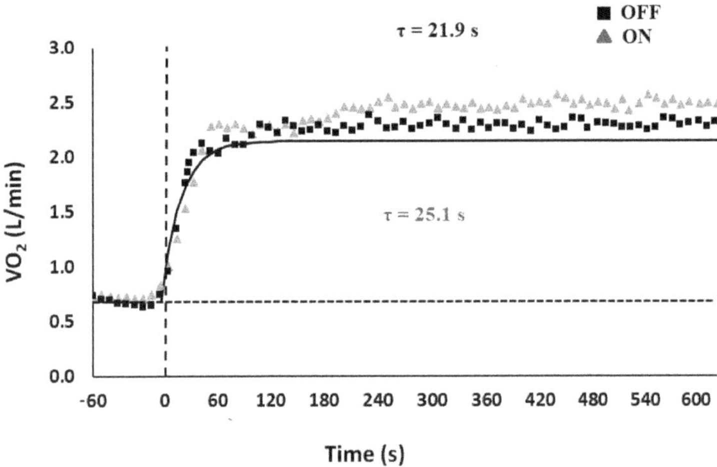

Figure 2. Characteristics of the two-component exponential model describing oxygen uptake (VO_2) during the on-transient of heavy intensity. Data refer to transition (at time 0, vertical dashed line) from unloaded pedaling to constant-load exercise during OFF (■) and ON (▲) experimental condition. Data points are average values calculated over 1 s. Horizontal dashed line represents baseline. Data obtained during the first 20 s of the transition were excluded from analysis.

Table 2. Pulmonary VO_2 on-kinetics parameters from unloaded pedaling to constant-load exercise across conditions (OFF; ON).

	VO_2(b) (L/min)	VO_2(ss) (L/min)	Ap (L/min)	TDp (s)	τp (s)	As (L/min)	TDs (s)	τs (s)	MRTp (s)	MRTs (s)	Sc (L/min)
OFF	0.70 ± 0.10	**2.44 ± 0.20**	1.80 ± 0.10	17.58 ± 0.90	**21.94 ± 1.30**	0.20 ± 0.04	**171.70 ± 11.30**	140.85 ± 19.40	**39.52 ± 1.80**	312.55 ± 27.03	0.12 ± 0.02
ON	0.79 ± 0.05	**2.58 ± 0.20 ***	1.79 ± 0.10	19.21 ± 0.90	**25.19 ± 2.10 ***	0.19 ± 0.04	**141.87 ± 15.50 ***	159.61 ± 26.10	**44.40 ± 2.80 ***	301.48 ± 25.90	0.10 ± 0.03

Values are mean ± SD. VO_2(b), oxygen consumption at baseline level; VO_2(ss), oxygen consumption at steady-state level; Ap, amplitude of response for primary component; TDp, time delay for primary component; τp, time constant for primary component; As, amplitude of response for slow component; TDs, time delay for slow component; τs, time constant for slow component; MRTp and MRTs, mean reaction time for primary and slow component; Sc, slow component. Bold values with asterisk indicate significant differences between conditions at $p < 0.05$.

Muscle oxygenation parameters. HHb on-kinetics analysis, in both conditions, for a representative subject are shown in Figure 3. Time values of HHb were significantly lower during ON than OFF conditions both at the primary, with faster τp ($t(9) = 2.96$; $p = 0.008$; mean diff. = −1.37; d = −0.88) and mean response time ($t(9) = 2.65$; $p = 0.013$; mean diff. = −1.39; d = −0.82), and at the secondary component with faster mean response time ($t(9) = 2.35$; $p = 0.022$; mean diff. = −32.75; d = −0.63) from rest-to-exercise transition (see Table 3).

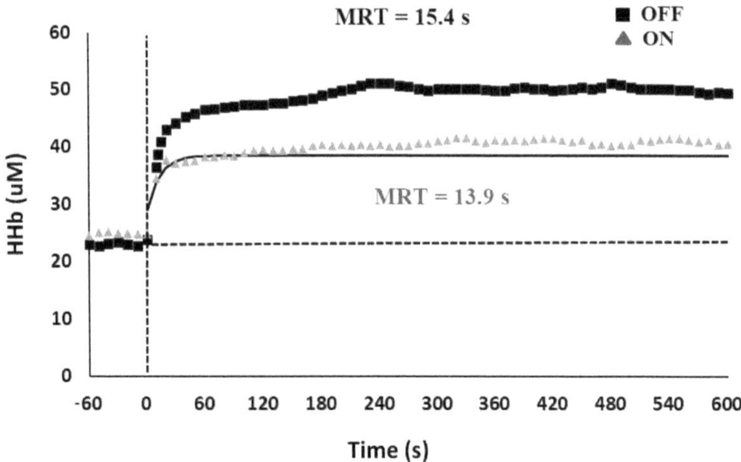

Figure 3. Characteristics of the two-component exponential model describing deoxyhemoglobin (HHb) during the on-transient of heavy intensity. Data refer to transition (at time 0, vertical dashed line) from unloaded pedaling to constant-load exercise during OFF (■) and ON (▲) experimental condition. Data points are average values calculated over 1 s. Data obtained during the first 20 s of the transition were excluded from analysis.

Table 3. Deoxygenated hemoglobin on-kinetics parameters from unloaded pedaling to constant-load exercise across conditions (OFF; ON).

	HHb(b) (μM)	HHb(ss) (μM)	Ap (μM)	TDp (s)	Tp (s)	As (μM)	TDs (s)	Ts (s)	MRTp (s)	MRTs (s)	Sc (μM)
OFF	23.99 ± 3.50	50.40 ± 6.10	24.54 ± 4.10	9.56 ± 0.20	**5.80 ± 0.50**	9.96 ± 2.20	113.22 ± 16.40	110.69 ± 17.60	**15.36 ± 0.50**	**223.91 ± 20.40**	2.47 ± 0.40
ON	23.95 ± 3.30	43.08 ± 5.90	19.13 ± 2.80	9.54 ± 0.20	**4.43 ± 0.50 ***	9.34 ± 2.20	90.61 ± 2.60	100.54 ± 11.70	**13.97 ± 0.50 ***	**191.15 ± 11.40 ***	2.11 ± 0.60

Values are mean ± SD. HHb(b), deoxyhemoglobin at baseline level; HHb(ss), deoxyhemoglobin at steady-state level; Ap, amplitude of response for primary component; TDp, time delay for primary component; τp, time constant for primary component; As, amplitude of response for slow component; TDs, time delay for slow component; τs, time constant for slow component; MRTp and MRTs, mean reaction time for primary and slow component; Sc, slow component. Bold values with asterisk indicate significant differences between conditions at $p < 0.05$.

Hemodynamic, cardiac autonomic variables. Table 4 shows all hemodynamic and autonomic variables investigated. Significant differences were found for systolic (t(9) = 2.67; p = 0.013; mean diff. = 7.48; d = 0.70) and mean arterial pressure (t(9) = 2.43; p = 0.024; mean diff. = 3.23; d = 0.51), with greater values during ON than OFF conditions. Time and frequency domain analysis showed significant differences for RMSSD (t(9) = 1.75; p = 0.047; mean diff. = 4.86; d = 0.28), HF (t(9) = 2.56; p = 0.015; mean diff. = 0.30; d = 0.37), and LF/HF ratio (t(9) = 1.95; p = 0.044; mean diff. = 0.42; d = 0.62) between ON and OFF conditions, respectively. It appeared that during stimulation (ON), participants recovered faster than during placebo condition (OFF).

Lactate and the rate of perceived exertion. Figure 4 shows lactate trends, analyzed at baseline; after the first MENS treatment that preceded the exercise; during the third minute of exercise; and after the second MENS treatment that has followed the exercise. A T-test revealed significant differences for comparison between ON and OFF conditions done before (t(9) = 1.65; p = 0.048; mean diff. = 0.16; d = 0.51), and after (t(9) = 1.89; p = 0.046; mean diff. = 0.36; d = 0.70) cycling. Rate of perceived exertion showed not significant difference between conditions (12.45 ± 2 vs. 11.25 ± 1.8 for ON and OFF, respectively).

Table 4. Hemodynamic and autonomic variables.

	OFF	*ON*
ΔSAP (mmHg)	**−4.98 ± 2.10**	**2.50 ± 2.00 ***
ΔDAP (mmHg)	0.01 ± 1.50	2.72 ± 2.10
ΔMAP (mmHg)	**−1.35 ± 1.50**	**1.88 ± 2.00 ***
ΔCO (L/min)	0.16 ± 0.20	0.31 ± 0.30
ΔSV (mL/min)	−8.35 ± 1.60	−5.41 ± 3.40
ΔHR (beat/min)	7.12 ± 1.70	7.00 ± 1.50
ΔEJT (s)	−0.02 ± 0.01	−0.02 ± 0.01
ΔTPR (mmHg s/mL)	−0.04 ± 0.05	−0.03 ± 0.10
ΔHRV (ms)	−140.30 ± 30.10	−132.25 ± 28.80
ΔSDRR (ms)	−4.32 ± 4.10	−1.60 ± 4.20
ΔRMSSD (ms)	**−12.97 ± 5.80**	**−8.10 ± 5.20 ***
ΔLF (Ln/ms^2)	0.02 ± 0.20	0.07 ± 0.30
ΔHF (Ln/ms^2)	**−0.43 ± 0.20**	**−0.13 ± 0.20 ***
ΔLF/HF	**0.59 ± 0.20**	**0.17 ± 0.10 ***
ΔBRS (ms/mmHg)	−2.75 ± 1.20	−2.15 ± 1.80

Delta values (mean ± SD) are obtained subtracting the recovery values from the baseline values. SAP, systolic arterial pressure; DAP, diastolic arterial pressure; MAP, mean arterial pressure; CO, cardiac output; SV, stroke volume; HR, heart rate; EJT, ejection time; TPR, total peripheral resistance; HRV, heart rate variability; SDRR standard deviation of the R-R intervals; RMSSD, root mean square of the successive differences; LF, low frequency; HF, high frequency; BRS, baroreflex sensitivity; Ln, logarithm. Bold values with asterisk represent significant differences at $p < 0.05$.

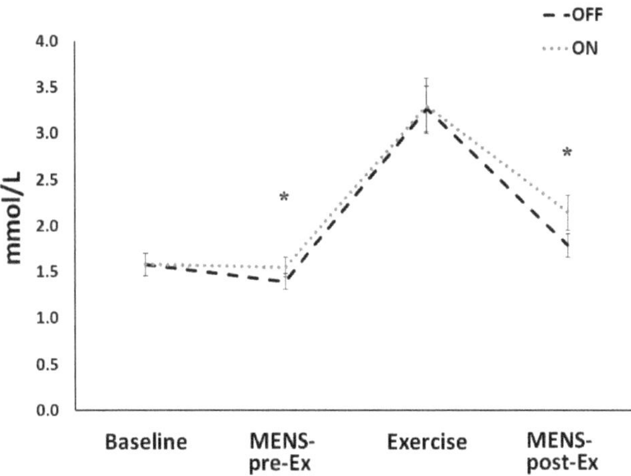

Figure 4. Mean (±SD) lactate values recorded at the baseline, after the first MENS treatment (MENS-pre-Ex) before exercise, at the third minute of the constant-load exercise (Exercise), and after the second MENS treatment (MENS-post-Ex), subsequently to exercise performance. Black dashed line represents OFF, the gray dashed line represents ON condition. Asterisks showed mean significant differences at $p < 0.05$.

4. Discussion

Different studies have discovered that the application of electric fields through the human body can significantly enhance cell metabolism [35] and injury restoration [36] when applied following exercise. The rationale behind the application of MENS is based on its efficacy to generate ATP at the cellular level and other health-related benefits, such as the increase in mitochondrial numbers [16], protein synthesis [7], and the activation of hormone-sensitive lipase, which increases the lipolysis process [16]. According to these different physiological mechanisms found in MENS treatment, our intention was to investigate the short-term effects of MENS on constant-load exercise at submaximal intensities and to the subsequent recovery process. The key results of our study are that MENS stimulation applied before exercise produced an increase in oxygen extraction at muscle microvasculature, while when applied after exercise, improved recovery through faster parasympathetic reactivation with respect to control condition. Moreover, electrical stimulation caused higher lactate levels, which may be due to the magnitude of the muscular stress by both manual treatment and electrical stimulation with respect to the control condition in which the right quadricep received only a manual treatment.

VO_2 and HHb kinetics. The main finding of the present study was the faster HHb on-transition kinetics during exercise executed after MENS stimulation, and surprisingly, by slower VO_2 on-transition kinetics. After the onset of exercise, a delay has been reported before an increase in muscle O_2 consumption [37], suggesting that the activation of mitochondrial respiration does not increase immediately, but rather, it has been delayed relative to the start of exercise. It could be argued that combining MENS with exercise might have been increased vasodilation and stimulated hyperemia, which could have, consequently, released nitric oxide, with the effect of accelerating O_2 availability at the muscle level. Nitric oxide represents an important component of the metabolic inertia to the VO_2 kinetics during supra-maximal exercise [38]. The precise mechanism by which nitric oxide contributes to the metabolic inertia at exercise onset is unclear but, in vitro, it has been demonstrated its role in inhibiting several mitochondrial enzymes, as it is a competitive inhibitor of oxygen consumption in the mitochondrial respiratory chain [39]. MENS treatment has received more widespread attention in the last years, as it not only relieves pain but also has a positive effect on reparative processes in the skin [1]. Microcurrents penetrate in the body's cells, normalize the biochemical processes, such as improving metabolism, increasing enzyme activity, ATP synthesis, proteins, lipids, and other vital substances [2]. In addition, microcurrents tone up the smooth muscles of blood vessels as well as improve skin turgor and tissue temperature, with an increase in blood flow through area treated [9]. They are associated with vasodilation, then stimulating the metabolism of waste and toxins from the blood, therefore increasing healing and decreasing pain [2]. Vasodilation and hyperemic processes might have stimulated NO release, even if, to the best of our knowledge, studies are lacking to support this hypothesis.

As shown in previous studies [40,41], the NIRS-derived HHb signal provides a continuous, noninvasive measurement of changes in muscle deoxygenation and reflects the balance between local muscle O_2 delivery and utilization. Our results have shown an immediate increment in muscle fractional O_2 extraction after a few seconds of delay ($\approx$10 s) following the onset of contraction. The rate of adaptation of muscle deoxygenation was faster than the adaptation of the primary phase of the VO_2, reflecting an accelerated O_2 extraction in the active muscle microvasculature as a consequence of microstimulation. Our results are in agreement with other studies that investigate increasing the availability of muscle O_2; through hyperoxia, adenosine, or drug administration to the O_2–hemoglobin dissociation curve, which facilitated O_2 release at the working muscle, the primary component of pulmonary VO_2 does not accelerates, even during high-intensity exercise [42,43]. This is in accordance with the hypothesis that VO_2 during the transition from rest-to-exercise is not managed by the rate of adjustment of convective oxygen delivery to the exercising muscles [42]. After the time delay, during the ON condition, HHb increased more rapidly toward a "steady-state" level, suggesting that oxygen delivery in the on-transition was more adequate to meet the metabolic demand of the muscle, thus requiring a rapid increase in O_2 extraction [41]. The slightly lower but not significantly different muscle HHb value exhibited by MENS stimulation at steady-state level, in concomitant with the significantly higher value of VO_2 consumption at the same working rate, suggests that our procedure may have improved oxygen availability/distribution within the muscle microvasculature. The cause of the slower phase II of VO_2 kinetics is unclear, although it is known that this parameter is sensitive to a number of factors, including the high percentage of type II fiber distribution in the working muscles [44]. However, it is difficult to see how MENS stimulation could alter muscle fiber recruitment patterns, although this should not, of course, be excluded yet.

Autonomic nervous system parameters. A second purpose of the present investigation was to examine the different physiological recovery responses to MENS exposure after exercise. The common physiological variable used to evaluate recovery time is the heart rate variability. At the end of exercise, HRV returns exponentially to control value, and its increment is functionally related to the athlete's training status and the exercise intensity previously executed. HRV is the tool used to investigate the cardiac autonomic responses in combination with the baroreflex sensitivity, which is a reflex that adapts the heart period in response to variations in systolic blood pressure. These parameters have been used to evaluate the different adaptations to exercise and the recovery times after exercise [11–14]. With a transition from exercise to passive recovery, there is a loss of central command and activation of the arterial baroreflex, resulting in a decrease in heart rate toward its pre-exercise level [45]. The vagal system plays a main role in reducing heart rate immediately after the cessation of exercise, and its further decrease is mediated by both the vagal and sympathetic system [13]. In the present study, we found significantly different effects on the autonomous nervous system parameters, with higher increase in vagal reactivation (RMSSD and HF band of the HRV frequency spectrum) after MENS compared to sham-exposure. Moreover, sympathovagal balance, assessed by LF/HF ratio, was shifted toward a state of parasympathetic predominance, revealing a faster recovery after stimulation treatment than in the control condition. A possible explanation of the microcurrent effects on faster recovery after exercise could be related to its effect on muscle metaboreflex. Until now, no study has investigated the effect of MENS on metaboreflex activity. One study found that the transcutaneous electric nerve stimulation, a technique similar to MENS (both are accepted mode of electrotherapy) [1], augments peripheral blood flow by reduction of the muscle metaboreflex, increasing oxygen supply to stimulated muscles, with a decrease in sympathetic activity evaluated with the heart rate variability [46]. These findings support the idea that the acute application of electrotherapy improves sympathovagal balance, which could be linked to an intense peripheral vasodilatation response, contributing to a faster recovery process.

Lactate and rate of perceived exertion. Hyperlactatemia is observed during exercise and severe inflammation [47], as well as in muscle cells subjected in vitro to electrical pulse stimulation [48]. In the present study, lactate levels were significantly higher after MENS treatments, both before and after exercise, whereas during constant-load cycling, participants produced the same lactate values in both experimental conditions. We can speculate that higher lactate values could have caused vasodilation at muscle level, through the changes in osmolarity and acidity, which are necessary to speed-up HHb on-transition kinetics but not higher enough to accelerate VO_2 on-kinetics. The finding that the lactate values increased with respect to inactive stimulation is somewhat surprising. We could assume that MENS increased the magnitude of muscular activity, which may be due to both manual and electrical stimulation with respect to a sham condition in which the right quadricep received only a manual treatment. Moreover, the lactate is crucial for muscle to make cytosolic NAD+, which is necessary to ATP regeneration from glycolysis, protecting muscles from acidosis. Lactate utilizes two protons, which is necessary to promoting proton elimination from muscles. Moreover, MENS efficacy on blood lactate values could be influenced by parameters used (pulse duration, frequency, amplitude, and muscles stimulated) [1,9], target population [49], and the type of fatiguing exercise or duration of recovery [9,10]. Finally, the RPE was not significantly different between conditions, and this result is similar to that of Barcala-Furelos et al. [10], in which electrical stimulation did not alter the RPE values when compared with the passive recovery in lifeguards following a water rescue.

Limitations of the study. Some limitations to the current investigation warrant discussion. Although we know that NIRS has several limitations, most of them have been prevailed by recent technological developments. For example, when the probe is applied on the skin overlying the muscle that we want to investigate, NIRS can measure only a relatively small and superficial volume of skeletal muscle tissue. However, the method has also important strengths and can give valuable and noninvasive useful insights into skeletal muscle oxidative metabolism in vivo during exercise [50].

Furthermore, the main limitation of the whole-body pulmonary oxygen consumption measurements is the difficulty to differentiate between the exercising muscles and the rest of the body, or between different muscles involved in the exercise. Moreover, the presence of O_2 stores between the location of measurement (the mouth) and the sites of gas exchange at the skeletal muscle level complicate data interpretation during metabolic transitions [50].

The sample size was in line with the most important articles published in this area, in which the number of participants is under 10 units and unbalanced between male and female. Although we have applied it on athletes, it could be important to direct future studies on patients and aged, in which their agility and life quality are limited for impairment in oxygen delivery and utilization.

We have also highlighted the fact that exercise duration, rate of increased in work rate, blood sampling location, instrument utilized, and measurement error are all potential sources of variability in measuring lactate values. However, the reliability of our portable blood lactate analyzer was <0.5 mM for concentrations in the range of $\approx$1.0–10 mM, with a measurement error of $\leq$3% [26]. Further investigations are needed, in terms of stimulation parameters (e.g., time, frequency, amplitude, duration) and with different exercise protocols.

5. Conclusions

In summary, we found that MENS stimulation causes a faster HHb and a slower VO_2 on-transition kinetics during exercise, with higher lactate levels immediately after the treatments. Sympathovagal balance was shifted toward a state of parasympathetic predominance, revealing a faster recovery after stimulation executed following cycling. These results could be due to the increased vasodilation and hyperemia, which are a consequence of stimulation. It seems plausible to consider MENS as an electrotherapy useful for improving recovery through faster parasympathetic reactivation following exercise. The absence of any positive effect on VO_2 on-transition kinetics could be partly imputed to methodological procedures such as the arbitrary choice of stimulation intensity and duration. Nevertheless, additional studies are needed to approve or discard this hypothesis and to shed light on the correlation of these consequences with a short period of training program with concurrent MENS stimulation.

Author Contributions: A.P. conceived and designed research; A.P. and L.Z. performed experiments; A.P. analyzed data; A.P., L.Z., A.T., F.C., and M.R. interpreted results of experiments; A.P. prepared figures; A.P. drafted manuscript; A.P., L.Z., A.T., F.C., and M.R. edited and revised manuscript; A.P. approved final version of manuscript. All authors have read and agreed to the published version of the manuscript.

Funding: Supported by University of Bologna, RFO 2018.

Institutional Review Board Statement: This study was approved by the Bioethics committee of the University of Bologna in accordance with the Declaration of Helsinki. The protocol n° 0021246 was approved on 03-02-2020.

Informed Consent Statement: Written informed consent was obtained from all subjects involved in the study.

Data Availability Statement: Data sharing is not applicable to this article because of the consent provided by participants on the use of confidential data.

Acknowledgments: The authors would like to thanks all subjects involved in this study.

Conflicts of Interest: The authors declare no conflict of interest.

References

1. Babault, N.; Cometti, C.; Maffiuletti, N.A.; Deley, G. Does electrical stimulation enhance post-exercise performance recovery? *Eur. J. Appl. Physiol.* **2011**, *111*, 2501–2507. [CrossRef] [PubMed]
2. Cheng, N.; van Hoof, H.; Bockx, E.; Hoogmartens, M.J.; Mulier, J.C.; De Dijcker, F.J.; Sansen, W.M.; De Loecker, W. The effects of electric currents on ATP generation, protein synthesis, and membrane transport in rat skin. *Clin. Orthop. Relat. Res.* **1982**, *171*, 264–272. [CrossRef]
3. Zizic, T.M.; Hoffman, K.C.; Holt, P.A.; Hungerford, D.S.; O'Dell, J.R.; Jacobs, M.A.; Lewis, C.G.; Deal, C.L.; Caldwell, J.R.; Cholewczynski, J.G.; et al. The treatment of osteoarthritis of the knee with pulsed electrical stimulation. *J. Rheumatol.* **1995**, *22*, 1757–1761.
4. Cook, H.A.; Morales, M.; La Rosa, E.M.; Dean, J.; Donnelly, M.K.; McHugh, P.; Otradovec, A.; Wright, K.S.; Kula, T.; Tepper, S.H. Effects of electrical stimulation on lymphatic flow and limb volume in the rat. *Phys. Ther.* **1994**, *74*, 1040–1046. [CrossRef]
5. Mitchell, P. Chemiosmotic coupling in oxidative and photosynthetic phosphorylation. *Biochim. Biophys. Acta Bioenerg.* **2011**, *1807*, 1507–1538. [CrossRef] [PubMed]
6. Pall, M.L. Electromagnetic fields act via activation of voltage-gated calcium channels to produce beneficial or adverse effects. *J. Cell. Mol. Med.* **2013**, *17*, 958–965. [CrossRef]
7. Cortis, C.; Tessitore, A.; Dartibale, E.; Meeusen, R.; Capranica, L. Effects of post-exercise recovery interventions on physiological, psychological, and performance parameters. *Int. J. Sports Med.* **2010**, *31*, 327–335. [CrossRef]
8. Ohno, Y.; Fujiya, H.; Goto, A.; Nakamura, A.; Nishiura, Y.; Sugiura, T.; Ohira, Y.; Yoshioka, T.; Goto, K. Microcurrent electrical nerve stimulation facilitates regrowth of mouse soleus muscle. *Int. J. Med. Sci.* **2013**, *10*, 1286–1294. [CrossRef]
9. Malone, J.K.; Blake, C.; Caulfield, B.M. Neuromuscular electrical stimulation during recovery from exercise: A systematic review. *J. Strength Cond. Res.* **2014**, *28*, 2478–2506. [CrossRef]
10. Barcala-Furelos, R.; González-Represas, A.; Rey, E.; Martínez-Rodríguez, A.; Kalén, A.; Marques, O.; Rama, L. Is low-frequency electrical stimulation a tool for recovery after a water rescue? A cross-over study with lifeguards. *Int. J. Environ. Res. Public Health* **2020**, *17*, 5854. [CrossRef]
11. Piras, A.; Campa, F.; Toselli, S.; Di Michele, R.; Raffi, M. Physiological responses to partial-body cryotherapy performed during a concurrent strength and endurance session. *Appl. Physiol. Nutr. Metab.* **2019**, *44*, 59–65. [CrossRef]
12. Piras, A.; Persiani, M.; Damiani, N.; Perazzolo, M.; Raffi, M. Peripheral heart action (PHA) training as a valid substitute to high intensity interval training to improve resting cardiovascular changes and autonomic adaptation. *Eur. J. Appl. Physiol.* **2015**, *115*, 763–773. [CrossRef] [PubMed]
13. Piras, A.; Gatta, G. Evaluation of the effectiveness of compression garments on autonomic nervous system recovery after exercise. *J. Strength Cond. Res.* **2017**, *31*, 1636–1643. [CrossRef]
14. Piras, A.; Cortesi, M.; Campa, F.; Perazzolo, M.; Gatta, G. Recovery time profiling after short-, middle-and long-distance swimming performance. *J. Strength Cond. Res.* **2019**, *33*, 1408–1415. [CrossRef] [PubMed]
15. Toni, G.; Belvederi Murri, M.; Piepoli, M.; Zanetidou, S.; Cabassi, A.; Squatrito, S.; Bagnoli, L.; Piras, A.; Mussi, C.; Senaldi, R.; et al. Physical Exercise for Late-Life Depression: Effects on Heart Rate Variability. *Am. J. Geriatr. Psychiatry* **2016**, *24*, 989–997. [CrossRef] [PubMed]
16. Noites, A.; Nunes, R.; Gouveia, A.I.; Mota, A.; Melo, C.; Viera, Á.; Adubeiro, N.; Bastos, J.M. Effects of Aerobic Exercise Associated with Abdominal Microcurrent: A Preliminary Study. *J. Altern. Complement. Med.* **2015**, *21*, 229–236. [CrossRef]
17. Lambert, M.I.; Marcus, P.; Burgess, T.; Noakes, T.D. Electro-membrane microcurrent therapy reduces signs and symptoms of muscle damage. *Med. Sci. Sport. Exerc.* **2002**, *34*, 602–607.
18. Naclerio, F.; Seijo, M.; Karsten, B.; Brooker, G.; Carbone, L.; Thirkell, J.; Larumbe-Zabala, E. Effectiveness of combining microcurrent with resistance training in trained males. *Eur. J. Appl. Physiol.* **2019**, *119*, 2641–2653. [CrossRef] [PubMed]
19. Kwon, D.R.; Kim, J.; Kim, Y.; An, S.; Kwak, J.; Lee, S.; Park, S.; Choi, Y.H.; Lee, Y.K.; Park, J.W. Short-term microcurrent electrical neuromuscular stimulation to improve muscle function in the elderly: A randomized, double-blinded, sham-controlled clinical trial. *Medicine* **2017**, *96*. [CrossRef]
20. Hirai, D.M.; Copp, S.W.; Holdsworth, C.T.; Ferguson, S.K.; McCullough, D.J.; Behnke, B.J.; Musch, T.I.; Poole, D.C. Skeletal muscle microvascular oxygenation dynamics in heart failure: Exercise training and nitric oxide-mediated function. *Am. J. Physiol. Hear. Circ. Physiol.* **2014**, *306*. [CrossRef]
21. Burnley, M.; Jones, A.M. Oxygen uptake kinetics as a determinant of sports performance. *Eur. J. Sport Sci.* **2007**, *7*, 63–79. [CrossRef]

22. Koppo, K.; Bouckaert, J.; Jones, A.M. Effects of Training Status and Exercise Intensity on Phase II VO_2 Kinetics. *Med. Sci. Sport. Exerc.* **2004**, *36*, 225–232. [CrossRef] [PubMed]
23. Campa, F.; Piras, A.; Raffi, M.; Trofè, A.; Perazzolo, M.; Mascherini, G.; Toselli, S. The effects of dehydration on metabolic and neuromuscular functionality during cycling. *Int. J. Environ. Res. Public Health* **2020**, *17*, 1161. [CrossRef]
24. Beaver, W.L.; Wasserman, K.; Whipp, B.J. A new method for detecting anaerobic threshold by gas exchange. *J. Appl. Physiol.* **1986**, *60*, 2020–2027. [CrossRef]
25. Camm, A.J.; Malik, M.; Bigger, J.T.; Günter, B. Task force of the European Society of Cardiology and the North American Society of Pacing and Electrophysiology. Heart rate variability: Standards of measurement, physiological interpretation and clinical use. *Circulation* **1996**, *93*, 1043–1065.
26. Bonaventura, J.M.; Sharpe, K.; Knight, E.; Fuller, K.L.; Tanner, R.K.; Gore, C.J. Reliability and accuracy of six hand-held blood lactate analysers. *J. Sport. Sci. Med.* **2014**, *14*, 203–214.
27. Ferrari, M.; Muthalib, M.; Quaresima, V. The use of near-infrared spectroscopy in understanding skeletal muscle physiology: Recent developments. *Trans. R. Soc. A* **2011**, *369*, 4577–4590. [CrossRef]
28. Ferrari, M.; Binzoni, T.; Quaresima, V. Oxidative metabolism in muscle. *Philos. Trans. R. Soc. B Biol. Sci.* **1997**, *352*, 677–683. [CrossRef]
29. Grassi, B.; Pogliaghi, S.; Rampichini, S.; Quaresima, V.; Ferrari, M.; Marconi, C.; Cerretelli, P. Muscle oxygenation and pulmonary gas exchange kinetics during cycling exercise on-transitions in humans. *J. Appl. Physiol.* **2003**, *95*, 149–158. [CrossRef] [PubMed]
30. Borg, G.; Löllgen, H. Borg's perceived exertion and pain scales. *Dtsch. Z. Sportmed.* **2001**, *52*, 252. [CrossRef]
31. Whipp, B.J.; Ward, S.A.; Lamarra, N.; Davis, J.A.; Wasserman, K. Parameters of ventilatory and gas exchange dynamics during exercise. *J. Appl. Physiol. Respir. Environ. Exerc. Physiol.* **1982**, *52*, 1506–1513. [CrossRef]
32. Bos, W.J.; van Goudoever, J.; van Montfrans, G.A.; van den Meiracker, A.H.; Wesseling, K.H. Reconstruction of brachial artery pressure from noninvasive finger pressure measurements. *Circulation* **1996**, *94*, 1870–1875. [CrossRef]
33. Hopkins, W.G.; Marshall, S.W.; Batterham, A.M.; Hanin, J. Progressive statistics for studies in sports medicine and exercise science. *Med. Sci. Sport. Exerc.* **2009**, *41*, 3–12. [CrossRef]
34. Fylstra, D.; Lasdon, L.; Watson, J.; Waren, A. Design and Use of the Microsoft Excel Solver. *INFORMS J. Appl. Anal.* **1998**, *28*, 29. [CrossRef]
35. Huckfeldt, R.; Flick, A.B.; Mikkelson, D.; Lowe, C.; Finley, P.J. Wound closure after split-thickness skin grafting is accelerated with the use of continuous direct anodal microcurrent applied to silver nylon wound contact dressings. *J. Burn Care Res.* **2007**, *28*, 703–707. [CrossRef]
36. Ahmed, A.F.; Elgayed, S.S.A.; Ibrahim, I.M. Polarity effect of microcurrent electrical stimulation on tendon healing: Biomechanical and histopathological studies. *J. Adv. Res.* **2012**, *3*, 109–117. [CrossRef]
37. Grassi, B.; Poole, D.C.; Richardson, R.S.; Knight, D.R.; Erickson, B.K.; Wagner, P.D. Muscle O_2 uptake kinetics in humans: Implications for metabolic control. *J. Appl. Physiol.* **1996**, *80*, 988–998. [CrossRef] [PubMed]
38. Wilkerson, D.P.; Campbell, I.T.; Jones, A.M. Influence of nitric oxide synthase inhibition on pulmonary O_2 uptake kinetics during supra-maximal exercise in humans. *J. Physiol.* **2004**, *561*, 623–635. [CrossRef] [PubMed]
39. Brown, G.C. Nitric oxide as a competitive inhibitor of oxygen consumption in the mitochondrial respiratory chain. *Acta Physiol. Scand.* **2000**, *168*, 667–674. [CrossRef] [PubMed]
40. Ferreira, L.F.; Townsend, D.K.; Lutjemeier, B.J.; Barstow, T.J. Muscle capillary blood flow kinetics estimated from pulmonary O_2 uptake and near-infrared spectroscopy. *J. Appl. Physiol.* **2005**, *98*, 1820–1828. [CrossRef] [PubMed]
41. DeLorey, D.S.; Kowalchuk, J.M.; Paterson, D.H. Relationship between pulmonary O_2 uptake kinetics and muscle deoxygenation during moderate-intensity exercise. *J. Appl. Physiol.* **2003**, *95*, 113–120. [CrossRef]
42. Grassi, B.; Gladden, L.B.; Stary, C.M.; Wagner, P.D.; Hogan, M.C. Peripheral O_2 diffusion does not affect $V'O_2$ on-kinetics in isolated in situ canine muscle. *J. Appl. Physiol.* **1998**, *85*, 1404–1412. [CrossRef]
43. MacDonald, M.; Pedersen, P.K.; Hughson, R.L. Acceleration of $V'O_2$ kinetics in heavy submaximal exercise by hyperoxia and prior high-intensity exercise. *J. Appl. Physiol.* **1997**, *83*, 1318–1325. [CrossRef]
44. Barstow, T.J.; Jones, A.M.; Nguyen, P.H.; Casaburi, R. Influence of muscle fiber type and pedal frequency on oxygen uptake kinetics of heavy exercise. *J. Appl. Physiol.* **1996**, *81*, 1642–1650. [CrossRef]
45. O'Leary, D.D.; Kimmerly, D.S.; Cechetto, A.D.; Shoemaker, J.K. Differential effect of head-up tilt on cardiovagal and sympathetic baroreflex sensitivity in humans. *Exp. Physiol.* **2003**, *88*, 769–774. [CrossRef] [PubMed]
46. Vieira, P.J.C.; Ribeiro, J.P.; Cipriano, G.; Umpierre, D.; Cahalin, L.P.; Moraes, R.S.; Chiappa, G.R. Effect of transcutaneous electrical nerve stimulation on muscle metaboreflex in healthy young and older subjects. *Eur. J. Appl. Physiol.* **2012**, *112*, 1327–1334. [CrossRef] [PubMed]
47. Alamdari, N.; Constantin-Teodosiu, D.; Murton, A.J.; Gardiner, S.M.; Bennett, T.; Layfield, R.; Greenhaff, P.L. Temporal changes in the involvement of pyruvate dehydrogenase complex in muscle lactate accumulation during lipopolysaccharide infusion in rats. *J. Physiol.* **2008**, *586*, 1767–1775. [CrossRef] [PubMed]
48. Nikolić, N.; Skaret Bakke, S.; Tranheim Kase, E.; Rudberg, I.; Flo Halle, I.; Rustan, A.C.; Thoresen, G.H.; Aas, V. Electrical pulse stimulation of cultured human skeletal muscle cells as an in vitro model of exercise. *PLoS ONE* **2012**, *7*, e33203. [CrossRef] [PubMed]

49. Malone, J.K.; Coughlan, G.F.; Crowe, L.; Gissane, G.C.; Caulfield, B. The physiological effects of low-intensity neuromuscular electrical stimulation (NMES) on short-term recovery from supra-maximal exercise bouts in male triathletes. *Eur. J. Appl. Physiol.* **2012**, *112*, 2421–2432. [CrossRef] [PubMed]
50. Grassi, B.; Quaresima, V. Near-infrared spectroscopy and skeletal muscle oxidative function in vivo in health and disease: A review from an exercise physiology perspective. *J. Biomed. Opt.* **2016**, *21*, 091313. [CrossRef]

International Journal of
Environmental Research and Public Health

MDPI

Article

Effect of an Endurance and Strength Mixed Circuit Training on Regional Fat Thickness: The Quest for the "Spot Reduction"

Antonio Paoli [1], Andrea Casolo [1], Matteo Saoncella [1], Carlo Bertaggia [1], Marco Fantin [1], Antonino Bianco [2], Giuseppe Marcolin [1] and Tatiana Moro [2,*]

[1] Department of Biomedical Sciences, University of Padua, 35131 Padua, Italy; antonio.paoli@unipd.it (A.P.); andrea.casolo@unipd.it (A.C.); matteo.saoncella@gmail.com (M.S.); carlo.bertaggia@gmail.com (C.B.); marco.fantin.3@studenti.unipd.it (M.F.); giuseppe.marcolin@unipd.it (G.M.)

[2] Sport and Exercise Sciences Research Unit, University of Palermo, 90128 Palermo, Italy; antonino.bianco@unipa.it

* Correspondence: tatiana.moro@unipd.it

Abstract: Accumulation of adipose tissue in specific body areas is related to many physiological and hormonal variables. Spot reduction (SR) is a training protocol aimed to stimulate lipolysis locally, even though this training protocol has not been extensively studied in recent years. Thus, the present study sought to investigate the effect of a circuit-training SR on subcutaneous adipose tissue in healthy adults. Methods: Fourteen volunteers were randomly assigned to spot reduction (SR) or to a traditional resistance training (RT) protocol. Body composition via bioimpedance analysis (BIA) and subcutaneous adipose tissue via skinfold and ultrasound were measured before and after eight weeks of training. Results: SR significantly reduced body mass ($p < 0.05$) and subcutaneous abdominal adipose tissue ($p < 0.05$). Conclusions: circuit-training SR may be an efficient strategy to reduce in a localized manner abdominal subcutaneous fat tissue depot.

Keywords: spot reduction; body composition; resistance training; adipose tissue

Citation: Paoli, A.; Casolo, A.; Saoncella, M.; Bertaggia, C.; Fantin, M.; Bianco, A.; Marcolin, G.; Moro, T. Effect of an Endurance and Strength Mixed Circuit Training on Regional Fat Thickness: The Quest for the "Spot Reduction". *Int. J. Environ. Res. Public Health* **2021**, *18*, 3845. https://doi.org/10.3390/ijerph18073845

Academic Editors: António Carlos Sousa and Pantelis T. Nikolaidis

Received: 11 March 2021
Accepted: 5 April 2021
Published: 6 April 2021

1. Introduction

Regular physical activity can impact body composition, reducing fat mass and therefore positively improving health status. Accumulation of adipose tissue (AT) in specific areas of the body can be influenced by lifestyle behavior, such as working for most of the time in a sitting position or using only the upper body. Adipose tissue does not develop regularly but normally spread in distinctive anatomical depots [1]. Approximately 10–20% of total fat mass is contained in the visceral adipose tissue (VAT), located centrally and surrounds the internal organs [2]. The majority of total body fat is represented by the subcutaneous AT (SAT) positioned immediately below the skin: SAT normally accumulates in the gluteal, femoral and abdominal region and its distribution are regulated by different physiological and/or hormonal variables [3]. A small portion of AT consists in the ectopic AT and is localized around vital organs, such as the liver, heart and kidney.

The total amount of fat mass is considered a risk factor for several cardiometabolic diseases [4,5]; however, the location of lipids storage seems also to be critical for cardiometabolic consequences [6–8]. If central obesity is associated with metabolic dysfunction and hypertension [9,10], lower-body fat accumulation appears to have a protective effect and seems to be negatively correlated with cardiovascular disease and type 2 diabetes mellitus development [11,12]. The reduction of total body fat can be achieved through diet and/or exercise intervention [13,14]. While it has been widely demonstrated that an adequate amount of physical activity can have a favorable impact on the weight loss process [15], the existence of a "localized fat loss" is still on debate. As a matter of fact, for more than 60 years, the possibility of a localized removal of AT has raised interest in the scientific and social community. Even the "father" of the Mediterranean diet, Ancel Keys, admitted the possibility of a localized fat reduction, although not in a scientific journal, but rather on Vogue in 1956 [13]. Later, in the '50s, some researchers reported that certain sports, such as gymnast [16], basketball [17] or running [18], promoted greater loss of fat mass in those parts of the bodies that were vigorously exercised. Since then, different strategies have been developed to advance the localized loss of SAT with exercise, and all those protocols have been termed "spot reduction". More recently, Stallknecht and coll. [19], hypothesized that exercise on specific muscles may induce "spot lipolysis" via an increased blood flow and release of fatty acids in the SAT nearby the contracting muscle regardless of exercise intensity. However, most of the studies found conflicting conclusions: some authors found a positive effect of spot reduction on localized lipolysis [20–22], while others were inconclusive [23–26]. The discrepancy between results can be found on the several exercise modalities employed, on the different body areas examined and, on the technique used for measuring SAT [27]. On the latter, most of the studies used skinfolds to evaluate changes after training [20–25]; recent comparative studies indicated that skinfold measurements do not permit accurate evaluation of SAT thickness because it is operator-dependent and influenced by anatomical site and skin thickness [28,29]. Regarding training modalities, it is well known that combining in the same training session endurance and strength exercises may exert a greater effect on total body fat loss [30] and, as recently demonstrated, and it might also have some positive effects on regional fat loss [31]. However, the effects of an alternation of strength and endurance training (mixed circuit training: MCT) has, until now, not been investigated. As a matter of fact, we demonstrated that a MCT induces greater total body fat loss and an improvement of metabolic variables compared to endurance training [32,33] but, except for our pilot trial in the 90′s [34] no one analyzed the effects of a MCT on regional fat loss.

In the light of the above, the purpose of the present study was to reconsider the spot reduction approach using a modified MCT protocol. MCT protocols usually alternate various total body strength exercises with short bouts of aerobic training. In the present study, we aimed to emphasize the positive effect of MCT by streamlining the order of proposed exercise to focus the major metabolic stress on the target body areas. To our best knowledge, this strategy has not been explored yet. We hypothesized that an MCT protocol concentrated on specific muscle would have exerted a great local lipolytic effect compare to a non-circuit mixed training. We tested our hypothesis on a group of healthy adults, using skinfolds and a modern ultrasound technique to measure SAT before and after eight weeks of training targeting abdominal and triceps.

2. Materials and Methods

This study is a randomized controlled parallel study. The study was approved by the ethical committee of the Department of Biomedical Sciences, University of Padova (HEC-DSB 05/17, 22 March 2017), according to the current Declaration of Helsinki. All participants read and signed a written informed consent form before enrollment.

Subjects were evaluated in a single visit before and after eight weeks of intervention. During the visit, height and body mass were measured, body composition was assessed via bioimpedance analysis (BIA) and skinfolds, and ultrasound was used to quantify the thickness of the adipose panicle. All measurements were taken by the same operator before and after the study. After the first screening visit, participants were randomized into two different groups: spot reduction (SR) or traditional resistance training (RT) and started the supervised training program.

2.1. Subjects

Eighteen volunteers (9 female and 9 male) aged between 20 and 46 years took part in the study. To be included in the protocol, subjects had to pass a medical interview and be aged 18–50 years old. Exclusion criteria were more than 1 year of training experience, chronic use of medication, metabolic disorders or any other clinical problems that could be aggravated by the study procedures or engagement in weight loss dietetic regimen.

During the intervention, participants were allowed to continue their recreational physical activity, but were instructed not to perform any structured, high-impact training. Four subjects (two from each group) were excluded from the final analysis due to noncompliance with the training schedule. To be considered for the analysis, subjects had to complete all sessions and maintain a frequency of 3 training per week. Table 1 shows the anthropometric characteristics at baseline.

Table 1. Baseline characteristics of spot reduction (SR) and resistance training (RT) groups.

	SR (N = 7)	RT (N = 7)	Difference between Group (p-Value)
Age (y)	23.29 ± 1.89	26.57 ± 9.14	0.37
Weight (kg)	69.24 ± 6.90	75.93 ± 12.47	0.24
Height (cm)	170.64 ± 5.73	175.57 ± 12.02	0.35
BMI (kg/m^2)	23.78 ± 2.11	24.67 ± 3.54	0.58
Body fat (%)	24.69 ± 10.32	28.03 ± 7.44	0.50

All values are means ± SD.

2.2. Measurements

Body mass index (BMI) was calculated in kg/m^2, obtained from body mass and height measurement using a Wunder stadiometer (Holtain Ltd., Crymych, UK) with a precision of 0.1 kg and 1 cm, respectively.

Before proceeding with the measurements of the adipose panniculus by skinfolds and ultrasound, specific detection points were traced on the right portion of the body. For skinfolds, the 4 points described by the Durnin protocol [35] were used: bicipital, triceps, suprailiac and subscapularis. With regard to ultrasound scans, to standardize the procedure and detection points between subjects, we employed the protocol described by Muller et al. [36], and 8 regions were selected for the analysis: upper abdomen, lower abdomen, spinal erectors, distal triceps, brachioradialis, front thigh, medial calf, lateral thigh.

A mechanical caliper (GIMA, Gessate MI, Italy) was used to determined skinfolds to the nearest 1 mm. Each skinfold was measured 3 times, and the arithmetic means were used as the final value. Test–retest reliability for skinfold analysis was ICC = 0.96. Test–retest intra-observer reliability for fat adipose tissue thickness in our lab was ICC = 0.96, similar to previous findings [28,37].

Using the points previously marked, the skin fold was "pinched" between thumb and forefinger one centimeter above the measurement site, perpendicularly for the triceps and bicipital folds, and at a 45° angle to the longitudinal axis for the suprailiac and subscapular fold. All measurements were taken with the subject in an upright position and with the arms relaxed at the sides, while for the suprailiac point, the subject's right arm was placed over the operator's shoulder. Successively, body density was determined with Durnin–Womersley method [38], according to the methods Equation (1) was used to estimate body dansiy in male, whilst Equation (2) was used in female volounteers:

$$\text{Male BD} = 1.1765 - (0.744 \times \log_{10} \Sigma \text{ skinfolds}); \tag{1}$$

$$\text{Female BD} = 1.1567 - (0.0717 \times \log_{10} \Sigma \text{ skinfolds}). \tag{2}$$

Ultimately, body density from Equation (1) and (2) was used in Equation (3) to estimate body fat percentage using the Siri formula [39]:

$$\text{FAT\%} = ((4.95/\text{BD}) - 4.5) \times 100 \tag{3}$$

Ultrasound measurements were performed using Xario 100 ultrasound (TOSHIBA, Tustin, CA, US) with a surface probe set on MSK 1. The probe was equipped with a spirit level to maintain the same inclination in all readings. During acquisition, no pressure was ever exerted on the probe placed on the subject's skin, except for the natural one resulting from the weight of the probe itself, by keeping the probe by the cable. For each detection point 3 snapshots were taken, following Muller's procedures:

- Upper abdomen and Lower abdomen: subjects were positioned supine and asked to inhale and then to stop breathing at mid-exhalation to take the 3 photographs without movement of the abdominal wall;
- Brachioradialis: subjects were positioned supine, arms at the sides, right-hand with thumb up;
- Front thigh: subjects were positioned supine and asked to stay relaxed;
- Spinal erectors: subjects were prone, with the chin resting on the edge of the bed and arms extended at the sides;
- Distal triceps: subjects were prone, arms at the sides with right palm upwards;
- Lateral thigh: subject in lateral decubitus on the left side and legs at a 90° angle at the knee;
- Medial calf: subject in lateral decubitus on the right side, right leg with a 90° angle at the knee.

Body composition, total body water (TBW), extracellular (ECW) and intracellular (ICW) water, body cell mass (BCM) and phase angle (PA) were measured via bioelectrical impendence analysis (Akern, Body Pro, Pontassieve, Italy). Subjects were asked to empty their bladder and rest for ~3–5 min in a supine position, while four electrodes were placed on their hands and feet to start the analysis. Test–retest reliability for body composition analysis using bioelectrical impendence was ICC = 0.99.

2.3. Training Protocols

Subjects were required to perform the prescribed exercise protocol 3 times per week for a total of 8 weeks. The training was supervised by a certified trainer whose task was to check adherence to the study and the correct execution of training protocols.

Training protocols were comparable for the type of exercise, intensity, and volume but differed for the order of execution. The SR protocol was an alternation between endurance strength exercise (MCT) in which, specifically, abdominal, triceps and aerobic exercise were performed in a circuit (as described in Table 2) while the other muscular area (back, shoulders, arms, and lower limbs) were trained through RT exercises at the end of the circuit. RT group completed first all the aerobic exercises, and in the second part, the resistance exercises. Aerobic exercise intensity was settled at 65% of max HR using Cooper formula, while resistance load was assessed based on the previous training schedule and during a preliminary familiarization session.

2.4. Statistical Analysis

Results are presented as mean $\pm$ SD. The sample size was calculated based on preliminary data from our laboratory, assuming within-subject variability of 25% and a fixed power of 0.8 and an alpha risk of 0.05 for the main variables (skinfolds). Initially, the analysis revealed that 9 subjects per group were needed to achieve the above parameters. However, only 7 participants were included in the final evaluation; we thus perform a post hoc analysis and the achieved power with the real sample size was 0.75. An independent t-test was performed on baseline characteristics to ensure no difference between groups. After checking for normal distribution via the Shapiro–Wilk W test, a two-way ANOVA for repeated measures was performed to compare the effect of training modalities through a "time x training" analysis. The post hoc Bonferroni test was used to identify specific intragroup differences when suitable. For each group, Cohen's d effect size was assessed by dividing the difference between mean values by the pooled SD. The p-value was set at 0.05. Data analysis was performed using GraphPad Prism software version 8.4.3 (GraphPad Software, San Diego, CA, USA).

Table 2. Training protocols.

SR			RT		
Exercise	**Set × Reps/Time**	**Rest**	**Exercise**	**Set × Reps**	**Rest**
Treadmill	5 min	-	Treadmill	5 min	-
Crunches	20 reps	-	Bike	5 min	-
Dumbbell overhead extension	15 reps	-	Step	5 min	-
Bike	5 min	-	Treadmill	5 min	-
Crunches	20 reps	-	Bike	5 min	-
Dumbbell overhead extension	15 reps	-	Dumbbell bench press	3 × 10	1 min
Step	5 min	-	Lat pulldown	3 × 10	1 min
Crunches	20 reps	-	Shoulder press	3 × 10	1 min
Dumbbell overhead extension	15 reps	-	Arm curl	3 × 12	45 s
Treadmill	5 min	-	Dumbbell overhead extension	4 × 15	45 s
Crunches	20 reps	-	Leg press	3 × 10	1 min 30 s
Dumbbell overhead extension	15 reps	-	Leg extension	3 × 12	1 min
Bike	5 min	-	Crunches	4 × 20	45 s
Dumbbell bench press	3 × 10	1 min	-	-	-
Lat pulldown	3 × 10	1 min	-	-	-
Shoulder press	3 × 10	1 min	-	-	-
Arm curl	3 × 12	45 s	-	-	-
Leg press	3 × 10	1 min 30 s	-	-	-
Leg extension	3 × 12	1 min	-	-	-

SR, Spot Reduction group; RT, Resostance Training group.

3. Results

Body mass significantly decreased (F(1,12) = 14.304; p = 0.003) in the SR group (from 69.24 ± 6.90 kg to 67.74 ± 6.34 kg; p = 0.01, d = −0.32), but not in the RT group (from 75.93 ± 12.47 kg to 74.96 ± 12.08 kg, p > 0.05, d = −0.11). As a consequence, also the BMI was significantly reduced (F(1,12) = 14.605; p = 0.002) only in the SR group (from 23.78 ± 2.11 kg/m^2 to 23.27 ± 1.93 kg/m^2, p = 0.01, d = −0.36) compare to RT group (from 24.67 ± 3.54 kg/m^2 to 24.36 ± 3.44 kg/m^2, p > 0.05, d = −0.13) as shown in Figure 1.

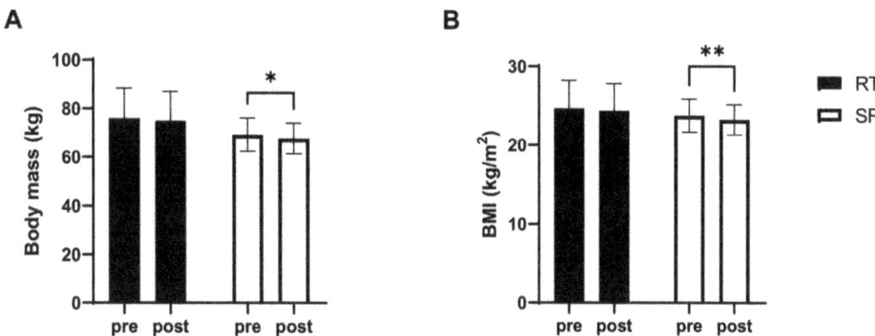

Figure 1. (**A**) body mass and (**B**) body mass index (BMI). RT resistance training group; SR, spot reduction group. * significantly different from pre-value (p < 0.05); ** significantly different from pre-value (p < 0.01).

No differences in skinfolds were detected after RT protocol; while SR resulted particularly effective on suprailiac (−13.29%; p = 0.02, d = −0.56) and subscapularis (−7.59%, p = 0.04, d = −0.44) skinfold. In both sites, the two-way ANOVA analysis revealed a significant main Time effect (suprailiac: F(1,12) = 6.993; p = 0.01; subscapularis: F(1,12) = 5.822; p = 0.01). Furthermore, body fat percentage estimated with Siri equation presented a significant main effect of time (F(1,12) = 7.776; p = 0.02), with a significant decrease observed only in the SR group (p = 0.01, d = −0.21). Data on skinfolds results are shown in Table 3.

Table 3. Skinfold results of spot reduction (SR) and resistance training (RT) groups.

	SR (N = 7)		RT (N = 7)	
	Pre	**Post**	**Pre**	**Post**
Bicipital (mm)	11.66 ± 9.04	9.68 ± 5.09	8.50 ± 4.85	9.02 ± 6.14
Triceps (mm)	18.75 ± 9.31	18.65 ± 9.31	19.87 ± 5.36	19.81 ± 5.14
Suprailiac (mm) §	20.43 ± 6.01	17.99 ± 6.37 *	19.33 ± 6.39	18.85 ± 6.36
Subscapularis (mm) §	16.94 ± 5.74	15.36 ± 4.25 *	18.25 ± 6.47	17.17 ± 5.61
Body fat (%) §	28.83 ± 8.92	27.52 ± 8.97 *	27.74 ± 6.81	27.49 ± 6.91

All values are means ± SD. * significantly different from pre-value ($p < 0.05$); § significant time effect ($p < 0.05$).

We observed a significant main time effect on ultrasound measurements of adipose panicle for upper abdomen (F(1,12) = 6.888; p = 0.02), spinal erectors (F(1,12) = 10.209; p = 0.01) and front thigh (F(1,12) = 5.855; p = 0.03) (Table 4). Post hoc test revealed a significant reduction only in the SR group for upper abdomen (−18.89%, p = 0.05, d = −0.66), spinal erector site (−19.45%, p = 0.04, d = −0.55).

Table 4. Ultrasound results of spot reduction (SR) and resistance training (RT) groups.

	SR (N = 7)		RT (N = 7)	
	Pre	**Post**	**Pre**	**Post**
Upper abdomen (mm) §	15.09 ± 6.08	12.35 ± 5.65 *	19.43 ± 10.34	18.13 ± 10.11
Lower abdomen (mm)	17.88 ± 7.44	17.30 ± 7.06	22.99 ± 9.23	22.33 ± 8.07
Spinal erectors (mm) §	9.96 ± 6.25	7.79 ± 4.83 *	10.77 ± 2.83	9.31 ± 3.28
Distal triceps (mm)	5.40 ± 3.95	6.21 ± 4.71	5.72 ± 2.13	5.19 ± 2.05
Brachioradialis (mm)	2.72 ± 1.37	3.24 ± 2.59	3.59 ± 2.12	3.66 ± 2.00
Front thigh (mm) §	8.46 ± 4.35	9.40 ± 5.17 *	11.56 ± 4.26	12.40 ± 3.29
Medial calf (mm)	4.87 ± 3.17	4.73 ± 2.90	5.90 ± 2.99	6.05 ± 2.49
Lateral thigh (mm)	18.44 ± 9.09	15.64 ± 7.08	24.10 ± 11.97	23.92 ± 11.08

All values are means ± SD. * significantly different from pre-value ($p < 0.05$). § significant time effect ($p < 0.05$).

Results from the body composition analysis via BIA are shown in Table 5. A significant time effect (F(1,12) = 5.776; p = 0.03) was observed only in the Intracellular body water, where RT resulted in a significant reduction (−7.89%, p = 0.04, d = −0.76). No other detectable differences were found.

Table 5. Bioimpedance analysis (BIA) results of spot reduction (SR) and resistance training (RT) groups.

	SR (N = 7)		RT (N = 7)	
	Pre	**Post**	**Pre**	**Post**
Total body water (L)	38.15 ± 6.09	36.74 ± 5.33	39.22 ± 6.22	38.36 ± 5.52
Extracellular water (L)	14.76 ± 1.97	13.93 ± 2.49	14.51 ± 2.76	14.41 ± 1.76
Intracellular water (L) §	23.38 ± 4.39	22.82 ± 3.79	24.71 ± 4.10	22.65 ± 3.63 *
Fat mass (kg)	17.13 ± 7.82	16.26 ± 6.57	21.42 ± 7.43	20.46 ± 8.12
Fat-free mass (kg)	52.15 ± 8.64	51.17 ± 7.27	54.58 ± 9.66	54.39 ± 9.45
Body cellular mass (kg)	28.85 ± 6.04	28.15 ± 5.53	30.91 ± 8.26	29.40 ± 6.85
Phase angle (°)	5.94 ± 0.56	5.91 ± 0.57	5.94 ± 1.17	5.66 ± 0.75

All values are means ± SD. * significantly different from pre-value ($p < 0.05$); § significant time effect ($p < 0.05$).

4. Discussion

The study aimed to revisit the spot reduction training in the light of new methods to analyze SAT. We observed a significant general reduction of body mass and abdominal SAT, measured both with skinfold and ultrasound after 12 weeks of spot reduction training. Skinfold measurements also showed a reduction on the subscapularis site, while ultrasound revealed a decrease in the spinal erectors SAT.

Compared to ultrasounds, the skinfolds' technique measures SAT within a compressed double layer of skin. Skin thickness may vary substantially among body area, for example, is lower in the upper arm compared to the abdomen [28,36], reducing the accuracy between measurements. A recent analysis revealed that ultrasound might overcome the problem linked to the compressibility and viscoelasticity of adipose tissue and thus represents a better tool to estimate SAT [36,40]. Despite the limitation mentioned, we decided to include skinfolds measurements to be consistent with most of the studies that have analyzed spot reduction protocols. It is also worth mentioning that the ultrasound technique relies on the operator performing the measurements as much as skinfolds. Despite trying to comply with all the standard procedures to reduce variability during measurements, we observed relatively large standard deviations of up to 5–6 times the observed difference. Overall, the pre-post analysis revealed a relative medium effect size (as suggested by the observed Cohen's d > 0.5 for most measures) which may slightly weaken the validity of our findings.

Spot reduction is a training protocol aimed to reduce subcutaneous fat on a particular part of the body. The first protocols of spot reduction were created based on the assumption that the accumulation of fat in a specific body area is related to the activity of the adjacent muscles [20,22,41]. However, a better understanding of the mechanism of adipose accumulation/oxidation has revealed that this assumption might not be completely correct. Fatty acids taken from the diet are deposited in the adipose tissue based on hormonal and receptor action [42,43], like energy storage. During physical activity, muscle contraction demands energy; if the energy request is not solved with glycogen store, fats are mobilized from adipose tissue, released into the bloodstream, and carried to target cells to be oxidized. Lipolysis is mediated by hormonal fluxes (catecholamines, insulin and autocrine/paracrine factors), which reach adipose tissue passing through the circulatory network [42,43]. Circulating fatty acids can be provided from any body district, which does not necessarily must be involved with muscular effort; therefore, performing countless series only of a specific exercise may not be sufficient to promote lipolysis in that specific site. However, it was recently observed that lipolytic activity is associated with an increase of blood flow in the adipose tissue and, thus, to the oxygenation of the adipocyte, suggesting that "blood flow and lipolysis are generally higher in subcutaneous adipose tissue adjacent to contracting than adjacent to resting muscle irrespective of exercise intensity. Thus, specific exercises can induce "spot lipolysis" in adipose tissue" [19]. Based on these premises, the goal of spot reduction training should be to increase blood perfusion in the areas where it is most needed, which are where the adipose tissue is located; and sequentially promote fat oxidation. For this reason, the SR protocol we have employed in the present study was composed by a circuit training, in which the localized SAT mobilization was stimulated by target exercises (crunches for the abdomen and dumbbell overhead extension for the triceps), while fat oxidation was induced by the aerobic phases. Apart from our previous pilot study [34], this is the first attempt to use an MCT in an SR protocol.

We observed a significant reduction in the suprailiac skinfold and in the upper abdomen measure via ultrasound. These data are in accordance with others [20,21] and support the idea that spot reduction protocol can improve local lipolysis in the abdomen. We also observed a significant SAT reduction in the spinal erector site, which was adjacent to the subscapularis skinfold site, while we did not observe any direct effect on the triceps measurements. This was an unexpected result, as our hypothesis was that the specific triceps exercise included in the circuit training would have reduced the local SAT. We included tricipital exercise into our protocol because this is one of the areas in which subcutaneous adipose tissue can concentrate and also because triceps brachii can be exercised with several specific exercises. It is possible that, due to their inexperience, participants had involved more the shoulder and scapula-stabilizing muscles than the triceps brachialis during the execution of the dumbbell overhead extension exercise, reducing the effect on the specific site. It is also worth noting that, although not significantly, the front thigh SAT increased in the SR group. This result may be explained with a greater effort expressed from the participant during the first MCT part, which might have tired them out before facing the second part of the workout. It is, therefore, possible that lower limbs were not successfully trained. Although this is only speculation, and these results raise interesting future questions on the SR approach.

Overall, we found that SR reduced total body mass, while any significant difference was obtained after a traditional resistance training protocol. However, body compositional analysis via BIA was unable to detect any significant changes in total fat mass or lean body mass. We observed a significant decrease in the intracellular water after RT, which normally indicates alteration of the number and size of muscle cells; however, this did not reflect on lean body mass value. Using the Siri equation to estimate body fat percentage, we found a significant decrease in the SR group compared to the RT group. However, the Siri formula is dependent on the precision of skin-fold measurements, and generally, the error of this method is approximately 5% [44].

The two protocols contained the same exercises and were comparable for duration and volume. This implicates that to reduce body mass, training intensity is a more important variable than the type of exercise. Training intensity could be manipulated in several ways: by increasing loads or oxygen consumption level, but also reducing rest intervals or altering movement velocity. In the initial MCT part of the spot reduction protocol, subjects did not rest between exercises, which increased the overall training metabolic demand. This is in contrast with other studies comparing the general and localized type of exercise training [21,45], which found similar effects on body composition. However, in the mentioned studies, the two protocols did not match the intensity of training modalities; for example, Noland and coll. Compared a general aerobic training with a localized calisthenic-type exercise [21]; while Schade concentrated one protocol only in the hip and abdominal areas and expanded the generalized training adding exercises on the upper and lower body [45]. Finally, we hypothesized that the alternation of endurance and strength exercises, or put another way, the insertion of a strength training exercise for specific muscles inside an endurance training might enhance the reduction of the fat tissue adjacent to the exercising muscles.

A limitation of the present study was the reduced number of participants. Due to the variability between subjects, it would be important to increase the sample size in future studies to determine the effectiveness of spot reduction in a larger population.

5. Conclusions

Spot reduction training, conducted in a mixed circuit-training format (triceps and abdomen inside an endurance training), seems to be efficient in promoting adipose tissue reduction in the subcutaneous abdominal region, but was not efficient on the triceps site.

Author Contributions: Conceptualization, A.P.; methodology, A.P., G.M. and T.M.; data curation: M.S., C.B., G.M. and M.F.; formal analysis: A.P., A.C., M.S., C.B., M.F., A.B., G.M. and T.M.; supervision: A.P., and T.M.; writing—original draft preparation, T.M.; writing—review and editing, A.P., and A.B.; All authors have read and agreed to the published version of the manuscript.

Funding: This research received no external funding.

Institutional Review Board Statement: The study was conducted according to the guidelines of the Declaration of Helsinki and approved by the Ethics Committee of the Department of Biomedical Sciences, University of Padova (HEC-DSB 05/17, 22 March 2017).

Informed Consent Statement: Informed consent was obtained from all subjects involved in the study.

Data Availability Statement: Original data are available upon request to the corresponding author.

Acknowledgments: We would like to thank all the subjects involved in the experiment for their patience and availability.

Conflicts of Interest: The authors declare no conflict of interest.

References

1. Walker, G.E.; Marzullo, P.; Ricotti, R.; Bona, G.; Prodam, F. The pathophysiology of abdominal adipose tissue depots in health and disease. *Horm. Mol. Biol. Clin. Investig.* **2014**, *19*, 57–74. [CrossRef] [PubMed]
2. Despres, J.P.; Lemieux, I. Abdominal obesity and metabolic syndrome. *Nature* **2006**, *444*, 881–887. [CrossRef]
3. Smith, S.R.; Lovejoy, J.C.; Greenway, F.; Ryan, D.; de Jonge, L.; de la Bretonne, J.; Volafova, J.; Bray, G.A. Contributions of total body fat, abdominal subcutaneous adipose tissue compartments, and visceral adipose tissue to the metabolic complications of obesity. *Metabolism* **2001**, *50*, 425–435. [CrossRef] [PubMed]
4. Kivimaki, M.; Kuosma, E.; Ferrie, J.E.; Luukkonen, R.; Nyberg, S.T.; Alfredsson, L.; Batty, G.D.; Brunner, E.J.; Fransson, E.; Goldberg, M.; et al. Overweight, obesity, and risk of cardiometabolic multimorbidity: Pooled analysis of individual-level data for 120 813 adults from 16 cohort studies from the USA and Europe. *Lancet Public Health* **2017**, *2*, e277–e285. [CrossRef]
5. Global Burden of Metabolic Risk Factors for Chronic Diseases Collaboration; Lu, Y.; Hajifathalian, K.; Ezzati, M.; Woodward, M.; Rimm, E.B.; Danaei, G. Metabolic mediators of the effects of body-mass index, overweight, and obesity on coronary heart disease and stroke: A pooled analysis of 97 prospective cohorts with 1.8 million participants. *Lancet* **2014**, *383*, 970–983. [CrossRef] [PubMed]
6. Jensen, M.D.; Haymond, M.W.; Rizza, R.A.; Cryer, P.E.; Miles, J.M. Influence of body fat distribution on free fatty acid metabolism in obesity. *J. Clin. Investig.* **1989**, *83*, 1168–1173. [CrossRef]
7. Alligier, M.; Meugnier, E.; Debard, C.; Lambert-Porcheron, S.; Chanseaume, E.; Sothier, M.; Loizon, E.; Hssain, A.A.; Brozek, J.; Scoazec, J.Y.; et al. Subcutaneous adipose tissue remodeling during the initial phase of weight gain induced by overfeeding in humans. *J. Clin. Endocrinol. Metab.* **2012**, *97*, E183–E192. [CrossRef]
8. Vogel, M.A.A.; Wang, P.; Bouwman, F.G.; Hoebers, N.; Blaak, E.E.; Renes, J.; Mariman, E.C.; Goossens, G.H. A comparison between the abdominal and femoral adipose tissue proteome of overweight and obese women. *Sci. Rep.* **2019**, *9*, 4202. [CrossRef] [PubMed]
9. Bosy-Westphal, A.; Geisler, C.; Onur, S.; Korth, O.; Selberg, O.; Schrezenmeir, J.; Muller, M.J. Value of body fat mass vs anthropometric obesity indices in the assessment of metabolic risk factors. *Int. J. Obes.* **2006**, *30*, 475–483. [CrossRef]
10. Selvaraj, S.; Martinez, E.E.; Aguilar, F.G.; Kim, K.Y.; Peng, J.; Sha, J.; Irvin, M.R.; Lewis, C.E.; Hunt, S.C.; Arnett, D.K.; et al. Association of Central Adiposity With Adverse Cardiac Mechanics: Findings From the Hypertension Genetic Epidemiology Network Study. *Circ. Cardiovasc. Imaging* **2016**, *9*, e004396. [CrossRef]
11. Seidell, J.C.; Perusse, L.; Despres, J.P.; Bouchard, C. Waist and hip circumferences have independent and opposite effects on cardiovascular disease risk factors: The Quebec Family Study. *Am. J. Clin. Nutr.* **2001**, *74*, 315–321. [CrossRef]
12. Snijder, M.B.; Dekker, J.M.; Visser, M.; Bouter, L.M.; Stehouwer, C.D.; Kostense, P.J.; Yudkin, J.S.; Heine, R.J.; Nijpels, G.; Seidell, J.C. Associations of hip and thigh circumferences independent of waist circumference with the incidence of type 2 diabetes: The Hoorn Study. *Am. J. Clin. Nutr.* **2003**, *77*, 1192–1197. [CrossRef]
13. Swift, D.L.; McGee, J.E.; Earnest, C.P.; Carlisle, E.; Nygard, M.; Johannsen, N.M. The Effects of Exercise and Physical Activity on Weight Loss and Maintenance. *Prog. Cardiovasc. Dis.* **2018**, *61*, 206–213. [CrossRef] [PubMed]
14. Elagizi, A.; Kachur, S.; Carbone, S.; Lavie, C.J.; Blair, S.N. A Review of Obesity, Physical Activity, and Cardiovascular Disease. *Curr. Obes. Rep.* **2020**, *9*, 571–581. [CrossRef] [PubMed]
15. Donnelly, J.E.; Blair, S.N.; Jakicic, J.M.; Manore, M.M.; Rankin, J.W.; Smith, B.K.; American College of Sports Medicine. American College of Sports Medicine Position Stand. Appropriate physical activity intervention strategies for weight loss and prevention of weight regain for adults. *Med. Sci. Sports Exerc.* **2009**, *41*, 459–471. [CrossRef] [PubMed]
16. Cureton, T. *The Effect of Gymnastics upon Boys*; Unpublished Paper; College Coaches Gymnastic Clinic: Sarasota, FL, USA, 1954.
17. Yuhasz, M.S. *The Effects of Sports Training on Body Fat in Man with Predictions of Optimal Body Weight*; University of Illinois at Urbana-Champaign: Champaign, IL, USA, 1962.
18. Kireilis, R.W.; Cureton, T.K. The relationships of external fat to physical education activities and fitness tests. *Res. Q. Am. Assoc. Healthphys. Educ. Recreat.* **1947**, *18*, 123–134. [CrossRef]
19. Stallknecht, B.; Dela, F.; Helge, J.W. Are blood flow and lipolysis in subcutaneous adipose tissue influenced by contractions in adjacent muscles in humans? *Am. J. Physiol. Endocrinol. Metab.* **2007**, *292*, E394–E399. [CrossRef] [PubMed]
20. Mohr, D.R. Changes in Waistline and Abdominal Girth and Subcutaneous Fat Following Isometric Exercises. *Res. Q.* **1965**, *36*, 168–173. [CrossRef] [PubMed]
21. Noland, M.; Kearney, J.T. Anthropometric and densitometric responses of women to specific and general exercise. *Res. Q.* **1978**, *49*, 322–328. [CrossRef]
22. Olson, A.L.; Edelstein, E. Spot reduction of subcutaneous adipose tissue. *Res. Q.* **1968**, *39*, 647–652. [CrossRef]
23. Gwinup, G.; Chelvam, R.; Steinberg, T. Thickness of subcutaneous fat and activity of underlying muscles. *Ann. Intern. Med.* **1971**, *74*, 408–411. [CrossRef] [PubMed]
24. Krotkiewski, M.; Aniansson, A.; Grimby, G.; Bjorntorp, P.; Sjostrom, L. The effect of unilateral isokinetic strength training on local adipose and muscle tissue morphology, thickness, and enzymes. *Eur. J. Appl. Physiol. Occup. Physiol.* **1979**, *42*, 271–281. [CrossRef] [PubMed]
25. Roby, F.B. Effect of exercise on regional subcutaneous fat accumulations. *Res. Q. Am. Assoc. Healthphys. Educ. Recreat.* **1962**, *33*, 273–278. [CrossRef]

26. Katch, F.I.; Clarkson, P.M.; Kroll, W.; McBride, T.; Wilcox, A. Effects of sit up exercise training on adipose cell size and adiposity. *Res. Q. Exerc. Sport* **1984**, *55*, 242–247. [CrossRef]
27. Kostek, M.A.; Pescatello, L.S.; Seip, R.L.; Angelopoulos, T.J.; Clarkson, P.M.; Gordon, P.M.; Moyna, N.M.; Visich, P.S.; Zoeller, R.F.; Thompson, P.D.; et al. Subcutaneous fat alterations resulting from an upper-body resistance training program. *Med. Sci. Sports Exerc.* **2007**, *39*, 1177–1185. [CrossRef] [PubMed]
28. Muller, W.; Horn, M.; Furhapter-Rieger, A.; Kainz, P.; Kropfl, J.M.; Maughan, R.J.; Ahammer, H. Body composition in sport: A comparison of a novel ultrasound imaging technique to measure subcutaneous fat tissue compared with skinfold measurement. *Br. J. Sports Med.* **2013**, *47*, 1028–1035. [CrossRef] [PubMed]
29. Bellisari, A.; Roche, A.F.; Siervogel, R.M. Reliability of B-mode ultrasonic measurements of subcutaneous adipose tissue and intra-abdominal depth: Comparisons with skinfold thicknesses. *Int. J. Obes. Relat. Metab. Disord.* **1993**, *17*, 475–480.
30. Villareal, D.T.; Aguirre, L.; Gurney, A.B.; Waters, D.L.; Sinacore, D.R.; Colombo, E.; Armamento-Villareal, R.; Qualls, C. Aerobic or Resistance Exercise, or Both, in Dieting Obese Older Adults. *N. Engl. J. Med.* **2017**, *376*, 1943–1955. [CrossRef]
31. Scotto di Palumbo, A.; Guerra, E.; Orlandi, C.; Bazzucchi, I.; Sacchetti, M. Effect of combined resistance and endurance exercise training on regional fat loss. *J. Sports Med. Phys. Fit.* **2017**, *57*, 794–801. [CrossRef]
32. Paoli, A.; Pacelli, F.; Bargossi, A.M.; Marcolin, G.; Guzzinati, S.; Neri, M.; Bianco, A.; Palma, A. Effects of three distinct protocols of fitness training on body composition, strength and blood lactate. *J. Sports Med. Phys. Fit.* **2010**, *50*, 43–51.
33. Paoli, A.; Pacelli, Q.F.; Moro, T.; Marcolin, G.; Neri, M.; Battaglia, G.; Sergi, G.; Bolzetta, F.; Bianco, A. Effects of high-intensity circuit training, low-intensity circuit training and endurance training on blood pressure and lipoproteins in middle-aged overweight men. *Lipids Health Dis.* **2013**, *12*, 131. [CrossRef]
34. Dussini, N.; Martino, R.; Neri, M.; Paoli, A.; Velussi, C. Multifactorial analysis of circuit-training induced regional fat reduction. *Pflügers Arch. Eur. J. Physiol.* **1994**, *426*, R186. [CrossRef]
35. Durnin, J.; Rahaman, M.M. The assessment of the amount of fat in the human body from measurements of skinfold thickness. *Br. J. Nutr.* **1967**, *21*, 681–689. [CrossRef]
36. Muller, W.; Horn, M.; Furhapter-Rieger, A.; Kainz, P.; Kropfl, J.M.; Ackland, T.R.; Lohman, T.G.; Maughan, R.J.; Meyer, N.L.; Sundgot-Borgen, J.; et al. Body composition in sport: Interobserver reliability of a novel ultrasound measure of subcutaneous fat tissue. *Br. J. Sports Med.* **2013**, *47*, 1036–1043. [CrossRef] [PubMed]
37. Chirita-Emandi, A.; Dobrescu, A.; Papa, M.; Puiu, M. Reliability of Measuring Subcutaneous Fat Tissue Thickness Using Ultrasound in Non-Athletic Young Adults. *Maedica* **2015**, *10*, 204–209.
38. Durnin, J.V.; Womersley, J. Body fat assessed from total body density and its estimation from skinfold thickness: Measurements on 481 men and women aged from 16 to 72 years. *Br. J. Nutr.* **1974**, *32*, 77–97. [CrossRef]
39. Siri, W.E. Body composition from fluid spaces and density: Analysis of methods. *Tech. Meas. Body Compos.* **1961**, *61*, 223–244.
40. Ramirez, M.E. Measurement of subcutaneous adipose tissue using ultrasound images. *Am. J. Phys. Anthr.* **1992**, *89*, 347–357. [CrossRef]
41. Checkley, E. *A Natural Material Method of Physical Education Training Making Muscle and Reducing Flesh without Dieting or Apparatus*; William C. Bryant and Co: New York, NY, USA, 1895.
42. Duncan, R.E.; Ahmadian, M.; Jaworski, K.; Sarkadi-Nagy, E.; Sul, H.S. Regulation of lipolysis in adipocytes. *Annu. Rev. Nutr.* **2007**, *27*, 79–101. [CrossRef]
43. Jaworski, K.; Sarkadi-Nagy, E.; Duncan, R.E.; Ahmadian, M.; Sul, H.S. Regulation of triglyceride metabolism. IV. Hormonal regulation of lipolysis in adipose tissue. *Am. J. Physiol. Gastrointest. Liver Physiol.* **2007**, *293*, G1–G4. [CrossRef] [PubMed]
44. Lustig, J.; Strauss, B. *Nutritional Assessment. Anthropometry and Clinical Examination*; Encyclopedia of Foood Sciences and Nutrition; Academic Press: Oxford, UK, 2003.
45. Schade, M.; Helledrandt, F.; Waterland, J.C.; Carns, M.L. Spot reducing in overweight college women: Its influence on fat distribution as determined by photography. *Res. Q. Am. Assoc. Healthphys. Educ. Recreat.* **1962**, *33*, 461–471. [CrossRef]

International Journal of
Environmental Research and Public Health

MDPI

Article

The Activation of Gluteal, Thigh, and Lower Back Muscles in Different Squat Variations Performed by Competitive Bodybuilders: Implications for Resistance Training

Giuseppe Coratella [1,*], Gianpaolo Tornatore [1], Francesca Caccavale [1], Stefano Longo [1], Fabio Esposito [1,2] and Emiliano Cè [1,2]

[1] Department of Biomedical Sciences for Health, Università degli Studi di Milano, via Giuseppe Colombo 71, 20133 Milano, Italy; gianpaolo.tornatore@email.it (G.T.); francesca.caccavale@studenti.unimi.it (F.C.); stefano.longo@unimi.it (S.L.); fabio.esposito@unimi.it (F.E.); emiliano.ce@unimi.it (E.C.)
[2] IRCCS Galeazzi Orthopaedic Institute, 20122 Milano, Italy
* Correspondence: giuseppe.coratella@unimi.it

Citation: Coratella, G.; Tornatore, G.; Caccavale, F.; Longo, S.; Esposito, F.; Cè, E. The Activation of Gluteal, Thigh, and Lower Back Muscles in Different Squat Variations Performed by Competitive Bodybuilders: Implications for Resistance Training. *Int. J. Environ. Res. Public Health* **2021**, *18*, 772. https://doi.org/10.3390/ijerph18020772

Received: 3 December 2020
Accepted: 14 January 2021
Published: 18 January 2021

Publisher's Note: MDPI stays neutral with regard to jurisdictional claims in published maps and institutional affiliations.

Abstract: The present study investigated the activation of gluteal, thigh, and lower back muscles in different squat variations. Ten male competitive bodybuilders perform back-squat at full (full-BS) or parallel (parallel-BS) depth, using large feet-stance (sumo-BS), and enhancing the feet external rotation (external-rotated-sumo-BS) and front-squat (FS) at 80% 1-RM. The normalized surface electromyographic root-mean-square (sEMG RMS) amplitude of *gluteus maximus, gluteus medius, rectus femoris, vastus lateralis, vastus medialis, adductor longus, longissimus,* and *iliocostalis* was recorded during both the ascending and descending phase of each exercise. During the descending phase, greater sEMG RMS amplitude of *gluteus maximus* and *gluteus medius* was found in FS vs. all other exercises ($p < 0.05$). Additionally, FS elicited *iliocostalis* more than all other exercises. During the ascending phase, both sumo-BS and external-rotated-sumo-BS showed greater *vastus lateralis* and *adductor longus* activation compared to all other exercises ($p < 0.05$). Moreover, *rectus femoris* activation was greater in FS compared to full-BS ($p < 0.05$). No between-exercise difference was found in *vastus medialis* and *longissimus* showed no between-exercise difference. FS needs more backward stabilization during the descending phase. Larger feet-stance increases thigh muscles activity, possibly because of their longer length. These findings show how bodybuilders uniquely recruit muscles when performing different squat variations.

Keywords: EMG; quadriceps; gluteus maximus; adductor longus; weight training; strength training; front squat; back squat; feet stance

1. Introduction

The squat is one of the most popular exercises used to elicit lower-limb strength, hypertrophy, and power [1–3]. It consists of a simultaneous flexion-extension of the hip, knee, and ankle joints, with the important role of the lower back muscles that stabilize the upper body, and consequently the whole movement [4,5]. Particularly, the gluteal, thigh, and lower back muscles are strongly activated during both the ascending and descending phase [6–9].

The squat can be performed in a multitude of variations, depending for example on the place where the barbell is located, the squatting depth, the feet stance, and/or feet rotation. Consequently, we may have back-squat (BS) or front-squat (FS) when the barbell is placed over the shoulders or in front of the clavicular line, respectively [10]. Alternatively, the squatting depth may lead to parallel or full squat, where the descending phase ends when the thighs are parallel to the ground or below this line, respectively [7]. Furthermore, the feet stance may be regular or wide, leading in the latter case to a so-called sumo-squat, where the direction of the feet can be parallel or rotated externally [11]. Obviously, all these independent parameters can be miscellaneously used to create many combinations of squatting techniques and exercises. Among these, full-BS, parallel-BS, FS, sumo-BS, and external-rotated-sumo-BS are widely performed in practice.

A number of studies have investigated the difference in muscle activation when different squat variations are performed. Overall, squatting depth was shown to affect *gluteus maximus* activation with inconsistent results, with greater activation recorded in partial vs. full squat performed by young resistance-trained men [12], greater activation in full vs. partial performed by experienced lifters [7], or no difference when performed by resistance-trained women [13]. Additionally, quadriceps activation was overall greater in full vs. partial squat [14]. Interestingly, no difference in muscle activation was found comparing BS vs. FS performed with 70% 1-RM by healthy men [10], while larger stance specifically activates medial thigh muscles in experienced lifters [15], although no difference was found in gluteal muscles activation [11].

Bodybuilders have a unique capacity to perform exercises with a profound consistency of their technique, and were recently used to investigate the differences in muscle activation when bench press [16] or shoulder raise variations [17] are performed. Additionally, examining the muscle activation during both the concentric and the descending phase may help practitioners to characterize the strength and the hypertrophic stimuli, given both the short-term [18,19] and long-term unique responses following traditional or eccentric-based exercise training [2,20–22]. Therefore, the present study investigated the differences in the gluteal, thigh, and lower back muscles' activation in bodybuilders when varying the squatting technique. Particularly, the exercises selected were full-BS, parallel-BS, sumo-BS, external-rotated-sumo-BS, and FS, and the gluteal, thigh, and lower back muscles' activation were recorded during both the ascending and descending phase.

2. Materials and Methods

2.1. Study Design

The present investigation was designed as a cross-over, repeated-measures, within-subject study. The participants were involved in seven different sessions. In the first five sessions, the 1-RM was measured in full-BS, parallel-BS, sumo-BS, external-rotation-sumo-BS, and FS in random order. In the sixth session, the participants were familiarized with the selected loads and the electrodes placement. In the seventh session, the muscles' maximum activation was first measured, i.e., the activation during a maximum voluntary contraction. Then, after a minimum of 30 min of passive recovery, the participant performed a non-exhausting set for each exercise performed in a random order, with an inter-set pause of 10 min. Each session was separated by at least three days, and the participants were instructed to avoid any further form of resistance training for the entire duration of the investigation.

2.2. Participants

The present investigation was advertised by the investigators during some regional and national competitions, and to be included in the study, the participants had to compete in regional competitions for a minimum of 5 years. Additionally, they had to be clinically healthy, without any reported history of upper-limb and lower back muscle injury and neurological or cardiovascular disease in the previous 12 months. To avoid possible confounding factors, the participants competed in the same weight category (Men's Classic Bodybuilding <80 kg, <1.70 m), according to the International Federation of Body Building Pro-League. The use of drugs or steroids was continuously monitored by a dedicated authority under its regulations, although we could have not checked for it. Thereafter, 10 male competitive bodybuilders (age 29.8 $\pm$ 3.0 years; body mass 77.9 $\pm$ 1.0 kg; stature 1.68 $\pm$ 0.01 m; training seniority 10.6 $\pm$ 1.8 years) were recruited for the present procedures. The participants were asked to abstain from alcohol, caffeine, or similar beverages in the 24 h preceding the test. After a full explanation of the aims of the study and the experimental procedures, the participants signed a written informed consent. They were also free to withdraw at any time. The current design was approved by the Ethical Committee of the Università degli Studi di Milano (CE 27/17) and performed following the Declaration of Helsinki (1975) for studies involving human subjects.

2.3. Maximum Voluntary Isometric Activation

The maximum voluntary isometric activation of gluteus maximus, gluteus medius, rectus femoris, vastus lateralis, vastus medialis, erector spinae longissimus, and erector spinae iliocostal was assessed in random order. The electrodes were placed on the dominant limb, defined as the one preferred to kick a ball [2]. The participants were required to exert their maximum force against manual resistance. Each attempt lasted 5 s and three attempts were completed for each movement separated by 3 min of passive recovery [16,17]. The operators provided strong standardized verbal encouragements to push as hard as possible against the resistance exerted. The surface electromyography (sEMG) electrodes were placed following the SENIAM recommendations [23]. To check for appropriate electrodes placement previous procedures were followed [17]. For example, if the electrode shifted over the innervation zone during part of the movement, the EMG amplitude was underestimated. Therefore, to check for any consequence due to a possible shift of the surface electrode over the innervation zone, a Fast-Fourier Transform approach was used, as suggested in a previous investigation [24]. Briefly, the electrode placement on each muscle was checked during the warm-up phase of each exercise, analyzing the power spectrum profile of the sEMG signal recorded at the starting-, middle-, and endpoint of each exercise in all muscles. The correct electrode placement results in a typical belly-shaped power spectrum profile of the EMG signal, while noise, motion artifacts, power lines, and electrodes placed on the innervation zone or myotendinous junction generate a different power spectrum profile [24]. If the power spectrum did not match with the typical belly-shaped power spectrum profile in any of the temporal points, the electrodes were repositioned, and the procedures repeated so to have a clear EMG signal from all the muscles throughout the movement. The same experienced operator placed the electrodes and checked the power-spectrum profile. This approach was shown to provide very high reliability in sEMG data [16,17].

For *gluteus maximus*, the participants laid prone with the flexed knee and the electrode was placed below the line between the posterior-superior iliac spine and the trochanter major [13]. The participants were then asked to extend the hip against a manual resistance on the distal thigh [13]. For *gluteus medius*, the participant laid on a side and the electrodes were placed at 50% on the line from the crista iliaca to the trochanter. The participant was then asked to abduct the limb against manual resistance [23]. For *rectus femoris, vastus lateralis*, and *vastus medialis*, the participants sat on a table with the knees in slight flexion and the trunk slightly bent backward. The electrode were respectively placed at 50% and 2/3 on the line between the anterior-superior iliac spine and the lateral side of the patella and at 90% on the line between the anterior-superior iliac spine and the joint space in front of the anterior border of the medial ligament [23]. The participant were then asked to extend the knee against manual resistance [23]. The *adductor longus* belly was found midway between the origin at the pubic tubercle and the insertion at the medial linea aspera of the femur [25]. To ensure electrode placement, the test leg was passively abducted and the adductor longus muscle belly was palpated just distal to the muscle's tendon, traced from the pubic tubercle on the medial side of the leg, and the participant was then asked to actively adduct the leg against resistance [25]. For *erector spinae longissimus* and *iliocostalis*, the participant laid prone and the electrodes were respectively placed at 2-finger width lateral from the processus spinalis of L1 and 1-finger width medial from the line from the posterior-superior iliac spine to the lowest point of the lower rib, at the level of L2 [23]. The participant was then asked to extend the trunk against manual resistance [23].

The electrodes were equipped with a probe (probe mass: 8.5 g, BTS Inc., Milano, Italy) that permitted the detection and the transfer of the sEMG signal by wireless modality. sEMG signal was acquired at 1000 Hz, amplified (gain: 2000, impedance and the common rejection mode ratio of the equipment are >1015 Ω//0.2 pF and 60/10 Hz 92 dB, respectively), and driven to a wireless electromyographic system (FREEEMG 300, BTS Inc., Milano, Italy) that digitized (1000 Hz) and filtered (filter type: IV-order Butterworth filter; bandwidth: 10–500 Hz) the raw sEMG signals.

2.4. 1-RM Protocol

The squat 1-RM was assessed following previous procedures [26] using an Olympic bar (Vulcan Standard 20 kg, Vulcan Strength Training System, Charlotte, NC, USA). Briefly, after a standardized warm-up consisting of 30 weight-free squats, the 1-RM attempts started from 80% of the self-declared 1-RM and additional 5% or less was added until failure [27]. Each attempt was separated by at least 3 min of passive recovery. A standard time under tension (2 s for the ascending and descending phase, 0.5 for the isometric phase) was used and the participants had to lower the bar until the thighs were parallel to the ground. A metronome was used to pace the intended duty cycle and a camera was used to provide a feedback about the squatting technique and depth. Strong standardized encouragements were provided to the participants to maximally perform each trial.

2.5. Exercises' Technique Description

The selected exercises are shown in Figure 1, and described here from left to right, first the upper and then the lower row. In parallel-BS, the bar was placed over the shoulder and the participants were required to descent until the thighs were parallel to the ground, with a regular feet stance. In full-BS, the bar was placed over the shoulder and the participants were required to descent below the parallel thighs, with a regular feet stance. In FS, the bar was placed in front of the clavicular line and sternum, and the participants were required to descent until the thighs were parallel to the ground, with a regular feet stance. In sumo-BS, the bar was placed over the shoulder, and the participants were required to descent until the thighs were parallel to the ground, with a two-fold feet stance compared to the previous exercises. In external-rotated-sumo-BS, the participants received the same instructions as for sumo-BS, with the exception of the feet that were rotated externally. Six non-exhaustive repetitions were performed for each exercise.

Figure 1. The squat variations are shown. From the left to the right, in the upper row: full-back squat (BS), parallel-BS, and front squat (FS). In the lower row: sumo-BS and external-rotated-sumo-BS.

2.6. Data Analysis

The sEMG signals from both the peak value recorded during the maximum voluntary isometric activation and from the ascending and descending phases of each exercise were analyzed in time-domain, using a 25-ms mobile window for the computation of the root mean square (RMS). For the maximum voluntary isometric activation, the average of the RMS corresponding to the central 2 s was considered. During each exercise, the RMS was calculated and averaged over the 2 s of the ascending and descending phase. To identify the ascending and the descending phase, the sEMG was synchronized with an integrated camera (VixtaCam 30 Hz, BTS Inc., Milano, Italy) that provided the duration of each phase. Such a duration was used to mark the start and the end of each phase while analyzing the sEMG signal. The sEMG data were averaged excluding the first repetition of each set, to possibly have more consistent technique during the following repetitions. After, the sEMG RMS of each muscle during each exercise was normalized for its respective maximum voluntary isometric activation [16,17,27] and inserted into the data analysis.

2.7. Statistical Analysis

The statistical analysis was performed using a statistical software (SPSS 22.0, IBM, Armonk, NY, USA). The normality of data was checked using the Shapiro–Wilk test and all distributions were normal. Descriptive statistics are reported as mean (SD). The differences in the normalized EMG RMS were separately calculated for each exercise (5 levels) and phase (2 levels) using a two-way repeated-measures ANOVA. Multiple comparisons were adjusted using the Bonferroni's correction. Significance was set at $p < 0.05$. The differences are reported as mean with 95% of confidence interval (95%CI). Cohen's d effect size (ES) with 95% confidence interval (CI) was reported and interpreted according to the Hopkins' recommendations: 0.00–0.19: trivial; 0.20–0.59: small: 0.60–1.19: moderate; 1.20–1.99: large; ≥2.00: very large [28].

3. Results

The 1-RM were as follows: 215(28) kg for full-BS, 238(31) kg for parallel-BS, 255(36) kg for sumo-BS, 258(41) kg for external-rotated-sumo-BS, and 176(33) kg for FS.

The results for *gluteus maximus* are shown in Figure 2. No phase x exercise interaction ($p = 0.197$) was found for the normalized RMS of *gluteus maximus*. A main effect was found for factor phase ($p < 0.001$), but not exercise ($p = 0.097$). With the exception of FS (11.1%, −6.5% to 28.8%, $p = 0.11$; ES: 0.48, −0.43 to 1.48), greater normalized RMS was found during the ascending vs. descending phase in all exercises (16.0% to 41.1%, $p < 0.05$; ES: 1.55 to 3.99). During the ascending phase, no between-exercise difference was observed. During the descending phase, greater normalized RMS was found in FS vs. full-BS (46.6%, 8.4% to 84.8%, $p = 0.017$; ES: 2.94, 1.58 to 4.05), parallel-BS (40.9%, 14.0% to 67.9%, $p = 0.005$; ES: 2.58, 1.31 to 3.63), sumo-BS (40.1%, 7.1% to 73.0%, $p = 0.017$; ES: 2.38, 1.16 to 3.41), and external-rotated-sumo-BS (44.9%, 10.3% to 79.5%, $p = 0.012$; ES: 2.83, 1.49 to 3.92).

The results for *gluteus medius* are shown in Figure 2. No phase x exercise interaction ($p = 0.157$) was found for the normalized RMS of *gluteus medius*. A main effect was found for factor phase ($p = 0.002$), but not exercise ($p = 0.125$). Greater normalized RMS was found during the ascending phase in full-BS (12.0%, 9.1% to 15%, $p < 0.001$; ES: 2.92, 1.56 to 4.02) and external-rotated-sumo-BS (12.9%, 4.2% to 21.7%, $p = 0.010$; ES: 1.57, 0.51 to 2.49). During the ascending phase, no between-exercise difference was observed. During the descending phase, greater normalized RMS was found in FS vs. full-BS (19.0%, 4.9% to 33.1%, $p = 0.010$; ES: 2.16, 0.98 to 3.16), parallel-BS (13.6%, 0.4% to 26.8%, $p = 0.016$; ES: 1.35, 0.10 to 2.70), sumo-BS (17.5%, 5.4% to 29.6%, $p = 0.006$; ES: 1.90, 0.78 to 2.86), and external-rotated-sumo-BS (19.4%, 7.3% to 31.5%, $p = 0.003$; ES: 2.10, 0.93 to 3.08).

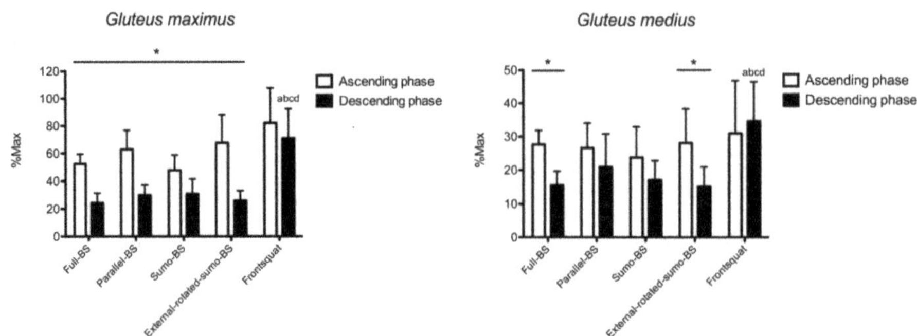

Figure 2. The surface electromyographic root-mean-square (sEMG) RMS amplitude of *gluteus maximus* and *gluteus medius* is shown. BS: back squat. *: $p < 0.05$ ascending vs. descending phase. a: $p < 0.05$ vs. full-BS. b: $p < 0.05$ vs. parallel-BS. c: $p < 0.05$ vs. sumo-BS. d: $p < 0.05$ vs. external-rotated-sumo-BS.

The results for *rectus femoris* are shown in Figure 3. Phase x exercise interaction ($p = 0.038$) was found for the normalized RMS, and no main effect was found for factor phase ($p = 0.417$) and exercise ($p = 0.231$). Greater normalized RMS was found during the ascending compared to the descending phase in FS (30.1%, 7.8% to 52.3%, $p = 0.015$; ES: 1.35, 0.33 to 2.25). During the ascending phase FS showed greater normalized RMS than full-BS (24.0%, 1.9% to 46.0%, $p = 0.032$; ES: 1.21, 0.21 to 2.11). No between-exercise difference was found during the descending phase.

The results for *vastus lateralis* are shown in Figure 3. Phase x exercise interaction ($p = 0.026$) was found for the normalized RMS, and a main effect was found for factor phase ($p = 0.011$), but not exercise ($p = 0.457$). Compared to the descending phase, greater normalized RMS was found during the ascending phase in full-BS (22.1%, 6.1% to 38.1%, $p = 0.013$; ES: 1.60, 0.54 to 2.53), sumo-BS (28.8%, 8.4% to 49.1%, $p = 0.012$; ES: 1.64, 0.57 to 2.58), and external-rotated-sumo-BS (30.0%, 14.6% to 45.5%, $p = 0.002$; ES: 1.26, 0.25 to 2.16). During the ascending phase, both sumo-BS (19.8%, 0.8% to 38.8%, $p = 0.040$; ES: 0.97, 0.01 to 1.85) and external-rotated-sumo-BS (23.0%, 3.8% to 42.1%, $p = 0.019$; ES: 0.88, −0.07 to 1.76) had greater normalized RMS than FS. No between-exercise difference was found during the descending phase.

The results for *vastus medialis* are shown in Figure 3. No phase x exercise interaction ($p = 0.133$) was found for the normalized RMS, and a main effect was found for factor phase ($p < 0.001$), but not exercise ($p = 0.102$). Compared to the descending phase, greater normalized RMS was found during the ascending phase in full-BS (25.2%, 11.8% to 38.5%, $p = 0.003$; ES: 1.06, 0.08 to 1.94) and sumo-BS (25.9%, 10.6% to 41.2%, $p = 0.005$; ES: 1.27, 0.26 to 2.17). No between-exercise difference was observed during both the ascending and descending phase.

The results for *adductor longus* are shown in Figure 3. Phase x exercise interaction ($p = 0.032$) was found for the normalized RMS, and a main effect was found for factor phase ($p < 0.001$) and exercise ($p = 0.021$). Compared to the descending phase, greater normalized RMS was observed during the ascending phase in all exercises (ES: 2.25 to 5.39). During the ascending phase, greater normalized RMS was found in external-rotated-sumo-BS than full-BS (17.9%, 1.7% to 34.0%, $p = 0.029$; ES: 2.01, 0.86 to 2.98), parallel-BS (16.7%, 3.0% to 30.3%, $p = 0.017$; ES: 1.47, 0.43 to 2.39), and FS [26.9%, 7.3% to 46.5%, $p = 0.009$; ES: 2.64, 1.35 to 3.70). Greater normalized RMS was also found for sumo-BS than FS (19.7%, 5.6% to 33.9%, $p = 0.008$; ES: 2.15, 0.98 to 3.14).

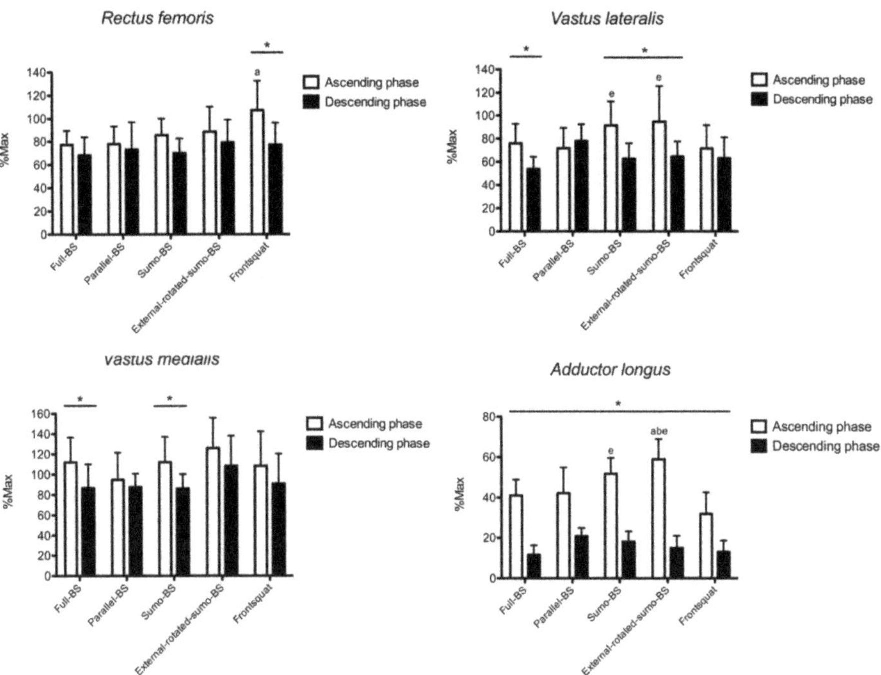

Figure 3. The surface electromyographic root-mean-square (sEMG) RMS amplitude of *rectus femoris, vastus lateralis, vastus medialis* and *adductor longus* is shown. BS: back squat. *: $p < 0.05$ ascending vs. descending phase. a: $p < 0.05$ vs. full-BS. b: $p < 0.05$ vs. parallel-BS. e: $p < 0.05$ vs. parallel front squat.

The results for *erector spinae longissimus* are shown in Figure 4. Phase x exercise interaction ($p = 0.004$) was found for the normalized RMS, and a main effect was found for factor phase ($p = 0.015$), but not exercise ($p = 0.477$). Compared to the descending phase, greater normalized RMS was found during the ascending phase in full-BS (39.6%, 16.0% to 63.1%, $p = 0.005$; ES: 1.76, 0.67 to 2.71). No between-exercise difference was observed during both ascending and descending phase.

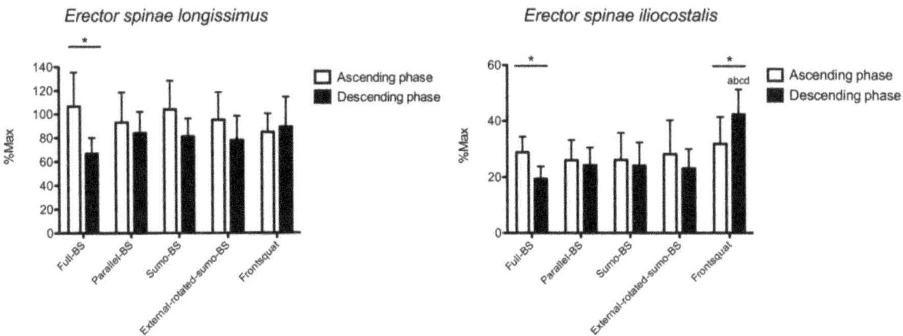

Figure 4. The surface electromyographic root-mean-square (sEMG) RMS amplitude of *erector spinae longissimus* and *erector spinae iliocostalis* is shown. BS: back squat. *: $p < 0.05$ ascending vs. descending phase. a: $p < 0.05$ vs. full-BS. b: $p < 0.05$ vs. parallel-BS. c: $p < 0.05$ vs. sumo-BS. d: $p < 0.05$ vs. external-rotated-sumo-BS.

The results for *erector spinae iliocostalis* are shown in Figure 4. Phase x exercise interaction ($p = 0.020$) was found for the normalized RMS, and a main effect was found for factor exercise ($p = 0.040$), but not phase ($p = 0.431$). Compared to the descending phase, the normalized RMS was greater during the ascending phase in full-BS (9.4%, 4.5% to 14.3%, $p = 0.003$; ES: 1.91, 0.78 to 2.87) and lower in FS (−10.4%, −18.6 to −2.3, $p = 0.019$; ES: −1.14, −2.03 to −0.15). During the descending phase, FS showed greater normalized RMS than full-BS (22.9%, 14.6% to 31.2%, $p < 0.001$; ES: 3.29, 1.84 to 4.46), parallel-BS (18.1%, 0.5% to 35.7%, $p = 0.043$; ES: 2.37, 1.14 to 3.39), sumo-BS (18.2%, 12.1% to 24.3%, $p < 0.001$; ES: 2.14, 0.97 to 3.13), and external-rotated-sumo-BS (19.2%, 7.1% to 31.2%, $p = 0.004$; ES: 2.43, 1.19 to 3.46). No between-exercise difference was found during the ascending phase.

4. Discussion

The current study examined how different squat variations influence the activation of the main muscles involved in these exercises. Both *gluteus maximus* and *gluteus medius* were more active during the descending phase of FS compared to all other exercises. *Rectus femoris* was more active during the ascending phase of FS compared to full-BS compared to all other exercises, while no between-exercise difference was visible for *vastus medialis*. *Vastus lateralis* and *adductor longus* were more active during the ascending phase of sumo-BS and external-rotation-sumo-BS compared to all other exercises. Lastly, while no between-exercise difference was observed for *erector spinae longissimus*, *erector spinae iliocostalis* was more active during the descending phase of FS elicited compared to all other exercises. As such, varying the squatting technique seems to affect selectively the muscle activation.

4.1. Gluteal Muscles

FS showed very large increases in the *gluteus maximus* activation compared with all other exercises during the descending phase, with no between-exercise difference recorded during the ascending phase. A direct comparison with the literature is challenging, since few previous studies used similar design. When recording the sEMG RMS amplitude of *gluteus maximus* and distinguishing the ascending from the descending phase, no difference in FS vs. BS was found [29]. However, the load was maximal and performed by healthy men that limits the inference towards the present population. Additionally, we found that the *gluteus maximus* activation recorded here is much greater compared to the aforementioned study (e.g., 70% vs. 30% of the maximum activation during the descending phase of FS), which underlines the capacity of bodybuilders to increase muscle activation while training [30]. Moreover, no difference in gluteus maximus activation was found comparing FS, full-BS, and parallel-BS in trained women [13]. However, the authors did not specifically state which phase (ascending or descending or both) was examined, since it leads to argue that these findings are consistent with the no between-exercise difference recorded here during the ascending phase. Additionally, FS vs. BS was previously investigated, but no gluteal muscle was examined [10]. Lastly, the effect of stance does not seem to play a key role in *gluteus maximus* activation, which contrasts with the greater activation reported at greater stance [11,15]. Again, it is possible that the present bodybuilders population may have cancelled such a difference, since they were able to recruit the gluteus maximus more than just experienced lifters irrespectively of the stance. Similarly, *gluteus medius* resulted in greater activation during the descending phase of FS compared to all other exercise, with no between-exercise difference during the ascending phase. In a previous study, no difference in *gluteus medius* activation was observed when increasing the feet stance, confirming the present findings [11]. Taking all together, gluteal muscles seem to be particularly involved during the descending phase of FS. This may derive from the need to maintain an adequate trunk extension to avoid the barbell slipping forward (i.e., *gluteus maximus*), and to avoid a medial collapsing of the knees (i.e., *gluteus medius*), particularly when controlling the descending phase. As such, a frontal barbell placement seems to be a good option to increase the stimuli towards gluteal muscles while squatting.

4.2. Thigh Muscles

Rectus femoris showed greater activation in FS compared to full-BS during the ascending phase, with no other between-exercise differences. The lack of differences between full-BS and parallel-BS agrees with the no-difference found previously in powerlifters or weightlifters [31] or in healthy resistance-trained men [12]. Similarly with previous results, no difference in rectus femoris activation was reported when varying the squatting stance [11]. The reduced activation in full-BS vs. FS can be possibly explained by the greater *rectus femoris* length forced by the more vertical trunk in FS, which agrees with the greater work performed by the aforementioned gluteal muscles. Indeed, since *rectus femoris* acts as hip flexor, a more extended trunk corresponds to a longer length throughout the whole movement, thus increasing its activation as previously shown for deltoids [17] and triceps [32]. Both the sumo squats showed greater activation in *vastus lateralis* vs. FS. As suggested previously, larger stance makes hip and knee joint to exert more force to lift the load due to the non-favorable less vertical lever, thus increasing their recruitment [15]. Indeed, larger stance was shown to increase the *vastus lateralis* activation [33], rather than an external feet rotation alone, as previously reported [34]. Moreover, *vastus medialis* showed no between-exercise difference, with all exercises highly recruiting it. This may depend by the role of profound stabilizer of the patella across all movements, that enhances its activation when high loads have to be lifted. Lastly, larger stance and feet external rotation increased the *adductor longus* activation. This may depend on the need to stabilize the thigh position and keep the trajectory as vertical as possible in conjunction with the thigh external rotators, and on the longer muscle length at which adductor longus act at larger squat stance [11,15,35]. Taking together, larger feet stance may be used as an effective stimulus to increase the thigh muscles activity and could be implemented in the training practice accordingly.

4.3. Lower Back Muscles

Erector spinae longissimus showed no between-exercise difference, displaying a great activation across all exercises and during both the ascending and descending phase. In line with our results, no difference was found between BS and FS in experienced lifters [10], not even at different squatting depth in resistance trained men [12]. The study that investigated the effects of feet stance did not examine any lower back muscle [11,15,36], so a direct comparison cannot be made. However, given the high load and the consistent squatting technique, it is possible that the feet stance does not play a role in the *erector spinae longissimus* activation. Intriguingly, the activation of *erector spinae iliocostalis* was greater in FS compared with all other exercises during the descending phase. This may imply that FS needs additional balance control by mean of the trunk extensors to avoid any possible forward unbalancing. However, it should be noted that the net activation was much lesser than what observed in *longissimus*, meaning that the whole trunk and not only the lower back is involved in stabilizing the body. Lastly, both erectors' activation was greater during the ascending vs. descending phase in full-BS. This may be accounted for the very closed joint angles that could need an additional backward action to start the movement from a non-favorable body position. In practice, in conjunction with the greater stimulus for the gluteal muscles, FS might be recommended to enhance the work of the lower back muscles.

4.4. Limitations

A number of limitations should be acknowledged. First, there is no information of any rear thigh muscle (e.g., *biceps femoris*) that could have deepened the between-exercise differences. Second, similarly, the stabilizer role of any anterior trunk muscle (e.g., *rectus abdominis*) was not examined. Third, we selected a group of squat variations among several possible different combinations, that cannot be examined in a single study, so further research is needed to widen these aspects. Fourth, adding kinematic data would deepen the knowledge and should be considered in future research. Last, it is acknowledged that the present results are specific for the present populations, and different sport background may result in different muscle activation.

5. Conclusions

In conclusion, the present study showed different muscle activation depending on the squat variation in competitive bodybuilders. A front vs. back bar position led to greater gluteal and lower back muscles activation compared to all other exercises. Additionally, larger feet stance increases the thigh muscles activation, particularly *rectus femoris, vastus lateralis,* and *adductor longus*. Lastly, squatting depth does not seem to promote any specific difference in muscle activation, with the exception of the greater *rectus femoris* activation in FS vs. full-BS. These findings could be used in resistance training practice to vary the training stimuli when performing the squat exercises depending on the muscle group needed to be highlighted. Additionally, the specific differences observed during the ascending or descending phase may increase the specificity of the training-induced effects.

Author Contributions: Conceptualization, G.C., F.E., and E.C.; methodology, G.T., F.C., and S.L.; formal analysis, G.C., G.T., F.C., and S.L.; investigation, G.T.; data curation, G.T.; writing—original draft preparation, G.C. and F.C.; writing—review and editing, G.C., G.T., F.C., S.L. F.E., and E.C.; All authors have read and agreed to the published version of the manuscript.

Funding: This research received no external funding.

Institutional Review Board Statement: The study was conducted according to the guidelines of the Declaration of Helsinki, and approved by the Ethics Committee of the Università degli Studi di Milano (protocol code CE 27/17, October 2017).

Informed Consent Statement: Informed consent was obtained from all subjects involved in the study.

Data Availability Statement: The data presented in this study are available on request from the corresponding author.

Acknowledgments: The Authors are grateful to the participants that volunteered for the present investigation.

Conflicts of Interest: The authors declare no conflict of interest.

References

1. Kubo, K.; Ikebukuro, T.; Yata, H. Effects of squat training with different depths on lower limb muscle volumes. *Eur. J. Appl. Physiol.* **2019**, *119*, 1933–1942. [CrossRef] [PubMed]
2. Coratella, G.; Beato, M.; Cè, E.; Scurati, R.; Milanese, C.; Schena, F.; Esposito, F. Effects of in-season enhanced negative work-based vs traditional weight training on change of direction and hamstrings-to-quadriceps ratio in soccer players. *Biol. Sport* **2019**, *36*, 241–248. [CrossRef] [PubMed]
3. Vanderka, M.; Longova, K.; Olasz, O.; Krčmár, M.; Walker, M. Improved Maximum Strength, Vertical Jump and Sprint Performance after 8 Weeks of Jump Squat Training with Individualized Loads. *J. Sports Sci. Med.* **2016**, *15*, 492–500.
4. Slater, L.V.; Hart, J.M. Muscle Activation Patterns during Different Squat Techniques. *J. Strength Cond. Res.* **2017**, *31*, 667–676. [CrossRef]
5. Clark, D.R.; Lambert, M.I.; Hunter, A.M. Muscle activation in the loaded free barbell squat: A brief review. *J. Strength Cond. Res.* **2012**, *26*, 1169–1178. [CrossRef]
6. Hamlyn, N.; Behm, D.G.; Young, W.B. Trunk muscle activation during dynamic weight-training exercises and isometric instability activities. *J. Strength Cond. Res.* **2007**, *21*, 1108–1112. [CrossRef]
7. Caterisano, A.; Moss, R.F.; Pellinger, T.K.; Woodruff, K.; Lewis, V.C.; Booth, W.; Khadra, T. The effect of back squat depth on the EMG activity of 4 superficial hip and thigh muscles. *J. Strength Cond. Res.* **2002**, *16*, 428–432. [CrossRef]
8. Glassbrook, D.J.; Helms, E.R.; Brown, S.R.; Storey, A.G. A Review of the Biomechanical Differences between the High-Bar and Low-Bar Back-Squat. *J. Strength Cond. Res.* **2017**, *31*, 2618–2634. [CrossRef]
9. Saeterbakken, A.H.; Stien, N.; Pedersen, H.; Andersen, V. Core Muscle Activation in Three Lower Extremity With Different Stability Requirements. *J. Strength Cond. Res.* **2019**. Publish Ahead of Print. [CrossRef]
10. Gullett, J.C.; Tillman, M.D.; Gutierrez, G.M.; Chow, J.W. A biomechanical comparison of back and front squats in healthy trained individuals. *J. Strength Cond. Res.* **2008**, *23*, 284–292. [CrossRef]

11. Paoli, A.; Marcolin, G.; Petrone, N. The effect of stance width on the electromyograhycal activity of eight superficial thigh muscles during back squat with different bar loads. *J. Strength Cond. Res.* **2009**, *23*, 246–250. [CrossRef] [PubMed]
12. Da Silva, J.J.; Schoenfeld, B.J.; Marchetti, P.N.; Pecoraro, S.L.; Greve, J.M.D.; Marchetti, P.H. Muscle activation differs between partial and full back squat exercise with external load equated. *J. Strength Cond. Res.* **2017**, *31*, 1688–1693. [CrossRef] [PubMed]
13. Contreras, B.; Vigotsky, A.D.; Schoenfeld, B.J.; Beardsley, C.; Cronin, J. A comparison of gluteus maximus, biceps femoris, and vastus lateralis electromyography amplitude for the barbell, band, and American hip thrust variations. *J. Appl. Biomech.* **2016**, *32*, 254–260. [CrossRef] [PubMed]
14. Trindade, T.B.; De Medeiros, J.A.; Dantas, P.M.S.; De Oliveira Neto, L.; Schwade, D.; De Brito Vieira, W.H.; Oliveira-Dantas, F.F. A comparison of muscle electromyographic activity during different angles of the back and front squat. *Isokinet. Exerc. Sci.* **2019**, *28*, 1–8. [CrossRef]
15. McCaw, S.T.; Melrose, D.R. Stance width and bar load effects on leg muscle activity during the parallel squat. *Med. Sci. Sports Exerc.* **1999**, *31*, 428–436. [CrossRef] [PubMed]
16. Coratella, G.; Tornatore, G.; Longo, S.; Esposito, F.; Cè, E. Specific prime movers' excitation during free-weight bench press variations and chest press machine in competitive bodybuilders. *Eur. J. Sport Sci.* **2019**, 1–22. [CrossRef]
17. Coratella, G.; Tornatore, G.; Longo, S.; Esposito, F. An electromyographic analysis of lateral raise variations and frontal raise in competitive bodybuilders. *Int. J. Environ. Res. Public Health* **2020**, *17*, E6015. [CrossRef]
18. Coratella, G.; Bertinato, L. Isoload vs isokinetic eccentric exercise: A direct comparison of exercise-induced muscle damage and repeated bout effect. *Sport Sci. Health* **2015**, *11*, 87–96. [CrossRef]
19. Coratella, G.; Chemello, A.; Schena, F. Muscle damage and repeated bout effect induced by enhanced eccentric squats. *J. Sports Med. Phys. Fitness* **2016**, *56*, 1540–1546.
20. Coratella, G.; Milanese, C.; Schena, F. Unilateral eccentric resistance training: A direct comparison between isokinetic and dynamic constant external resistance modalities. *Eur. J. Sport Sci.* **2015**, *15*, 720–726. [CrossRef]
21. Coratella, G.; Schena, F. Eccentric resistance training increases and retains maximal strength, muscle endurance, and hypertrophy in trained men. *Appl. Physiol. Nutr. Metab.* **2016**, *41*, 1184–1189. [CrossRef] [PubMed]
22. Coratella, G.; Milanese, C.; Schena, F. Cross-education effect after unilateral eccentric-only isokinetic vs dynamic constant external resistance training. *Sport Sci. Health* **2015**, *11*, 329–335. [CrossRef]
23. Hermens, H.J.; Freriks, B.; Disselhorst-Klug, C.; Rau, G. Development of recommendations for SEMG sensors and sensor placement procedures. *J. Electromyogr. Kinesiol.* **2000**, *10*, 361–374. [CrossRef]
24. Merlo, A.; Campanini, I. Technical Aspects of Surface Electromyography for Clinicians. *Open Rehabil. J.* **2010**, *3*, 98–109. [CrossRef]
25. Delmore, R.J.; Laudner, K.G.; Torry, M.R. Adductor longus activation during common hip exercises. *J. Sport Rehabil.* **2014**, *23*, 79–87. [CrossRef]
26. Coratella, G.; Beato, M.; Milanese, C.; Longo, S.; Limonta, E.; Rampichini, S.; Cè, E.; Bisconti, A.V.; Schena, F.; Esposito, F. Specific Adaptations in Performance and Muscle Architecture After Weighted Jump-Squat vs. Body Mass Squat Jump Training in Recreational Soccer Players. *J. Strength Cond. Res.* **2018**, *32*, 921–929. [CrossRef]
27. Coratella, G.; Tornatore, G.; Longo, S.; Borrelli, M.; Doria, C.; Esposito, F.; Cè, E.; Coratella, G.; Tornatore, G.; Longo, S.; et al. The Effects of Verbal Instructions on Lower Limb Muscles' Excitation in Back-Squat. *Res. Q. Exerc. Sport* **2020**, 1–7. [CrossRef]
28. Hopkins, W.G.; Marshall, S.W.; Batterham, A.M.; Hanin, J. Progressive Statistics for Studies in Sports Medicine and Exercise Science. *Med. Sci. Sport. Exerc.* **2009**, *41*, 3–13. [CrossRef]
29. Yavuz, H.U.; Erdağ, D.; Amca, A.M.; Aritan, S. Kinematic and EMG activities during front and back squat variations in maximum loads. *J. Sports Sci.* **2015**, *33*, 1058–1066. [CrossRef]
30. Alves, R.C.; Prestes, J.; Enes, A.; Moraes, W.M. Training Programs Designed for Muscle Hypertrophy in Bodybuilders: A Training Programs Designed for Muscle Hypertrophy in Bodybuilders: A Narrative Review. *Sports* **2020**, *8*, 149. [CrossRef]
31. Wretenberg, P.; Feng, Y.I.; Arborelius, U.P. High- and low-bar squatting techniques during weight-training. *Med. Sci. Sports Exerc.* **1996**, *28*, 218–224. [CrossRef] [PubMed]
32. Alves, D.; Matta, T.; Oliveira, L. Effect of shoulder position on triceps brachii heads activity in dumbbell elbow extension exercises. *J. Sports Med. Phys. Fitness* **2018**, *58*, 1247–1252. [CrossRef] [PubMed]
33. Signorile, J.F.; Rendos, N.K.; Heredia Vargas, H.H.; Alipio, T.C.; Regis, R.C.; Eltoukhy, M.M.; Nargund, R.S.; Romero, M.A. Differences in muscle activation and kinematics between cable-based and selectorized weight training. *J. Strength Cond. Res.* **2017**, *31*, 313–322. [CrossRef] [PubMed]
34. Boyden, G.; Kingman, J.; Dyson, R. A Comparison of Quadriceps Electromyographic Activity with the Position of the Foot during the Parallel Squat. *J. Strength Cond. Res.* **2000**, *14*, 379–382. [CrossRef]
35. Pereira, G.L.; Leporace, G.; Das Virgens Chabas, D.; Furtado, L.F.L.; Praxedes, J.; Batista, L.A. Influence of hip external rotation on hip adductor and rectus femoris myoelectric activity during a dynamic parallel squat. *J. Strength Cond. Res.* **2010**, *24*, 2749–2754. [CrossRef]
36. Signorile, J.F.; Kwiatkowski, K.; Caruso, J.F.; Robertson, B. Effect of foot position on the electromyographical activity of the superficial quadriceps muscles during the parallel squat and knee extension. *J. Strength Cond. Res.* **1995**, *9*, 182–187.

International Journal of
Environmental Research and Public Health

Article

The Impact of Wrist Percooling on Physiological and Perceptual Responses during a Running Time Trial Performance in the Heat

Kelsey Denby, Ronald Caruso, Emily Schlicht and Stephen J. Ives *

Department of Health and Human Physiological Sciences, Skidmore College, Saratoga Springs, NY 12866, USA; kdenby@skidmore.edu (K.D.); rcaruso@skidmore.edu (R.C.); eschlich@skidmore.edu (E.S.)

* Correspondence: sives@skidmore.edu; Tel.: +1-518-580-8366

Received: 23 September 2020; Accepted: 15 October 2020; Published: 17 October 2020

Abstract: Environmental heat stress poses significant physiological challenge and impairs exercise performance. We investigated the impact of wrist percooling on running performance and physiological and perceptual responses in the heat. In a counterbalanced design, 13 trained males (33 ± 9 years, 15 ± 7% body fat, and maximal oxygen consumption, VO_2max 59 ± 5 mL/kg/min) completed three 10 km running time trials (27 °C, 60% relative humidity) while wearing two cooling bands: (1) both bands were off (off/off), (2) one band on (off/on), (3) both bands on (on/on). Heart rate (HR), HR variability (HRV), mean arterial pressure (MAP), core temperature (T_{CO}), thermal sensation (TS), and fatigue (VAS) were recorded at baseline and recovery, while running speed (RS) and rating of perceived exertion (RPE) were collected during the 10 km. Wrist cooling had no effect ($p > 0.05$) at rest, except modestly increased HR (3–5 Δbeats/min, $p < 0.05$). Wrist percooling increased ($p < 0.05$) RS (0.25 Δmi/h) and HR (5 Δbeats/min), but not T_{CO} (Δ 0.3 °C), RPE, or TS. Given incomplete trials, the distance achieved at 16 min was not different between conditions (off/off 1.96 ± 0.16 vs. off/on 1.98 ± 0.19 vs. on/on 1.99 ± 0.24 miles, $p = 0.490$). During recovery HRV, MAP, or fatigue were unaffected ($p > 0.05$). We demonstrate that wrist percooling elicited a faster running speed, though this coincides with increased HR; although, interestingly, sensations of effort and thermal comfort were unaffected, despite the faster speed and higher HR.

Keywords: exercise; cooling; recovery; fatigue; thermal; environment; endurance

---

1. Introduction

Environmental stress, specifically heat stress, increases demand placed on the cardiovascular system [1,2]. Exercise also induces stress on the cardiovascular system, and the combination of heat stress with exercise can lead to a physiological challenge where demands for blood flow begin to challenge the maximal output of the heart, eventually leading to fatigue, exhaustion, and/or a decline in performance [1–7].

Accordingly, researchers have been developing strategies to prevent heat stress associated declines in exercise performance. One such approach has been the use of precooling, or reducing body temperature prior to exercise in the heat [4,8,9]. A review suggested that precooling via cold water immersion likely benefits performance, where ingestion of crushed ice/water ice slurry does not likely benefit performance [4]. Although the benefits of precooling, such as with cold water immersion, are not to be ignored, the issue of practicality raises concern over feasibility of implementation, and thus alternative methods ought to be explored. Strategies of attempting to cool during exercise, termed percooling, have demonstrated a positive effect on exercise performance, on par with precooling [8,9], though studies of percooling are far less abundant.

Recently, a company has developed a wearable, active cooling method (dhamaSPORT^{tm}, DhamaUSA, Scotts Valley, CA, USA) that is light weight (115 g, 6 cm wide) and can be worn on the wrist during activity while posing minimal disruption or burden to the athlete (e.g., ice vest). While we have demonstrated that this wrist cooling device improved physiological recovery and reduced fatigue from an occupationally relevant model of exercise-induced heat stress [10], it has yet to be determined whether wrist percooling is capable of improving endurance performance in the heat, and if this might impact post-exercise recovery. Aside from the obvious potential to provide cooling, mitigating exercise-induced elevations in core temperature, surface percooling might activate the transient receptor potential melastatin 8 (TRPM8) "cold receptor", which might alter thermal sensation and/or exercise performance [11]. Further, recent work by Phillips et al. [7] suggests that cooling can modulate prefrontal cortex activation, perceptions of muscle fatigue or effort, and partially mitigate declines in local muscle performance. However, the impact of wrist percooling on perceptual responses during exercise in hypothermia is unknown.

Therefore, the purpose of this study was to investigate whether the wrist cooling improves exercise performance in the heat or lessen physiological strain, and if this effect is "dose-dependent." We hypothesized that wrist percooling would reduce perceptions of effort and thermal stress, reduce heart rate, and/or improve performance on a 10 km running time trial, and these effects would be greater with the active cooling of both wrists. Second, use of the cooling bands would improve recovery as assessed by heart rate, heart rate variability, core temperature, and reduce fatigue and thermal sensations, all of which would also be greater with the active cooling of both wrists.

2. Materials and Methods

2.1. Subjects

Fourteen exercise-trained healthy male volunteers between the ages of 18 and 54 years were recruited for this study (Figure 1). To participate in this study, all participants must have been regularly exercise training for more than one hour at least three times a week for the past four months, have a maximal oxygen consumption (VO_2max) of >45 mL/kg/min, and >1 year of experience in competing in running events (e.g., 5 km, 10 km, half-, full-, ultra-marathon, half-, full-ironman, etc.). Participants were screened, via health history, and those with cardiovascular, pulmonary, musculoskeletal, or metabolic disease, those taking regular medication, or presenting contraindications to the ingestible telemetry pill (n = 1) were excluded. Methodologically, women were excluded to avoid the long periods of time that would be necessary (up to 3 months) to ensure adequate recovery with parallel desire for testing to occur in a singular phase of the menstrual cycle, thus reducing the impact of hormonal fluctuations. All participants provided written informed consent prior to any testing. The protocol was approved by the Institutional Review Board of Skidmore College (IRB # 1612-568) and was conducted in accordance with the most recent revisions to the Declaration of Helsinki.

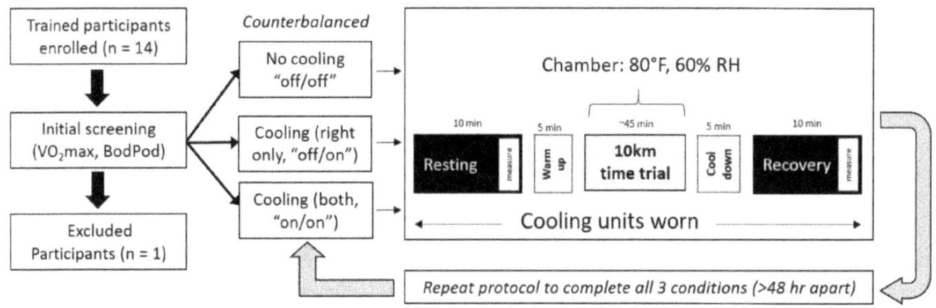

Figure 1. Experimental overview: RMR, resting metabolic rate; IET, incremental exercise test.

2.2. Study Overview

The current study was conducted in a single blind, counterbalanced, crossover design to investigate the potential impact of wrist cooling on performance in, and recovery from, exercise in the heat (Figure 1). As the number of participants did not equal or equally multiply by the number of possible sequences of three trials, we used a Latin square approach to counterbalance. All testing was conducted in the Environmental Physiology Laboratory at Skidmore College. For each visit, participants were asked to avoid strenuous exercise for 24 h prior and alcohol/caffeine use 12 h prior to each study visit. Participants were instructed to maintain a similar diet and sleep regimen throughout the duration of the study. Participants were asked to wear shorts and t-shirt and to dress similarly across trials. Finally, participants were instructed to arrive each day fueled and hydrated as if preparing for a race, which included drinking the proper amount of fluids prior to each experimental visit (e.g., ~500 mL 2–3 h prior and 250 mL within 15 min of the visit). All participants reported to the laboratory on four separate occasions: a screening day and three experimental trials. While wearing two cooling bands, the three trials were conducted as follows: (1) both bands were off (off/off), (2) one band on (off/on), (3) both bands on (on/on). In the off/on condition, the right wrist was always activated. All experimental trials for a subject were completed at the same time of day to reduce impact of diurnal variation. In a thermoneutral (21 ± 1 °C, 29 ± 12% relative humidity) and normobaric (~750 mmHg) environment, the first screening visit assessed participant characteristics, which included anthropometrics (height, weight), body composition using air displacement plethysmography (BodPod, Cosmed, Chicago, IL, USA), and aerobic fitness via graded exercise testing on a treadmill (PPS Med, Woodway, Waukesha, WI, USA). A running protocol (modified McConnell) [12,13] was used to determine maximal oxygen consumption (VO_2max) using open circuit spirometry and gas analysis (TrueOne 2400, Parvomedics, Sandy, UT, USA) [14]. Prior to each experimental trial, participants were given an FDA approved core temperature telemetry pill (HQ Inc, Palmetto, FL, USA), which was taken 8–12 h prior to the study visit [15,16].

2.3. Procedures

Upon arrival, a urine sample was collected and hydration status was confirmed via urine specific gravity (USG < 1.020) as described previously [17]. If USG was >1.020, participants were given 500 mL of water and USG was retested thereafter (though as the participants were familiar with race preparations, this only occurred once out of 39 total visits). Participants were then instrumented with a heart rate monitor (H7, Polar USA, Lake Success, NY, USA), and the presence of the core temperature telemetry pill was confirmed (CorTemp Recorder, HQ Inc, Palmetto, FL, USA). Participants were then seated and were outfitted with two wrist cooling bands (Dhama Sport Pro, Dhama USA, Scott's Valley, USA) (Figure 2). In the "on" condition, the bands were activated and set to the coolest setting (7.2 °C). While we attempted to avoid investigator cues and reduce possible anticipatory responses by single blinding and not making the participants' aware of which condition they were receiving, when the band was active, participants' were able to detect the cooling, but when the band was off (off/off condition) they were unsure. The device elicits cooling through a one-inch square ceramic cooling plate placed over the anterior vascular portion of the wrist, which dissipates heat via Peltier effect over a larger heat sink area on the exterior portion of the device. The heat transfer rate for this device ranges from 0.2 to 200 watts, with typical values of 0.5–50 watts, depending upon ambient conditions. After 10 min of quiet rest, a one minute [18], breathing frequency paced [19], recording of heart rate (HR) and R-R intervals were obtained via HR monitor, sent to a mobile device (IPad Pro, Apple, Cupertino, CA, USA) via Bluetooth™ and analyzed by a mobile device application (Elite HRV, Gloucester, MA, USA). The Elite HRV application performs artifact correction and has been shown to be valid [20] and has been used in previous studies [10,20–22]. Specifically, along with mean HR, R-R intervals were analyzed for the standard deviation of R-R intervals, SDNN; root mean square of successive differences, RMSSD; and the log transformed RMSSD, LnRMSSD. HRV was measured to assess potential impacts of wrist cooling on recovery as it is an increasingly recognized method to assess or monitor athlete acute and chronic physiological response to training, or recovery and readiness to train [22–26]. After HR and HRV were obtained, to further characterize potential impacts of wrist cooling on recovery, blood pressure (BP) was measured via oscillometric cuff method (Mobilograph, GmbH, Stolberg, Germany) [27–29], after which thermal sensation/comfort via thermal sensation (TS) scale (0 "unbearably cold" to 8 "unbearably hot"), and fatigue via a visual analog scale (0 "no fatigue" to 10 "severe fatigue") were recorded.

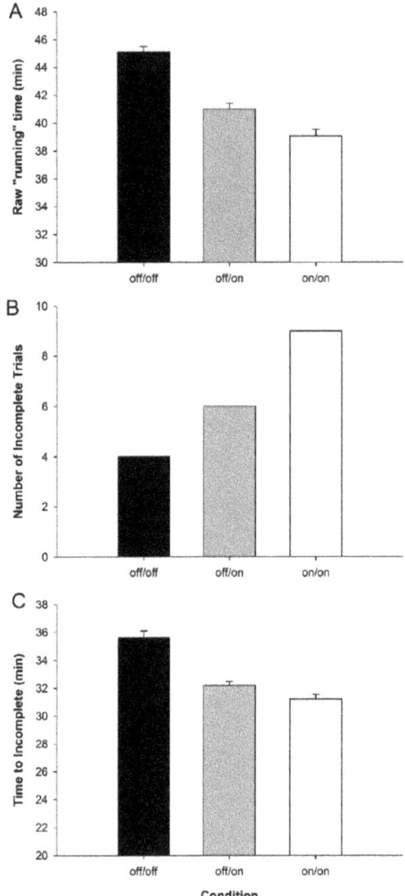

Figure 2. Raw performance data for all trials (n = 13): (**A**) average run time, including incomplete trials across condition; (**B**) number of incomplete trials across condition; and (**C**) average time to incompletion across conditions. Data are means ± Standard Error.

Participants were then allowed to warm up for a maximum of 5 min outside the chamber in the thermoneutral laboratory, typically followed by use of the restroom to void their bladder. Subsequently, participants entered the heated environmental chamber (26.7 °C, 60% relative humidity, heat index of 28 °C, "caution") and were instructed to complete the 10 km time trial (~6.2 miles, since the treadmill was in English units) as fast as possible at 0% grade. Thus, participants were able to see their speed and allowed direct control of the treadmill speed. Verbal encouragement was provided to all participants in a consistent manner between subjects and across trials. Participants were allowed to drink water, ad libitum, during all trials, but were asked to consume fluid in a similar volume and manner across trials, matched for their first trial completed. During exercise, participants were asked to report their thermal sensation, rating perceived exertion using standardized visual scales every 5 min, while HR and core temperature (T_{CO}) were monitored continuously and recorded every minute. Due to safety concerns, and institutional restrictions, in effort to avoid heat related injury, if core temperature reached two consecutive measures of 39.1 °C, or a single measure of 39.2 °C or higher, the trial was ended and the participant was immediately removed from the chamber and into a cool-down period in the thermoneutral laboratory. In such case, post-measures were obtained in an identical manner as if they had completed the trial.

Once the 10 km trial was completed, participants were escorted from the chamber and completed a 5 min cool-down, walking on a treadmill in the thermoneutral laboratory. HR and T_{CO} were continuously monitored for safety reasons. Fifteen minutes after the cessation of the exercise, a post-exercise assessment of the baseline measures, except USG, was performed, namely: VAS, thermal sensation, core temperature, HR, HRV, and BP. The timing of the post-exercise measurements was maintained for all trials, including those that were ended due to core temperature reaching our institutional safety threshold. Once post-exercise measures were obtained, the wrist cooling units were turned off and removed. Participants reported back to the laboratory to complete the other two trials in a randomized counterbalanced order as described above. Visits were completed with a minimum of 48 h in between (average time between visits ~72 h).

2.4. Statistical Analysis

Data were analyzed using commercially available software (SPSS v26, IBM, Armonk, NY, USA) As the number of athletes who reached our institutional safety cutoff turned out to be larger than anticipated, additional analyses were conducted to compare the number of incomplete trials between wrist percooling conditions using a chi square test, and pairwise comparisons were used to determine if the time to incompletion differed between conditions. Further, a Kaplan–Meier survival curve analysis was conducted to compare the rate and time of incompletion using a log rank test. Again, due to athletes' core temperatures reaching our institutional safety cutoff, to allow direct time-matched pairwise comparisons between trials, all trial data were analyzed to the point at which we had complete data for all participants (16 min), as well as the final data point for each participant. The final data point was either the final data at incompletion due to reaching the temperature cutoff or the final value at the end of the 10 km time trial. Thus, data were analyzed and plotted to the longest common time, plus each athlete's final data point, and only for one athlete in one condition were 16 min and final the same. Further, to estimate effects of wrist percooling on 10 km time trial performance, if the trial was incomplete for reaching core temperature cutoff (see Figure 2), trial performance was estimated using average running speed for the trial.

Prior to analysis, any anomalous individual data points presenting as an outlier (>2 SD) were removed from the data set, and where appropriate, interpolated using a linear function. Accordingly, heart rate, core temperature, and speed were analyzed using a 3 (condition) by 17 (time, 16 min + final) repeated measures ANOVA. For RPE and TS, a 3 (condition) by 4 (time) repeated measures ANOVA was completed. To compare the pre- and post-measurements, a 3 (condition) by 2 (time) repeated measures ANOVA was completed for HR, core temperature, RMSSD, MAP, SDNN, LnRMSSD, diastolic blood pressure (DBP), systolic blood pressure (SBP), VAS, RPE, and TS Significance was established at $p < 0.05$. Data are presented as means ± standard deviation (SD), unless indicated otherwise.

3. Results

3.1. Participant Characteristics

The participant characteristics are presented in Table 1. Most participants (n = 10 of 13) were active triathletes, having competed in half or full distance Ironman events as well as running events, but all had road and/or trail running racing experience. Participants were fit, with an average VO_2 max of 59 mL/kg/min (range 50–71), particularly considering their average age of 33 years.

Table 1. Subject characteristics (n = 13).

Variable	Means ± SD
Age (years)	32.6 ± 8.9
Height (cm)	178.6 ± 9.1
Weight (kg)	76.9 ± 8.0
Body Mass Index (kg/m^2)	24.1 ± 2.1
VO_2Max (mL/kg/min)	59.1 ± 5.2
Body Fat %	14.8 ± 6.8
Fat Free %	80.5 ± 15.3
Body Fat Mass (kg)	11.5 ± 5.4
Body Fat Free Mass (kg)	65.4 ± 7.4

3.2. Effects of Wrist Cooling on Baseline Parameters

No significant differences were found in resting core temperature, indicators of heart rate variability, blood pressure, thermal sensation, rating of perceived exertion, or in reported fatigue using a visual analog scale between conditions (Table 2). However, use of the bands tended to affect heart rate (p = 0.05), where HR was elevated by ~5 beats/min in the off/on condition (Table 2).

Table 2. Pre- and post-measurements for all three conditions (n = 13).

Variable		Off/Off	Off/On	On/On
HR (beats/min)	Pre	56.0 ± 7.0	61.0 ± 8.0 #	59.0 ± 6.5
	Post	86.0 ± 6.0	87.0 ± 6.0	87.0 ± 6.8 *,†
Core Temperature (°C)	Pre	37.0 ± 0.5	37.2 ± 0.6	37.1 ± 0.7
	Post	38.0 ± 1.0	37.3 ± 1.6	37.8 ± 1.0
MAP (mmHg)	Pre	106 ± 11	109 ± 11	106 ± 10
	Post	101 ± 6	104 ± 8	99 ± 11 *
SBP (mmHg)	Pre	124 ± 13	125 ±13	120 ±12
	Post	119 ± 10	120 ± 9	113 ± 16
DBP (mmHg)	Pre	85 ± 12	85 ± 12	86 ± 10
	Post	81 ± 8	80 ± 10	80 ± 10*
LnRMSSD (a.u.)	Pre	4.6 ± 0.4	4.5 ± 0.4	4.56 ± 0.5
	Post	3.4 ± 0.5	3.5 ± 0.7	3.4 ± 0.6 *
SDNN (ms)	Pre	159.0 ± 65.4	133.8 ± 36.3	146.5 ± 45.6
	Post	59.6 ± 28.2	64.1 ± 45.1	64.5 ± 29.1 *
TS (0–8)	Pre	3.4 ± 0.6	3.2 ± 0.4	3.5 ± 0.6
	Post	4.6 ± 0.9	4.2 ± 1.2	4.5 ± 0.8*
RPE (1–10)	Pre	2.0 ±1.0	1.5 ± 0.7	2.0 ± 0.6
	Post	4.5 ± 1.6	5.0 ± 2.0	4.0 ± 2.0 *
Fatigue (VAS 1–10)	Pre	1.4 ± 1.3	1.7 ± 1.1	1.2 ± 1.0
	Post	6.2 ± 1.8	6.4 ± 1.6	5.3 ± 2.5 *

HR, heart rate; MAP, mean arterial pressure; SBP, systolic blood pressure; DBP, diastolic blood pressure; LnRMSSD, natural log transformed root mean squared of successive differences; A.U., arbitrary units; SDNN, standard deviation of normal R-R intervals; Msec, milliseconds; TS, thermal sensation; RPE, rating of perceived exertion; VAS, visual analog scale. * main effect of time; † main effect of condition, # vs. off/off, $p < 0.05$. Note: all time effects were $p < 0.05$ pre vs. post. Means ± SD.

3.3. Ten km Time Trial Performance

Raw running time, including any incomplete trials, tended to decrease with the use of the bands (Figure 2A, off/on p = 0.14 and on/on p = 0.08 vs. off/off). However, due to participants reaching our institutionally mandated core temperature safety cut off, and importantly not of volitional means, this aforementioned time is tainted by a number of incomplete trials that tended to increase with the use of the bands (Figure 2B), and trended to an earlier incompletion time (Figure 2C). Chi squared analysis found that the proportion of incomplete trials did not significantly differ by condition ($p = 0.16$), nor were times to incompletion also not different between trials (pairwise comparisons, $p = 0.51$–0.81). Additionally, Kaplan–Meier survival curve analysis using a log rank test revealed no significant differences in survival distribution between wrist percooling conditions ($p = 0.14$, Figure 3).

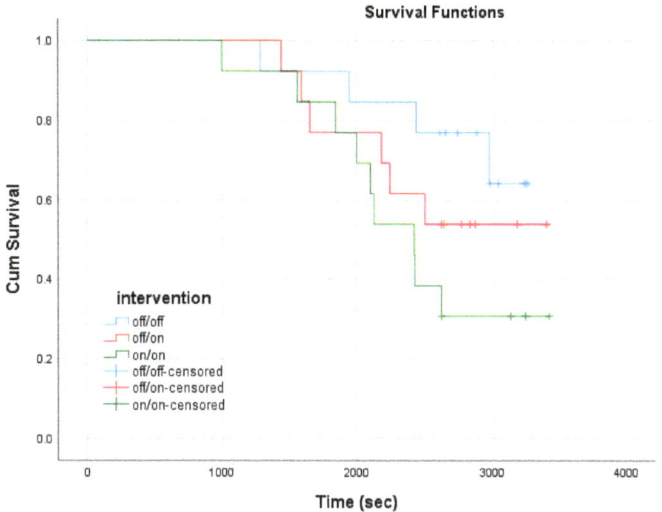

Figure 3. Kaplan–Meier survival curve across time (seconds) between wrist cooling conditions during the 10 km time trial.

There was a significant interaction of wrist percooling condition and time on running speed, where RS tended to increase more over time with wrist percooling (Figure 4A), though there was no main effect for condition ($p = 0.13$). Focusing on the time to which all participants had completed, the distance achieved at 16 min was not different between conditions (off/off 1.96 ± 0.16 vs. off/on 1.98 ± 0.19 vs. on/on 1.99 ± 0.24 miles, $p = 0.490$, Figure 4B). Using both actual or projected 10 km times, there was no statistically significant effect of the bands (off/on $p = 0.49$ on/on $p = 0.77$ vs. off/off), despite a tendency for an approximate 30 s improvement in 10 km time for the off/on condition and a 10 s improvement in 10 km time in the on/on condition (off/off: 50:14.6, off/on: 49:45.9, on/on: 50:04.2 min:s). Using only those with complete trials, within a condition, the trend is less clear (off/off: 49:17.4, off/on: 50:02.3, on/on: 48:51.3 min:s).

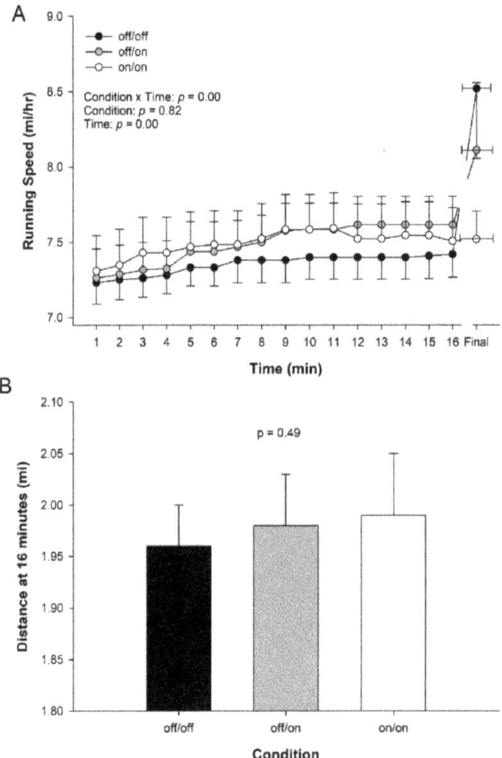

Figure 4. Running performance across wrist percooling condition. (**A**) Self-selected running speed during the 10 km time trial in the heat. Results of two-way ANOVA are presented (inset). Note: due to safety tolerance in core temperature trials were ended early and plotted to the shortest time, plus each athlete's final data point (with SE for time). (**B**) Distance to 16 min across wrist percooling condition ($n = 13$). This time was chosen as it was the longest point to which all participants completed at least 16 min for all 3 trials. Data are means ± Standard Error.

3.4. Physiological Response to 10 km Time Trial in the Heat

A significant interaction between condition and time was found for heart rate ($p = 0.00$, Figure 5A). Expectedly, a main effect for time was found for heart rate throughout the trial ($p = 0.00$). No significant differences were found for heart rate between conditions ($p = 0.39$).

There was no significant interaction of condition by time for core temperature during the 10 km time trial (TT) ($p = 0.15$, Figure 5B). No main effect of condition was found during the 10 km time trial for core temperature ($p = 0.88$). Expectedly, however, there was a significant effect of time on core temperature during the trial ($p < 0.001$, Figure 5B).

3.5. Perceptual Measures during 10 Km Time Trial in the Heat

There was no significant interaction between condition and time for thermal sensation during the 10 km ($p = 0.96$, Figure 6A). A significant main effect was found for time, where the participants' TS increased over time ($p = 0.00$), though no significant differences were observed between conditions for TS ($p = 0.47$).

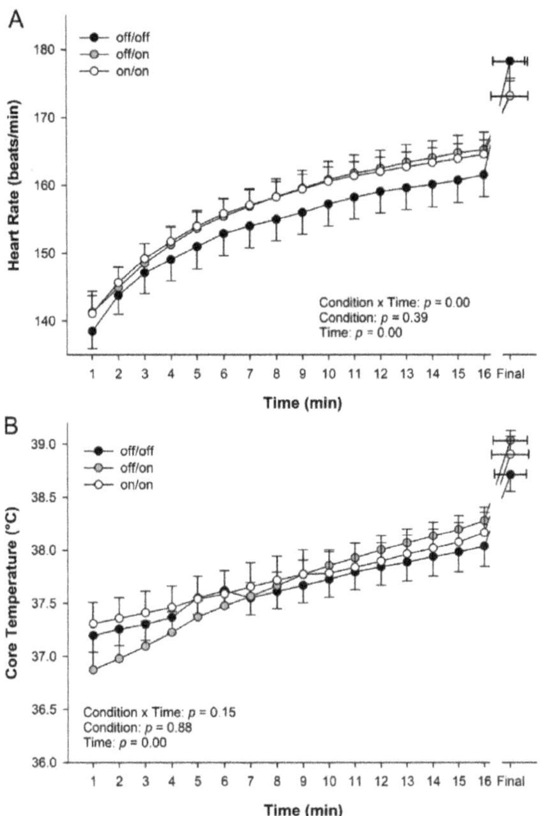

Figure 5. Physiological responses to 10 km time trial in the heat across wrist percooling condition. (**A**) Heart rate and (**B**) core temperature during 10 km time trial. Note: due to safety tolerance in core temperature, trials were ended early and plotted to the shortest time, plus each athlete's final data point. Data are means ± Standard Error (n = 13).

No significant interaction of condition and time was found for RPE ($p = 0.38$, Figure 6B). Naturally, a main effect for time was found ($p = 0.000$). However, RPE did not significantly differ between conditions ($p = 0.93$).

3.6. Impact of Wrist Cooling on Recovery

Pre- and post-measurements are shown in Table 2. No significant interaction ($p = 0.58$) was found between condition and time for core temperature, though core temperature was on average 0.2 to 0.7 °C cooler in recovery with use of the bands. Although core temperature approached significance, there was not a significant effect of time ($p = 0.05$); core temperature was not different from baseline and had recovered. However, there was no effect of condition on core temperature ($p = 0.64$).

There was no significant interaction between condition and time for heart rate ($p = 0.36$, Table 2). There was a significant main effect for time ($p = 0.00$) on heart rate with elevations post-exercise. Heart rate significantly differed between conditions, where heart rate tended to increase with the use of the bands (condition effect $p = 0.03$). To measure heart rate variability, the root mean square of the successive differences (RMSSD) was measured. No significant interaction effect for condition by time ($p = 0.23$) and no main effect of condition was found ($p = 0.97$). A significant main time effect was found for RMSSD ($p = 0.000$, Table 2) with a significantly reduced RMSSD post-exercise. In addition, there was no significant interaction effect for condition by time ($p = 0.43$) and no difference between conditions ($p = 0.96$) for log transformed RMSSD (LnRMSSD). Expectedly, similar to RMSSD, there was a main time effect for LnRMSSD ($p = 0.00$, Table 2) with lower HR variability post-exercise. Lastly, SDNN had no significant condition effect ($p = 0.41$) or condition by time effect ($p = 0.15$). Corresponding to the other heart rate variability variables, there was a main time effect for SDNN ($p = 0.00$) with lower HR variability post-exercise.

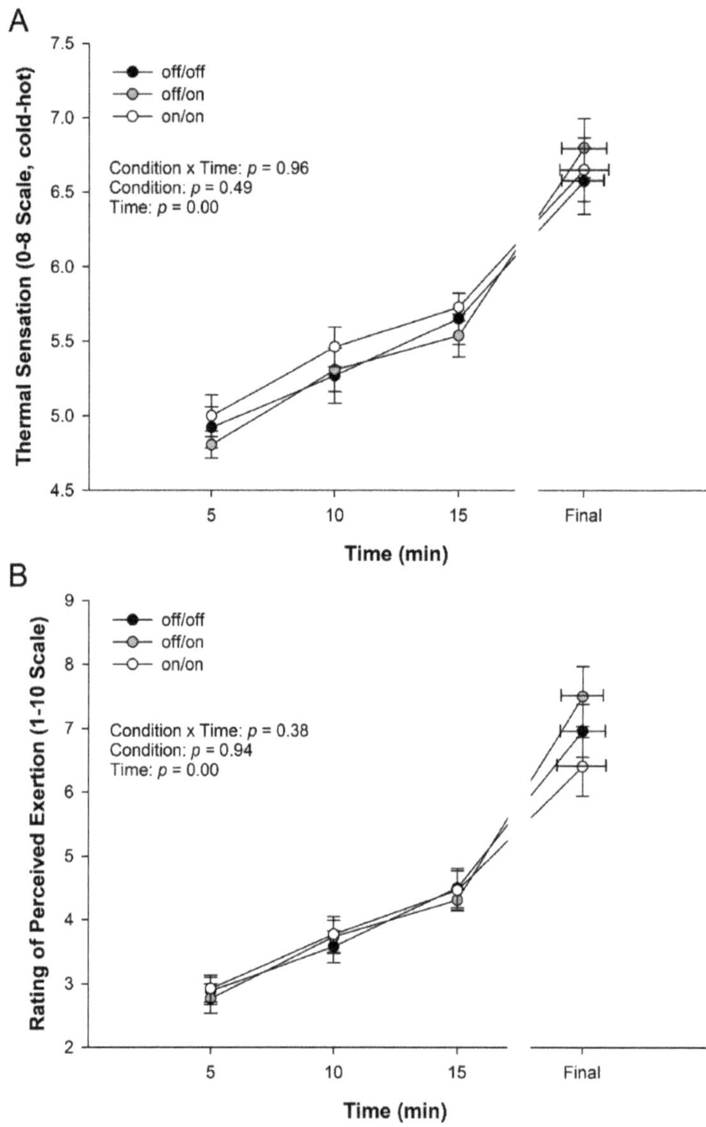

Figure 6. Perceptual measures during 10 km TT. (**A**) Thermal sensation (TS) and (**B**) Rating of perceived exertion. Note: due to safety tolerance in core temperature, trials were ended early and plotted to the shortest time, plus each athlete's final data point. Data are means ± Standard Error, ($n = 13$).

No significant interaction for condition by time was found for mean arterial pressure, MAP ($p = 0.90$, Table 2). During the recovery process, a main effect of time was found for MAP ($p = 0.01$) with lower MAP in recovery, though no significant differences between conditions were found ($p = 0.08$). Systolic blood pressure, SBP, showed no significance for main time effect, condition by time, or condition (all $p > 0.05$, Table 2). Diastolic blood pressure, DBP, had no significant interaction of condition by time ($p = 0.28$) or effect of condition ($p = 0.11$). In contrast to SBP, DBP had a significant main time effect ($p = 0.02$) with a reduction in diastolic pressure post-exercise.

During recovery, there was no significant interaction between condition and time for RPE ($p = 0.08$) and TS ($p = 0.72$). There was a significant main effect of time for RPE and TS (both $p = 0.000$, Table 2). No significant differences between conditions were present for RPE ($p = 0.43$) and TS ($p = 0.33$). There was no significant interaction between condition and time for the fatigue visual analog scale ($p = 0.47$, Table 2). Expectedly, a main effect of time was present during the recovery process for VAS ($p = 0.00$), showing higher reported levels of fatigue, but no significant difference was found between conditions in the recovery process ($p = 0.10$).

4. Discussion

The intent of this study was to ascertain whether percooling via wrist cooling bands would improve 10 km running time trial performance in the heat, or lessen the physiological strain, and enhance recovery. The main finding of this study was that the use of the bands seemed to promote the participants to run at a faster speed over time. Thus, when using the bands, there was a corresponding increase in heart rate over time as a result of this increased speed and energy demand. On average, core temperature, thermal sensation, and rating of perceived exertion were not different when using the bands. Use of the bands did not appear to alter baseline or enhance physiological or perceptual indicators of recovery from the 10 km running bout. There was also no clear evidence that two bands were more advantageous than one, in terms of the physiological and perceptual responses to exercise, performance, or recovery. Thus, athletes considering use of wrist percooling, wearing one band is likely sufficient and possibly optimal. In conclusion, the cooling bands elicited a faster running speed over time; however, this comes at a physiological cost, but surprisingly not a perceptual one. Further work in the field or in unrestricted settings are needed to ultimately demonstrate efficacy of wrist percooling.

4.1. Ten km Time Trial Performance

In the present study, we observed that wrist percooling through the use of wrist cooling bands seemed to elicit a faster running speed in the participants over time (interaction effect, Figure 4). The faster self-selected running speed over time could be interpreted as an increase in performance because, all else held constant, would be expected to lead to a faster 10 km time trial. Focusing on distance covered at a common time, the distance covered by the athletes was not statistically higher at 16 min, a critical time point in our study. Using completed and estimated 10 km times, wrist percooling via use of the band's lead to times that were on average ~20 s faster, but were not significantly different from control. The faster running speed, as an indicator of performance, agrees with prior research using precooling, which also observed an increase in performance [9,30–34].

Previous reviews indicated that the magnitude of the effect of precooling likely depends upon the modality of precooling (e.g., water immersion, and depth, ice slurry, cooling garment, etc.) and how performance is assessed (i.e., time to exhaustion, and prescribed intensity, time trial, distance trial, etc.) [4,9]. Specifically, cold water immersion elicits an effect size (Cohens d) of 0.4 to 2 on performance vs. control, whereas precooling garments elicit an effect size of 0.1 to 0.5 (small to medium) [4], the latter more reflecting the magnitude effect in the current study (Cohens d effect size of 0.2, small effect, in off/off vs. off/on). Although the trend for a positive effect on actual or estimated 10 km time was not statistically significant, it should not be interpreted necessarily as useless. A 10–30 s effect on 10 km performance could have practical implications. To put this modest effect into perspective, when looking at the professional men's results of the 2016 AJC Peachtree road race, one of the nation's largest

10 km events, run in Atlanta, GA, during July (similar environmental conditions at race start to those in our study), a 20 s boost in performance could mean being on the podium or not. For the Bolder Boulder 10 km race, one of the largest 10 km races in the world, first place through sixth place in the pro division was separated by 20 s. Again, further field testing is needed to support this notion, though as climatic temperatures rise and running events are held in hot environments (e.g., Tokyo 2021 Olympic games), developing viable methods of supporting athlete's performance is increasingly paramount.

4.2. Impact of Wrist Percooling on Physiological Responses to 10 km TT in the Heat

In the current study, core temperature during the 10 km time trial was unaffected by percooling via wrist cooling bands (Figure 5). The present findings are in contrast to previous work using precooling, which demonstrated reduced core temperature particularly during initial stages of exercise [31,32,35,36]. For example, in a study by Lee and Haymes [32], 30 min of precooling reduced core temperature at rest and during exercise, though final core temperatures were not different between precooled and control. In agreement, when using a mix-method of ice bags and a cooling vest, Duffield et al. [31] saw a decrease in core temperature during warm up and the first sprinting periods; however, there was no significant difference in the final core temperature. Precooling is thought to reduce core temperature, creating a larger reservoir or tolerance for core temperature increases, postponing increases in temperature to the latter stage of exercise. In the current study, it was hypothesized that the use of the bands would blunt the rise of core temperature and lower it during the 10 km time trial, but this was not observed. However, given the running speed was increased over time, further challenging the cooling capacity of the unit due to the additional metabolic heat load, in already loaded condition, observing such an effect may not be possible. The cooling power of the device, maximally 200 watts, is simply not capable of ablating human heat production, estimated at over 1000 watts [37], but whether it might attenuate the rise in core temperature in individuals warrants further study.

Concordant to the increased running speed over time seen with wrist percooling, heart rate also was elevated over time during the 10 km time trial (Figure 5). These results somewhat support one previous study done by Duffield et al. [31], where they found no significant difference in heart rate during the exercise. However, previous research found with various models of precooling that HR was suppressed [30,32,33,35,38], at least during the initial stages of exercise, as final HR often was not different between conditions. However, some of these protocols used steady state exercise models [35], others incremental [33,38], or time to exhaustion [30,32], where speed or work load were matched. Thus, the higher heart rate over the trial with the use of the bands, while perhaps in disagreement with previous studies and not supporting the hypothesis, makes physiological sense in the context of the increased running speed.

4.3. Impact of Wrist Percooling on Perceptual Measures during 10 km TT in the Heat

The current study found no significant difference between conditions and no interaction effect for RPE (Figure 6). In support of the present data, Duffield et al. [31,36] also found no significant difference between the precooling and control conditions for RPE during the performance trial. However, other studies showed that RPE decreased with the use of precooling methods during the exercise performance [35,39]. However, in both of these studies walking/running speed was fixed.

Similar to RPE, in the present study, thermal sensation or perception during the 10 km time trial was not different between conditions and did not see an interaction effect with the use of the bands (Figure 6). Only one previous study supports the present data that showed no significant difference between conditions during the performance [31]. Other research has proven that with the use of cooling, the TS decreases during exercise [35,39,40].

Thus, while the present data does not support the hypothesis that the use of the cooling bands would have lowered RPE and TS, the present data are to be considered in the context of altered running speed over time. Recent meta-analysis suggests that topical or ingestion of menthol, a known agonist of the "cold receptor," TRPM8, can alter thermal sensation and/or exercise performance [11], perhaps independent of core temperature [41]. Relatedly, work by Phillips et al. [7] suggests that precooling might modulate the prefrontal cortex and/or its processing of afferent feedback regarding perceptions of effort, fatigue, or skin temperature, which might support greater exercise tolerance, and ultimately an attenuation of muscle fatigue. Thus, while it is tempting to speculate that RPE and TS might have been expected to increase in response to the increased running speed, and that the cooling bands mitigated the expected increase in perception of effort and thermal strain, further work is needed to confirm this hypothesis and explore the potential psychophysiological effects of wrist percooling.

4.4. Impact of Wrist Percooling on Physiological Recovery

In contrast to our initial hypothesis, core temperature did not significantly differ between conditions during recovery (Table 2). A prior meta-analytical review demonstrated that more aggressive cooling methods, such as whole body cooling likely help to recover performance [42], though few studies have focused on recovery of core temperature. Much of this work has been done in occupational models, such as firefighting, using multiple interventions, some invasive, to induce cooling and recovery of heart rate after firefighting [43–47]. Accordingly, our previous work using this wrist cooling device to induce cooling after an occupational model of exercise-induced heat stress via exercise in encapsulating clothing, revealed a significant positive impact of wrist cooling on recovery of temperature and heart rate [10]. However, in that study the exercise was necessarily more modest (walking) and shorter, thus the rise in core temperature was lower, all only increasing by 1 °C or less, potentially creating multiple differentials between the present and the aforementioned study. Interestingly, heart rate was found to be significantly impacted by wrist cooling at rest, and agrees with previous work that demonstrated skin cooling to 7 °C resulted in a modest 5 beat/min increase, likely the result of activating sensory afferent neurons [48]. This resting difference, we believe, contributed to a main effect of condition when exploring rest and recovery, as post-exercise HR values were not actually different between conditions. In support of no difference in heart rate during recovery between conditions, Edmonds et al. [40] using the wrist cooling device also found no significant difference in heart rate after high intensity physical activity. Thus, wrist cooling may be insufficient to hasten recovery of HR after high intensity activity in the heat.

4.5. Impact of Wrist Percooling on Recovery of Perceptual Measurements

During the recovery period, there was no significant difference in TS. However, TS during the recovery period was recorded highest in the off/off condition (4.6) and the lowest in the on/on condition (4.2). In support of this, Edmonds et al. [40] found a significant decrease during the recovery period, at the 10 min marker. The present data do show a lower value with the use of the bands; however, due to the time to incompletion decreasing with the bands, and the number of incomplete trials increasing, perceptual cues could be altered due to the decrease in performance. In recovery, RPE, or perhaps more appropriately the fatigue visual analog scale values after the exercise did not significantly differ, which does disagree with our prior work [10] using wrist cooling. Although for the reasons mentioned above, this may be expected. Future studies should explore the potential impact of wrist cooling on recovery in the field or athletic setting where immediate cooling applications may not be readily available and thus wrist cooling could perhaps be a bridge from or to more powerful cooling methods, such as cold-water immersion.

4.6. Experimental Considerations

The 10 km time trial took place in a controlled environment during the winter months and therefore the athletes were not acclimated to running in hot and humid environments. Institutional safety mandated cessation of exercise just above 39 °C (Figure 2), impairing our ability to determine whether performance would have truly been affected; indeed, previous work has suggested that high-level athletes are capable of tolerating such core temperatures or higher, perhaps even to 41.5 °C [49]. Anecdotally, none of the participants exhibited any heat-related illness signs or symptoms, suggestive of a greater tolerance, and this may have underestimated potential performance effects as the athletes were unable to fully execute their individual race plan (e.g., negative splits or sprint at the end). Nonetheless due to this cutoff, in trials where the athlete reached this threshold we estimated or projected their performance. Future studies are needed to determine possible effects in a relatively unrestricted or field environment to observe fully the potential effects on performance. Although ingestible temperature telemetry pills have been demonstrated valid and able to track changes over time with heating or cooling [15,16], there was some variability in core temperature measurements with the telemetry pill and future studies might consider using more invasive measures such as esophageal or rectal thermistors. It was impossible to conduct the study in a fully blinded or placebo-controlled manner, though the research team sought to minimize eliciting any anticipatory responses, and participants wore both bands during all three trials. Measures of skin temperatures and/or localized thermal sensations would have enhanced the study, and future studies using this cooling method should include these measures, as well as pulmonary measures (e.g., VO_2, ventilation, respiratory exchange ratio).

5. Conclusions

In the present study, wrist percooling during a 10 km TT in the heat resulted in a faster self-selected running speed and higher heart rates, though thermal sensation or perceptions of effort were unaffected. The increased running speed over time with wrist percooling might be practically meaningful, but further work is needed to determine the potential impacts of wrist cooling on performance, particularly in the field.

Author Contributions: Conceptualization, S.J.I.; methodology, K.D., R.C., E.S. and S.J.I.; formal analysis, K.D., R.C., E.S. and S.J.I.; investigation, K.D., R.C., E.S. and S.J.I.; resources, S.J.I.; data curation, K.D., R.C., E.S. and S.J.I.; writing—original draft preparation, K.D. and S.J.I.; writing—review and editing, K.D., R.C., E.S. and S.J.I.; visualization, K.D. and S.J.I.; supervision, S.J.I.; project administration, S.J.I.; funding acquisition, S.J.I. All authors have read and agreed to the published version of the manuscript.

Funding: This research was supported, in part, by DhamaUSA.

Acknowledgments: The authors would like to thank the Empire Endurance triathlon team and those who graciously volunteered for the study. We would also like to thank Michael Lopez in the Department of Mathematics and Statistics at Skidmore for his assistance. We would like to thank DhamaUSA for providing cooling units and financial support.

Conflicts of Interest: The funders had no role in the design of the study; in the collection, analyses, or interpretation of data; in the writing of the manuscript, or in the decision to publish the results.

References

1. Crandall, C.G.; Gonzalez-Alonso, J. Cardiovascular function in the heat-stressed human. *Acta Physiol.* **2010**, *199*, 407–423. [CrossRef] [PubMed]
2. Rowell, L.B. Human cardiovascular adjustments to exercise and thermal stress. *Physiol. Rev.* **1974**, *54*, 75–159. [CrossRef] [PubMed]
3. Van Haitsma, T.A.; Light, A.R.; Light, K.C.; Hughen, R.W.; Yenchik, S.; White, A.T. Fatigue sensation and gene expression in trained cyclists following a 40 km time trial in the heat. *Eur. J. Appl. Physiol.* **2016**, *116*, 541–552. [CrossRef] [PubMed]
4. Jones, P.R.; Barton, C.; Morrissey, D.; Maffulli, N.; Hemmings, S. Pre-cooling for endurance exercise performance in the heat: A systematic review. *BMC Med.* **2012**, *10*, 166. [CrossRef] [PubMed]
5. Tatterson, A.J.; Hahn, A.G.; Martin, D.T.; Febbraio, M.A. Effects of heat stress on physiological responses and exercise performance in elite cyclists. *J. Sci. Med. Sport* **2000**, *3*, 186–193. [CrossRef]
6. Ely, B.R.; Cheuvront, S.N.; Kenefick, R.W.; Sawka, M.N. Aerobic performance is degraded, despite modest hyperthermia, in hot environments. *Med. Sci. Sports Exerc.* **2010**, *42*, 135–141. [CrossRef] [PubMed]
7. Phillips, K.C.; Verbrigghe, D.; Gabe, A.; Jauquet, B.; Eischer, C.; Yoon, T. The influence of thermal alterations on prefrontal cortex activation and neuromuscular function during a fatiguing task. *Int. J. Environ. Res. Public Health* **2020**, *17*, 7194. [CrossRef] [PubMed]

8. Bongers, C.C.; Thijssen, D.H.; Veltmeijer, M.T.; Hopman, M.T.; Eijsvogels, T.M. Precooling and percooling (cooling during exercise) both improve performance in the heat: A meta-analytical review. *Br. J. Sports Med.* **2015**, *49*, 377–384. [CrossRef]
9. Tyler, C.J.; Sunderland, C.; Cheung, S.S. The effect of cooling prior to and during exercise on exercise performance and capacity in the heat: A meta-analysis. *Br. J. Sports Med.* **2015**, *49*, 7–13. [CrossRef]
10. Schlicht, E.; Caruso, R.; Denby, K.; Matias, A.; Dudar, M.; Ives, S.J. Effects of wrist cooling on recovery from exercise-induced heat stress with firefighting personal protective equipment. *J. Occup. Environ. Med.* **2018**, *60*, 1049-00. [CrossRef]
11. Jeffries, O.; Waldron, M. The effects of menthol on exercise performance and thermal sensation: A meta-analysis. *J. Sci. Med. Sport* **2019**, *22*, 707–715. [CrossRef] [PubMed]
12. McConnell, T.R.; Clark, B.A. Treadmill protocols for determination of maximum oxygen uptake in runners. *Br. J. Sports Med.* **1988**, *22*, 3–5. [CrossRef] [PubMed]
13. Ives, S.J.; Blegen, M.; Coughlin, M.A.; Redmond, J.; Matthews, T.; Paolone, V. Salivary estradiol, interleukin-6 production, and the relationship to substrate metabolism during exercise in females. *Eur. J. Appl. Physiol.* **2011**, *111*, 1649–1658. [CrossRef]
14. Crouter, S.E.; Antczak, A.; Hudak, J.R.; DellaValle, D.M.; Haas, J.D. Accuracy and reliability of the parvomedics trueone 2400 and medgraphics vo2000 metabolic systems. *Eur. J. Appl. Physiol.* **2006**, *98*, 139–151. [CrossRef]
15. Byrne, C.; Lim, C.L. The ingestible telemetric body core temperature sensor: A review of validity and exercise applications. *Br. J. Sports Med.* **2007**, *41*, 126–133. [CrossRef] [PubMed]
16. O'Brien, C.; Hoyt, R.W.; Buller, M.J.; Castellani, J.W.; Young, A.J. Telemetry pill measurement of core temperature in humans during active heating and cooling. *Med. Sci. Sports Exerc.* **1998**, *30*, 468–472. [CrossRef]
17. Matias, A.; Dudar, M.; Kauzlaric, J.; Frederick, K.A.; Fitzpatrick, S.; Ives, S.J. Rehydrating efficacy of maple water after exercise-induced dehydration. *J. Int. Soc. Sports Nutr.* **2019**, *16*, 5. [CrossRef]
18. Esco, M.R.; Flatt, A.A. Ultra-short-term heart rate variability indexes at rest and post-exercise in athletes: Evaluating the agreement with accepted recommendations. *J. Sports Sci. Med.* **2014**, *13*, 535–541.
19. Aysin, B.; Aysin, E. Effect of respiration in heart rate variability (hrv) analysis. In Proceedings of the 2006 International Conference of the IEEE Engineering in Medicine and Biology Society, New York, NY, USA, 30 August–3 September 2006; pp. 1776–1779.
20. Perrotta, A.S.; Jeklin, A.T.; Hives, B.A.; Meanwell, L.E.; Warburton, D.E.R. Validity of the elite hrv smartphone application for examining heart rate variability in a field-based setting. *J. Strength Cond. Res.* **2017**, *31*, 2296–2302. [CrossRef]
21. Flatt, A.A.; Esco, M.R. Smartphone-derived heart-rate variability and training load in a women's soccer team. *Int. J. Sports Physiol. Perform.* **2015**, *10*, 994–1000. [CrossRef]
22. Flatt, A.A.; Esco, M.R. Evaluating individual training adaptation with smartphone-derived heart rate variability in a collegiate female soccer team. *J. Strength Cond. Res. Natl. Strength Cond. Assoc.* **2016**, *30*, 378–385. [CrossRef]
23. Edmonds, R.; Burkett, B.; Leicht, A.; McKean, M. Effect of chronic training on heart rate variability, salivary iga and salivary alpha-amylase in elite swimmers with a disability. *PLoS ONE* **2015**, *10*, e0127749. [CrossRef]
24. Egan-Shuttler, J.D.; Edmonds, R.; Ives, S.J. The efficacy of heart rate variability in tracking travel and training stress in youth female rowers: A preliminary study. *J. Strength Cond. Res. Natl. Strength Cond. Assoc.* **2018**. [CrossRef]
25. Flatt, A.A.; Esco, M.R.; Nakamura, F.Y. Individual heart rate variability responses to preseason training in high level female soccer players. *J. Strength Cond. Res. Natl. Strength Cond. Assoc.* **2017**, *31*, 531–538. [CrossRef]
26. Plews, D.J.; Laursen, P.B.; Stanley, J.; Kilding, A.E.; Buchheit, M. Training adaptation and heart rate variability in elite endurance athletes: Opening the door to effective monitoring. *Sports Med.* **2013**, *43*, 773–781. [CrossRef] [PubMed]
27. Izzo, J.; Saleem, O.; Khan, S.; Osmond, P. 7b.05: Differential effects nebivolol and valsartan alone and in combination on 24-h ambulatory rate-pressure product, stroke load, and blood pressure-heart rate variability. *J. Hypertens.* **2015**, *33* (Suppl. 1), e93. [CrossRef]

28. Izzo, J.L., Jr.; Khan, S.U.; Saleem, O.; Osmond, P.J. Ambulatory 24-hour cardiac oxygen consumption and blood pressure-heart rate variability: Effects of nebivolol and valsartan alone and in combination. *J. Am. Soc. Hypertens. JASH* **2015**, *9*, 526–535. [CrossRef]
29. Varrenti, M.; Meani, P.; Giupponi, L.; Vallerio, P.; Ferrari, E.; Stucchi, M.; Maloberti, A.; Bruno, J.; Turazza, F.; Parati, G.; et al. 1b.01: 24 h modulation of peripheral and central blood pressure, heart rate and arterial stiffness in heart transplant hypertensive individuals. *J. Hypertens.* **2015**, *33* (Suppl. 1), e5. [CrossRef]
30. Booth, J.; Marino, F.; Ward, J.J. Improved running performance in hot humid conditions following whole body precooling. *Med. Sci. Sports Exerc.* **1997**, *29*, 943–949. [CrossRef] [PubMed]
31. Duffield, R.; Steinbacher, G.; Fairchild, T.J. The use of mixed-method, part-body pre-cooling procedures for team-sport athletes training in the heat. *J. Strength Cond. Res. Natl. Strength Cond. Assoc.* **2009**, *23*, 2524–2532. [CrossRef]
32. Lee, D.T.; Haymes, E.M. Exercise duration and thermoregulatory responses after whole body precooling. *J. Appl. Physiol. (Bethesda Md. 1985)* **1995**, *79*, 1971–1976. [CrossRef] [PubMed]
33. Uckert, S.; Joch, W. Effects of warm-up and precooling on endurance performance in the heat. *Br. J. Sports Med.* **2007**, *41*, 380–384. [CrossRef] [PubMed]
34. Yeargin, S. Precooling improves endurance performance in the heat. *Clin. J. Sport Med.* **2008**, *18*, 177–178. [PubMed]
35. Kenny, G.P.; Schissler, A.R.; Stapleton, J.; Piamonte, M.; Binder, K.; Lynn, A.; Lan, C.Q.; Hardcastle, S.G. Ice cooling vest on tolerance for exercise under uncompensable heat stress. *J. Occup. Environ. Hyg.* **2011**, *8*, 484–491. [CrossRef]
36. Duffield, R.; Green, R.; Castle, P.; Maxwell, N. Precooling can prevent the reduction of self-paced exercise intensity in the heat. *Med. Sci. Sports Exerc.* **2010**, *42*, 577–584. [CrossRef]
37. Gleeson, M. Temperature regulation during exercise. *Int. J. Sports Med.* **1998**, *19* (Suppl. 2), S96–S99. [CrossRef]
38. Smith, D.L.; Fehling, P.C.; Hultquist, E.M.; Arena, L.; Lefferts, W.K.; Haller, J.M.; Storer, T.W.; Cooper, C.B. The effect of precooling on cardiovascular and metabolic strain during incremental exercise. *Appl. Physiol. Nutr. Metab. Physiol. Appl. Nutr. Metab.* **2013**, *38*, 935–940. [CrossRef]
39. Siegel, R.; Mate, J.; Brearley, M.B.; Watson, G.; Nosaka, K.; Laursen, P.B. Ice slurry ingestion increases core temperature capacity and running time in the heat. *Med. Sci. Sports Exerc.* **2010**, *42*, 717–725. [CrossRef]
40. Edmonds, R.C.; Wilkinson, A.F.; Fehling, P.C. Novel cooling device enhances autonomic nervous system recovery from live fire training: A pilot study. *Int. J. Innov. Res. Med Sci.* **2017**, *2*, 455–460. [CrossRef]
41. Hue, O.; Chabert, C.; Collado, A.; Hermand, E. Menthol as an adjuvant to help athletes cope with a tropical climate: Tracks from heat experiments with special focus on guadeloupe investigations. *Front. Physiol.* **2019**, *10*, 1360. [CrossRef]
42. Poppendieck, W.; Faude, O.; Wegmann, M.; Meyer, T. Cooling and performance recovery of trained athletes: A meta-analytical review. *Int. J. Sports Physiol. Perform.* **2013**, *8*, 227–242. [CrossRef] [PubMed]
43. Colburn, D.; Suyama, J.; Reis, S.E.; Morley, J.L.; Goss, F.L.; Chen, Y.-F.; Moore, C.G.; Hostler, D. A comparison of cooling techniques in firefighters after a live burn evolution. *Prehospital Emerg. Care* **2011**, *15*, 226–232. [CrossRef] [PubMed]
44. Hostler, D.; Reis, S.E.; Bednez, J.C.; Kerin, S.; Suyama, J. Comparison of active cooling devices with passive cooling for rehabilitation of firefighters performing exercise in thermal protective clothing: A report from the fireground rehab evaluation (fire) trial. *Prehospital Emerg. Care* **2010**, *14*, 300–309. [CrossRef]
45. Barr, D.; Reilly, T.; Gregson, W. The impact of different cooling modalities on the physiological responses in firefighters during strenuous work performed in high environmental temperatures. *Eur. J. Appl. Physiol.* **2011**, *111*, 959–967. [CrossRef] [PubMed]
46. McEntire, S.J.; Suyama, J.; Hostler, D. Mitigation and prevention of exertional heat stress in firefighters: A review of cooling strategies for structural firefighting and hazardous materials responders. *Prehospital Emerg. Care* **2013**, *17*, 241–260. [CrossRef]
47. Giesbrecht, G.G.; Jamieson, C.; Cahill, F. Cooling hyperthermic firefighters by immersing forearms and hands in 10 degrees c and 20 degrees c water. *Aviat. Space Environ. Med.* **2007**, *78*, 561–567. [PubMed]

48. Kregel, K.C.; Seals, D.R.; Callister, R. Sympathetic nervous system activity during skin cooling in humans: Relationship to stimulus intensity and pain sensation. *J. Physiol* **1992**, *454*, 359–371. [CrossRef] [PubMed]
49. Racinais, S.; Moussay, S.; Nichols, D.; Travers, G.; Belfekih, T.; Schumacher, Y.O.; Periard, J.D. Core temperature up to 41.5 °C during the uci road cycling world championships in the heat. *Br. J. Sports Med.* **2019**, *53*, 426–429. [CrossRef]

Publisher's Note: MDPI stays neutral with regard to jurisdictional claims in published maps and institutional affiliations.

International Journal of
Environmental Research and Public Health

Article

Sarcopenia as a Mediator of the Effect of a Gerontogymnastics Program on Cardiorespiratory Fitness of Overweight and Obese Older Women: A Randomized Controlled Trial

Pablo Jorge Marcos-Pardo [1,2], Noelia González-Gálvez [1,2,*], Gemma María Gea-García [1,2], Abraham López-Vivancos [1,2], Alejandro Espeso-García [1] and Rodrigo Gomes de Souza Vale [2,3]

1 Research Group on Health, Physical Activity, Fitness and Motor Behaviour (GISAFFCOM), Physical Activity and Sport Sciences Department, Faculty of Sport, Catholic University San Antonio of Murcia, 30107 Murcia, Spain; pmarcos@ucam.edu (P.J.M.-P.); gmgea@ucam.edu (G.M.G.-G.); alvivancos@ucam.edu (A.L.-V.); aespeso@ucam.edu (A.E.-G.)
2 Active Aging, Exercise and Health/HEALTHY-AGE Network, Consejo Superior de Deportes (CSD), Ministry of Culture and Sport of Spain, 28040 Madrid, Spain; rodrigovale@globo.com
3 Exercise Physiology Laboratory, Estacio de Sa University, 20261-063 Rio de Janeiro, Brazil
* Correspondence: ngonzalez@ucam.edu

Received: 4 August 2020; Accepted: 24 September 2020; Published: 27 September 2020

Abstract: The objectives were to analyze the effect of a gerontogymnastics program on functional ability and fitness on overweight and obese older woman and to understand if sarcopenia mediates its effect. This randomized controlled trial involved 216 overweight and obese women. The experimental group (EG) carried out 12 weeks of a gerontogymnastics program. The assessment was of gait speed, cardiorespiratory fitness, functional capacity, and muscle strength. EG showed significant improvements in almost every test. When the effect of training was adjusted by gait speed, the improvement of the 6 min walk test (MWT) for the trained group was no longer significant ($p = 0.127$). The improvement of the 6 MWT was significantly and positively associated with the 10 m test ($\beta = -10.087$). After including the 10-m test in the equations, the association between the 6MWT and carrying out the training program decreased but remained significant ($\beta = -19.904$). The mediation analysis showed a significant, direct and indirect effect with a significant Sobel test value ($z = 6.606 \pm 7.733$; $p = 0.000$). These results indicate that a gerontogymnastics program improves functional capacity and fitness; and the effect of a gerontogymnastics program on CRF is mediated by sarcopenia in older women who are overweight and obese.

Keywords: Sarcopenia; gait speed; cardiorespiratory responses; walking; physical fitness; older people

---

1. Introduction

Sarcopenia is a progressive disease that involves the loss of muscle mass and strength [1]. It is associated with aging and causes a decrease in functional capacity, increasing the risk of falls and negatively affecting the quality of life, and in many cases may require hospitalization or rehabilitation [2,3]. Sarcopenia affects about 5 to 13% of individuals in their 60s and 70s, and 11 to 50% of octogenarians [4]. From the age of 40, there is a decline of muscle mass of approximately 8% per decade, which around the age of 70 can reach up to 15% per decade [5]. Besides, older women are more susceptible to present sarcopenia, as opposed to young women and men [6].

Age is associated with loss of mass, strength, and muscle power [7–9]. The loss of muscle mass and strength increases the risk of falls and potential fractures and contributes to the loss of functionality, impedes the older individual's ability to live independently [10], and results in a worse quality of life of a person [11–13]. The ability of the leg muscles to produce strength is a key factor in maintaining the older person's balance and walking speed [14]. According to the definition by the European Working Group on Sarcopenia in Older People (EWGSOP) [3,15], the diagnosis of sarcopenia can be made by assessing the low muscle mass, plus low muscle strength or low physical performance. The most commonly used parameters to measure muscle mass loss are dual energy x-ray absorptiometry (DEXA) and bioelectrical impedance analysis (BIA), to measure muscle strength it's handgrip strength, and to measure physical performance it's short physical performance (SPPB) and gait speed (GS) (Table 1).

Table 1. Sarcopenia: measurable variables and cut-off points [15].

Criterion	Measurement Method	Cut-off Points by Gender
Muscle mass	DEXA	Skeletal muscle mass index (SMI) (Appendicular skeletal muscle mass/height2) Men: 7 kg/m^2 Women: 5.5 kg/m^2
	BIA	SMI using BIA predicted skeletal muscle mass (SM) equation (SM/$height^2$) Men: 8.87 kg/m^2 Women: 6.42 kg/m^2
Muscle strength	Handgrip strength	Men: <27 kg Women: <16 kg
	Chair stand	>15 s for five rises
Physical performance	SPPB SPPB score is a summation of scores on three tests: Balance, Gait Speed and Chair Stand. Each test is weighted equally with scores between 0 and 4—quartiles generated from Established Populations for Epidemiologic Studies of the Older people (EPESE) data (n = 6534). The maximum score on the SPPB is 12	SPPB ≤ 8 SPPB 0–6 Low performance SPPB 7–9 Intermediate performance SPPB 10–12 High Performance
	GS	GS ≤ 0.8 m/s

The EWGSOP suggests the gait speed test as an easy and valid method for assessing physical performance [3,15,16]. Gait speed has been performed to evaluate various health-related factors such as physical functions, health status [17–19], and quality of life [20]. A well-known meta-analysis of 2888 older people set the minimum health threshold for gait speed at ≥ 0.8 m/s [21]. The promotion of physical exercise programs can prevent or even reverse the loss of functional capacity associated with sarcopenia if they emphasize improving the gait speed of the older population [22].

The increased fat mass associated with sarcopenia has been linked with a higher incidence of chronic diseases in older people [5,23,24], and developing sarcopenia contributes to the development of cardiovascular and metabolic diseases [25]. Being sedentary or not very physically active contributes to sarcopenia, as shown by research result, which have also linked age to the development of sarcopenia. The ICD-10 code for sarcopenia in 2016 was established by the World Health Organization to promote effective therapeutic strategies that include physical exercise to prevent loss of muscle mass and function with aging [26].

Recent research suggests effective intervention strategies to combat sarcopenia that include physical exercise, and more specifically, strength training [27–31]. Also, maximum oxygen consumption (VO^2max) is a measurement of cardiorespiratory fitness (CRF), which can predict longevity in older adults [32]. People who have a higher risk of cardiovascular disease (CVD) and mortality are those who have a lower CRF [33]. For older individuals, regular aerobic exercise helps them to attain better VO^2 values [34]. With age, exercise that includes long-term aerobic exercise can help combat the effects of sarcopenia [35].

Being overweight and obese, coupled with a poor physical condition, are related to aging and are also associated with the risk of death from chronic diseases. Therefore, strategies are needed to encourage changes in body composition and physical condition [36,37]. Exercise programs, such as gerontogymnastics, which include resistance and aerobic training, are an optimal strategy for maintaining muscle mass and its protective effects against a variety of chronic diseases [38–42]. However, older adults with low functional capacity may not be able to develop resistance programs leading to improved CRF, due to their low fitness [43]. Besides, the improvement of CRF could also be influenced by improved strength [44].

Therefore, the objectives of the study were the following: a) to analyze the effect of a gerontogymnastics program with overweight and obese older women (≥65 years old) on functional ability and fitness, and b) to understand if sarcopenia mediates the effect of a gerontogymnastics program on cardiovascular fitness. We hypothesized that the older women who participate in the trained group will show improvements in all the tests, whereas the control group will not show changes, and that sarcopenia will mediate the effect of the program on cardiorespiratory capacity.

2. Materials and Methods

2.1. Design

The study was conducted from September to December 2019, with a total of 12 weeks of training. This is a randomized controlled trial study that investigates the effect of gerontogymnastics program with overweight and obese older women on functional ability and physical fitness. This study trial followed the Consolidated Standards of Reporting Trials (CONSORT) guidelines. Older women were informed of the study and signed an informed consent to participate, according to the process approved by the local ethics committee (CE111908) of the San Antonio Catholic University of Murcia (Spain) and in accordance with the Declaration of Helsinki. This study was conducted at a women's association and sport science laboratory from Region de Murcia (Spain).

2.2. Participants

Participants were recruited through advertisements in women's associations, senior centers, and presentations in the local community. Recruitment was done before registration for convenience and accessibility to the sample, but the intervention did not begin until the registration was completed. Inclusion criteria were: (a) at least 1 year not engaged in a structured exercise program, (b) having a body mass index between 25 and 29.9 (overweight) or between 30 and 34.9 (obese), (c) women aged between 65 and 90 years old, and (d) being physically independent. The exclusion criteria were: (a) having musculoskeletal injuries or limitations that could affect the person's health and physical performance; (b) being under a doctor's prescription for taking medication that could influence physical performance; (c) no regular attendance at the proposed sessions.

Sample size and power were established in connection with the 10-m walk test in a previous study [45]. An estimated error of 0.045 s and a significance level of $\alpha = 0.05$ were utilized. A valid sample size for a confidence interval of 95% was 207.84. Based on previous research, a dropout rate of 10% was assumed; therefore, 230 participants were recruited.

Participants were divided to a trained group (TG) and a control group (CG). Thus, the experimental sample of the study consisted of 230 women, being electronically randomized [46] into the TG (115 subjects) and the CG (115 subjects). A researcher who was not involved in participant recruitment performed the randomization.

Two-hundred-and-sixteen older women aged between 65 and 88 years old volunteered (mean ± standard deviation [SD]; age = 68.26 ± 4.19 years, body mass = 71.11 ± 10.66 kg, height = 1.55 ± 0.07 m; BMI = 29.97 ± 3.86) and completed the study (TG = 114; CG = 102). The CONSORT 2010 flow diagram is shown in Figure 1.

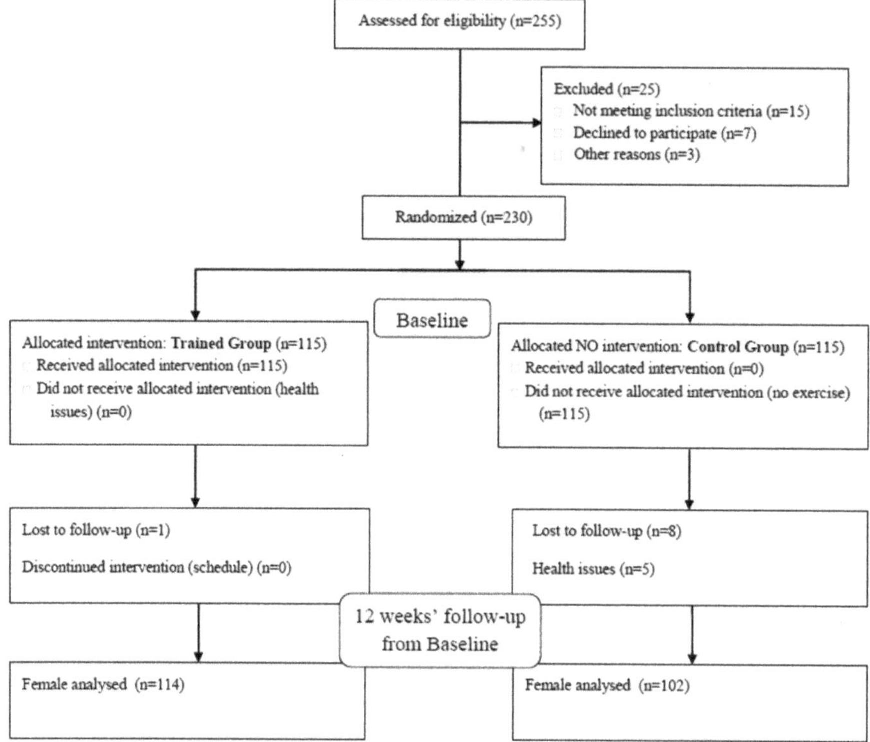

Figure 1. Flow diagram of the sample.

Anthropometric measurements were recorded. Weight (kg) was evaluated in light clothing without footwear to the nearest 0.1 kg by using an electronic scale, and height (cm) was measured using a stadiometer to the nearest millimeter (Seca 763 digital scale, Birmingham, UK). Body mass index (BMI) was calculated by dividing their weight in kilograms by their height in square meters (kg/m^2). All anthropometric measurements were completed by experienced and well-trained persons (ISAK level 1 and 2 certificate). The same researchers performed all the measurements in a single session between the hours of 9:00 and 13:00 without warming up and allowing for a 5-min break between tests.

2.3. Procedure

2.3.1. Trained Group

The gerontogymnastics program was implemented following the recommendations of the Otago Exercise Program (OEP) [47], as it is a renowned exercise program with widespread use at the international level that aims to improve strength and mobility to help in the prevention of falls. The gerontogymnastics program was planned principally to help prevent the risk of falls with exercise training that improves muscle strength, power, and balance in the lower extremities and cardiorespiratory endurance. A professional Sports Physical Educator directed the training. The total training group was divided into subgroups of a maximum of 20 subjects for security and the correct direction by the trainer. The participants in the TG trained for 1 h, three times a week for the 12 weeks of intervention (36 sessions). The training session consisted of: (a) 10 min of warm-up, consisting of joint mobility and stretching of the main muscle groups involved; (b) 30 min of an exercise circuit (three sets of 15 repetitions in each exercise, with 2 min rest between sets). The exercise circuit consisted of 12 exercises, of which, seven focused on strength: knee extension, squat, knee curl, leg press, elbow curl, chest press, and shoulder overhead press using the OMNI resistance exercise scale [48]; five focused on balance: walking on marked lines on the floor, walking on tiptoes, walking sideways, walking on heels, and walking from heel to toe; and (c) 20 min of cardiovascular exercises. The cardiovascular exercise consisted of walking at maximum speed without running to maintain a moderate to hard level of perception of exertion [49]. The training sessions were held on non-consecutive days to facilitate recovery.

2.3.2. Control Group

The CG participants were asked to carry out their normal life and not to alter their habits during the study period, and they did not practice any physical activity or exercise program.

2.3.3. Assessment of gait speed

Gait speed was assessed with a 10-m test. The time in seconds that the person took to walk 10 m was analyzed, with the person walking at their usual pace. This test has been widely used in large epidemiological studies, showing high concurrent and predictive validity [45,50–53]. The results of previous studies [45] indicate an excellent relationship between the 4-m test and the 10-m test (ICC = 0.959 and 0.976, respectively), with a good average between both tests (ICC = 0.867). The 10-m test proved to be somewhat better. This test is a valid method for predicting sarcopenia [54]. The reference value for gait speed in the 10-m test is 0.8 m/s [55]. In the present study, in order to obtain reliable measurements, two photocells were placed at the beginning and the end of a 10-m lane, and through a connection to a computer, recorded the time spent in carrying out the test (MuscleLab, Ergotest, Langesund, Norway). The older women were asked to stand at the starting line mark and walk at their usual pace at the sound signal. Two attempts were made and the average value between the two repetitions was recorded.

2.3.4. Assessment of Cardiorespiratory Fitness Level

Aerobic endurance was assessed using the 6-min walk test (6 MWT). The 6 MWT has been shown to be a valid, reliable, objective, inexpensive, and easy test used to evaluate cardiorespiratory capacity [56–60]. It is a simple test to perform and is better related to the person's daily life activities than other tests [58,59]. It is used to measure an individual's sub-maximum aerobic capacity while walking for 6 min.

It is suggested that this test should be performed on a flat surface that allows walking for 20 to 30 m. The subject should be relaxed and wear comfortable clothing and shoes and the heart rate of each subject was recorded with a POLAR 400 heart rate monitor just before the start of the test and just after the end. The route was marked every 5 m and cones were placed at the turning area. The subject was walking at a pace appropriate to his/her condition, being able to stop or slow down if he/she is fatigued and resume as soon as possible. The trainer can motivate the subjects with phrases such as "You are doing well", and the total meters walked is recorded [61]. This test has good reliability (ranging from 0.95 to 0.97) [62].

2.3.5. Assessment of Functional Capacity

The Latin American Group for Maturity (GDLAM) protocol is used to evaluate the functional capacity in older adults [27,63,64]. The battery consists of five tests: walking 10 m; rising from a sitting position; standing up from a prone position on the floor; getting up from a chair and moving around; and the putting on and taking off a T-shirt test. These tests were to calculate the GDLAM functionality index (GI) using a mathematical formula. The material needed for carrying out the tests consisted of a standard chair with a height of 48 cm from the seat to the floor, a digital chronometer, four cones, a sports mat, and a metal measuring tape. The magnitude of the statistical significance demonstrated high reliability ($r = 0.9$; $p < 0.001$) and validity [63].

2.3.6. Assessment of Muscle Strength

Two tests from the "Senior Fitness Tests" (SFT) battery [59,65] were used to assess strength variables: extension flexion elbow test and lift chair 30 s test. The extension flexion elbow test measures the muscle strength of the upper extremity. The subject, while sitting on a chair, was asked to perform the maximum number of repetitions for 30 s with a dumbbell (2.3 kg for women). The lift chair 30 s test reflects lower body strength. The participant was asked to sit on a chair with his arms across his chest and perform the most sitting and standing repetitions for 30 s. Reliability and validity indicators for the standards ranged between 0.79 and 0.97 [66].

2.4. Data Analysis

The normality of the data was evaluated using the Kolmogorov–Smirnov test, and Mauchly's W-test was used to analyze the normality and the sphericity of the data. The inter- and intra-groups differences and the interaction between groups and time were analyzed with a two-way ANOVA with repeated measurements of one factor (time). Also, an ANCOVA (adjusted for gait speed) with repeated measurement of one factor (time) was used. To check intra-groups change, the post-hoc Bonferroni test and the Wilcoxon signed-rank test were used to evaluate the statistical significance of parametric and non-parametric variables, respectively. The Mann–Whitney test was used to check for inter-group differences for non-parametric variables. The partial eta-squared (η2p) for variance analysis was used to calculate the size effect, and this was defined as small: ES ≥ 0.10; moderate: ES ≥ 0.30, large: ES ≥ 1.2; or very large: ES ≥ 2.0, with an error of $p \leq 0.05$ utilized [67].

To determine if the effect on the 6MWT test was mediated by the change in the 10-m test, the analysis of the mediation variables was performed using the Process macro for SPSS (SPSS Inc, Chicago, Illinois). A resample procedure of 10,000 bootstrap samples for non-parametric variables was utilized, [68] and the classical Baron and Kenny step regression method was used for parametric ones. [69]. In order to analyze the statistical significance of the mediation effect, the Sobel test was used [70]. If after the mediation, the independent variable was no longer associated with the dependent variable, it was considered complete mediation. However, if after the mediation, the independent variable was reduced but was still significant, it was considered partial mediation. The statistical analysis was performed using IBM SPSS Statistics (version 24.0), and an error of $p \leq 0.05$ was set for the analysis.

3. Results

The characteristics of the participants are shown in Table 2. The TG showed significant improvements in the 10-m test ($p < 0.000$), the 6 MWT ($p = 0.001$), stand from siting test ($p < 0.000$), the rising from sitting test ($p < 0.000$), the rise from the floor test ($p < 0.000$), the t-shirt test ($p < 0.000$), the GDLAM index ($p < 0.000$), the extension and flexion elbow test ($p < 0.000$), and the lift chair 30 s test ($p < 0.000$). TG did not show changes in the stand-up and go test ($p = 0.150$) and showed an increase of BMI ($p = 0.021$) but with a very low effect size (ES = 0.03).

The CG experienced a significant decrease in the 10-m test ($p < 0.000$), 6 MWT ($p = 0.011$), rise from the floor test ($p = 0.032$), stand-up and go test ($p < 0.000$), and extension flexion elbow test ($p < 0.000$), and showed a significant improvement in the rise from the floor test ($p = 0.032$), although they did not show changes in the rest of the tests. Although both groups showed a significant improvement in the rise from the floor test, the effect size was small for the CG (ES = 0.12), whereas the effect size for the TG was large (ES = 0.72) (Table 3).

Table 2. Characteristics of the participants.

	M ± SD
Age	68.03 ± 4.03
Weight (kg)	69.59 ± 11.28
Height (m)	1.55 ± 0.07
BMI (kg/m^2)	29.97 ± 3.86
10-m test (s)	7.29 ± 1.51
6MWT (m)	460.29 ± 78.02
Stand from siting (s)	13.13 ± 3.31
Rise from the floor (s)	8.47 ± 3.64
Stand and go (s)	42.83 ± 5.03
T-shirt (s)	15.92 ± 6.16
GDLAM index	33.11 ± 5.18
Ex Flex Elbow 30 s (rep)	14.54 ± 4.16
Lift chair 30 s (rep)	10.98 ± 2.78

Legend: s = seconds; m = meters; rep = repetitions; M = Mean; SD = Standard Deviation; 6 MWT = 6 min walk test.

Table 3. Differences pre- to post-test (intra-groups) for functional and fitness test.

		Pre-Test (M ± SD)	Post-Test (M ± SD)	Difference Post-Pre (M ± SD)	p	CI 95% (Mpost–Mpre)	ES
10-m test (s)	TG	6.93 ± 1.14	6.05 ± 0.94	0.878 ± 0.14	0.000	0.601; 1.155	0.77
	CG	7.59 ± 1.70	8.40 ± 1.68	−0.811 ± 0.15	0.000	−1.103; −0.518	0.47
6MWT (m)	TG	473.54 ± 70.42	491.06 ± 74.00	−17.528 ± 5.36	0.001	−28.083; −6.973	0.25
	CG	448.30 ± 79.33	433.70 ± 79.49	14.601 ± 5.66	0.011	3.442; 25.760	0.18
Stand from siting (s)	TG	13.56 ± 3.15	11.29 ± 2.69	2.272 ± 0.37	0.000	1.535; 3.010	0.72
	CG	12.96 ± 3.62	13.08 ± 3.70	−0.123 ± 0.4	0.756	−0.903; 0.657	0.03
Rise from the floor (s)	TG	8.92 ± 3.80	6.18 ± 3.27	2.740 ± 0.2	0.000	2.357; 3.124	0.72
	CG	8.44 ± 3.71	8.00 ± 3.92	0.444 ± 0.21	0.032	0.039; 0.850	0.12
Stand and go (s)	TG	41.07 ± 3.57	40.51 ± 4.04	0.566 ± 0.39	0.150	−0.206; 1.338	0.16
	CG	44.41 ± 5.44	46.36 ± 5.02	−1.950 ± 0.41	0.000	−2.766; −1.134	0.36
T-shirt (s)	TG	16.46 ± 6.94	11.73 ± 4.41	4.721 ± 0.37	0.000	3.999; 5.444	0.68
	CG	15.85 ± 5.66	15.36 ± 4.81	0.489 ± 0.39	0.208	−0.275; 1.252	0.09
GDLAM index	TG	33.20 ± 5.92	27.75 ± 4.50	5.448 ± 0.34	0.000	4.777; 6.118	0.91
	CG	33.52 ± 4.68	34.01 ± 3.83	−0.489 ± 0.36	0.176	−1.197; 0.220	0.10
Ex Flex Elbow 30 s (rep)	TG	14.59 ± 4.25	17.46 ± 4.55	−2.868 ± 0.28	0.000	−3.417; −2.320	0.67
	CG	14.63 ± 4.09	13.34 ± 3.64	1.284 ± 0.29	0.000	0.704; 1.865	0.31
Lift chair 30 s (rep)	TG	10.81 ± 2.46	13.09 ± 2.70	−2.281 ± 0.21	0.000	−2.699; −1.863	0.92
	CG	11.26 ± 3.17	11.15 ± 3.07	0.118 ± 0.22	0.600	−0.324; 0.560	0.04
BMI (kg/m^2)	TG	31.13 ± 4.15	31.26 ± 4.13	0.13 ± 0.54	0.021	0.020; 0.242	0.03
	CG	28.68 ± 3.04	28.59 ± 3.17	−0.09 ± 0.66	0.125	−0.209; 0.026	0.02

Legend: TG = trained group; CG = control group; M = Mean; SD = Standard Deviation; ES = Effect Size; s = seconds; m = meters; rep = repetitions.

highlighting its importance in the creation of strategies for the preservation of physical function with age [77]. These results support the evidence that a 12-week gerontogymnastics program that included endurance and strength training exercises improves functional capacity, CRF, and strength and endurance of musculature of overweight and obese older women; and could thus delay the harmful effects of aging.

The second objective of this study was to understand if sarcopenia mediated the effect of a gerontogymnastics program on cardiovascular fitness. The major finding of our study was that an improvement in CRF was associated with an improvement in gait speed, in consonance with the decrease in sarcopenia. Our results are in agreement with a previous study, showing a connection between CRF and gait speed and sarcopenia [78]. In our study, sarcopenia acted as a partial mediator on the association between carrying out a gerontogymnastics program and improved CRF. To the best of our knowledge, this is the first randomized controlled trial with an analysis of the mediation that assesses how sarcopenia influences the effect of an exercise program on CRF.

A recent study [79] assessed 527 women aged 75 years and older (79.7 ± 3.5) in a cross-sectional study. The objective of this study was to investigate if the connection between physical activity and gait speed was mediated by strength and weight. These authors reported that the association between physical activity and gait speed was partially mediated by the absolute and relative strength of the lower limbs and that muscle mass partially mediated the relationship between physical activity and muscle strength.

On the other hand, it has been demonstrated that there is a connection between walking balance and strength [80] and that sarcopenia influences walking balance [81]. A study with older adults with mild to moderate frailty improved their CRF but at a modest level [82]. This suggests that it will be necessary to increase leg strength to further increase walking speed, in order to improve CRF. In this sense, a study was performed to determine the mechanisms responsible for the effect of exercise training on CRF in older adults, utilizing a strength training program before an endurance training program, with a sample of 22 older adults, to improve their functional capacity [44].

In agreement with another study [43], older adults with declined functionality were not able to participate in endurance training until they improved their neuromuscular capacity. Therefore, endurance training for older women should be performed with previous strength and resistance training to achieve the highest CRF adaptations.

It has also been reported that gait speed, muscle mass, and sarcopenia are strongly associated with functional capacity [17–19]. However, our study expands this finding by showing that sarcopenia is not just a predictor, but also an important mediator of the effect of an exercise program on another important factor for the health such as CRF.

Strong research methodologies, such as a randomized clinical trial with a blinded examiner, is one of the strengths of the present study. Also, to minimize the risk of bias, a large sample size was utilized. However, our study is not without limits. This research was developed with older women who were overweight and obese, and thus, we are not able to generalize the result to other populations of interest.

5. Conclusions

A gerontogymnastics program improves the functional capacity and fitness of older women who are overweight and obese. Sarcopenia acts as a mediator of the effect of a gerontogymnastics program on CRF in overweight and obese older women.

In this sense, the results support the new interest in changing the type of intervention and could be used to suggest that the improvements in strength, gait speed, and reduction of sarcopenia at the start of the exercise program could be needed to secure or improve the effects of the program on CRF and help improve the health of overweight and obese older people.

Author Contributions: Conceptualization, P.J.M.-P., N.G.-G. and R.G.d.S.V; Formal analysis, P.J.M.-P. and N.G.-G.; Methodology, P.J.M.-P., N.G.-G., A.L.-V. and A.E.-G.; Resources, P.J.M.-P., N.G.-G., G.M.G.-G., A.L.-V., A.E.-G. and R.G.d.S.V.; Supervision, P.J.M.-P. Validation, P.J.M.-P., N.G.-G. and R.G.d.S.V; Writing—original draft, P.J.M.-P.,

N.G.-G., A.L.-V., A.E.-G. and R.G.d.S.V; Writing—review & editing, P.J.M.-P., N.G.-G., G.M.G.-G., A.L.-V., A.E.-G. and R.G.d.S.V. All authors have read and agreed to the published version of the manuscript.

Funding: The present research on active aging of members (GISAFFCOM) of HEALTHY-AGE Network (reference 08/UPR/20) is supported by a grant from the Spanish Ministry of Culture and Sport- Sports Sciences Networks and GISAFFCOM research group is supported by a grant from the Spanish Ministry of Science, Innovation and Universities- RETOS I+d+i 2018 (RTC-2017-6145-1).

Acknowledgments: The research team would like to thank the heads of the social and women's centers and all the older women for their participation in this research, and the San Antonio Catholic University of Murcia (UCAM) for its support to the line of research on healthy and active aging.

Conflicts of Interest: The authors declare no conflict of interest.

References

1. Santilli, V.; Bernetti, A.; Mangone, M.; Paoloni, M. Clinical definition of sarcopenia. *Clin. Cases Miner. Bone Metab.* **2014**, *11*, 177–180. [CrossRef] [PubMed]
2. Larsson, L.; Degens, H.; Li, M.; Salviati, L.; Lee, Y.; Thompson, W.; Kirkland, J.L.; Sandri, M. Sarcopenia: Aging-related loss of muscle mass and function. *Physiol. Rev.* **2019**, *99*, 427–511. [CrossRef] [PubMed]
3. Cruz-Jentoft, A.J.; Pierre Baeyens, J.; Bauer, J.M.; Boirie, Y.; Cederholm, T.; Landi, F.; Martin, F.C.; Michel, J.-P.; Rolland, Y.; Schneider, S.M.; et al. Sarcopenia: European consensus on definition and diagnosis Report of the European Working Group on Sarcopenia in Older People. *Age Ageing* **2010**, *39*, 412–423. [CrossRef] [PubMed]
4. Von Haehling, S.; Morley, J.E.; Anker, S.D. An overview of sarcopenia: Facts and numbers on prevalence and clinical impact. *J. Cachexia Sarcopenia Muscle* **2010**, *1*, 129–133. [CrossRef] [PubMed]
5. Melton, L.J.; Khosla, S.; Crowson, C.S.; O'Connor, M.K.; O'Fallon, W.M.; Riggs, B.L. Epidemiology of Sarcopenia. *J. Am. Geriatr. Soc.* **2000**, *48*, 625–630. [CrossRef]
6. Chen, L.; Xia, J.; Xu, Z.; Chen, Y.; Yang, Y. Evaluation of Sarcopenia in Elderly Women of China. *Int. J. Gerontol.* **2017**, *11*, 149–153. [CrossRef]
7. Aagaard, P.; Suetta, C.; Caserotti, P.; Magnusson, S.P.; Kjær, M. Role of the nervous system in sarcopenia and muscle atrophy with aging: Strength training as a countermeasure. *Scand. J. Med. Sci. Sport.* **2010**, *20*, 49–64. [CrossRef]
8. Guizelini, P.; de Aguiar, R.; Denadai, B.; Caputo, F.; Greco, C. Effect of resistance training on muscle strength and rate of force development in healthy older adults: A systematic review and meta-analysis. *Exp. Gerontol.* **2018**, *102*, 51–58. [CrossRef]
9. Coen, P.M.; Musci, R.V.; Hinkley, J.M.; Miller, B.F. Mitochondria as a target for mitigating sarcopenia. *Front. Physiol.* **2019**, *10*, 1883. [CrossRef]
10. Mthembu, T.G.; Brown, Z.; Cupido, A.; Razak, G.; Wassung, D. Family caregivers' perceptions and experiences regarding caring for older adults with chronic diseases. *South Afr. J. Occup. Ther.* **2016**, *46*, 83–88. [CrossRef]
11. Gadelha, A.B.; Neri, S.G.R.; de Oliveira, R.J.; Bottaro, M.; de David, A.C.; Vainshelboim, B.; Lima, R.M. Severity of sarcopenia is associated with postural balance and risk of falls in community-dwelling older women. *Exp. Aging Res.* **2018**, *44*, 1–12. [CrossRef] [PubMed]
12. Trombetti, A.; Reid, K.F.; Hars, M.; Herrmann, F.R.; Pasha, E.; Phillips, E.M.; Fielding, R.A. Age-associated declines in muscle mass, strength, power, and physical performance: Impact on fear of falling and quality of life. *Osteoporos. Int.* **2016**, *27*, 463–471. [CrossRef] [PubMed]
13. Ma, A.; Sun, Y.; Zhang, H.; Zhong, F.; Lin, S.; Gao, T.; Cai, J. Association between Sarcopenia and Metabolic Syndrome in Middle-Aged and Older Non-Obese Adults: A Systematic Review and Meta-Analysis. *Nutrients* **2018**, *10*, 364. [CrossRef]
14. Li, Z.; Wang, X.-X.; Liang, Y.-Y.; Chen, S.-Y.; Sheng, J.; Ma, S.-J. Effects of the visual-feedback-based force platform training with functional electric stimulation on the balance and prevention of falls in older adults: A randomized controlled trial. *PeerJ* **2018**, *6*, e4244. [CrossRef]
15. Cruz-Jentoft, A.J.; Bahat, G.; Bauer, J.; Boirie, Y.; Bruyère, O.; Cederholm, T.; Cooper, C.; Landi, F.; Rolland, Y.; Sayer, A.A.; et al. Sarcopenia: Revised European consensus on definition and diagnosis. *Age Ageing* **2019**, *48*, 16–31. [CrossRef]
16. Jung, H.J.; Lee, Y.M.; Kim, M.; Uhm, K.E.; Lee, J. Suggested assessments for sarcopenia in patients with stroke who can walk independently. *Ann. Rehabil. Med.* **2020**, *44*, 20–37. [CrossRef]

17. Peel, N.M.; Kuys, S.S.; Klein, K. Gait Speed as a Measure in Geriatric Assessment in Clinical Settings: A Systematic Review. *J. Gerontol. A Biol. Sci. Med. Sci.* **2012**, *68*, 39–46. [CrossRef]
18. Studenski, S.; Perera, S.; Patel, K.; Rosano, C.; Faulkner, K.; Inzitari, M.; Brach, J.; Chandler, J.; Cawthon, P.; Connor, E.B.; et al. Gait speed and survival in older adults. *JAMA* **2011**, *305*, 50–58. [CrossRef]
19. Noh, B.; Youm, C.; Lee, M.; Park, H. Age-specific differences in gait domains and global cognitive function in older women: Gait characteristics based on gait speed modification. *PeerJ* **2020**, *8*, e8820. [CrossRef]
20. Lee, T.; Lee, M.; Youm, C.; Noh, B.; Park, H. Association between Gait Variability and Gait-Ability Decline in Elderly Women with Subthreshold Insomnia Stage. *Int. J. Environ. Res. Public Health* **2020**, *17*, 5181. [CrossRef]
21. Kuys, S.S.; Peel, N.M.; Klein, K.; Slater, A.; Hubbard, R.E. Gait Speed in Ambulant Older People in Long Term Care: A Systematic Review and Meta-Analysis. *J. Am. Med. Dir. Assoc.* **2014**, *15*, 194–200. [CrossRef] [PubMed]
22. Perez-Sousa, M.A.; Venegas-Sanabria, L.C.; Chavarro-Carvajal, D.A.; Cano-Gutierrez, C.A.; Izquierdo, M.; Correa-Bautista, J.E.; Ramírez-Vélez, R. Gait speed as a mediator of the effect of sarcopenia on dependency in activities of daily living. *J. Cachexia Sarcopenia Muscle* **2019**, *10*, 1009–1015. [CrossRef] [PubMed]
23. Volpi, E.; Nazemi, R.; Fujita, S. Muscle tissue changes with aging. *Curr. Opin. Clin. Nutr. Metab. Care* **2004**, *7*, 405–410. [CrossRef]
24. Kim, T.N.; Choi, K.M. Sarcopenia: Definition, Epidemiology, and Pathophysiology. *J. Bone Metab.* **2013**, *20*, 1. [CrossRef] [PubMed]
25. Batsis, J.A.; Mackenzie, T.A.; Lopez-Jimenez, F.; Bartels, S.J. Sarcopenia, sarcopenic obesity, and functional impairments in older adults: National Health and Nutrition Examination Surveys 1999-2004. *Nutr. Res.* **2015**, *35*, 1031–1039. [CrossRef]
26. Anker, S.D.; Morley, J.E.; von Haehling, S. Welcome to the ICD-10 code for sarcopenia. *J. Cachexia Sarcopenia Muscle* **2016**, *7*, 512–514. [CrossRef]
27. Marcos-Pardo, P.J.; Orquin-Castrillón, F.J.; Gea-García, G.M.; Menayo-Antúnez, R.; González-Gálvez, N.; Vale, R.; Martínez-Rodríguez, A. Effects of a moderate-to-high intensity resistance circuit training on fat mass, functional capacity, muscular strength, and quality of life in elderly: A randomized controlled trial. *Sci. Rep.* **2019**, *9*, 7830. [CrossRef]
28. Marcos-Pardo, P.J.; González-Hernández, J.M.; García-Ramos, A.; López-Vivancos, A.; Jiménez-Reyes, P. Movement velocity can be used to estimate the relative load during the bench press and leg press exercises in older women. *PeerJ* **2019**, *7*, e7533. [CrossRef]
29. Landi, F.; Marzetti, E.; Martone, A.M.; Bernabei, R.; Onder, G. Exercise as a remedy for sarcopenia. *Curr. Opin. Clin. Nutr. Metab. Care* **2014**, *17*, 25–31. [CrossRef]
30. Singh, N.A.; Quine, S.; Clemson, L.M.; Williams, E.J.; Williamson, D.A.; Stavrinos, T.M.; Grady, J.N.; Perry, T.J.; Lloyd, B.D.; Smith, E.U.R.; et al. Effects of high-intensity progressive resistance training and targeted multidisciplinary treatment of frailty on mortality and nursing home admissions after hip fracture: A randomized controlled trial. *J. Am. Med. Dir. Assoc.* **2012**, *13*, 24–30. [CrossRef]
31. Lolascon, G.; Di Pietro, G.; Gimigliano, F.; Mauro, G.L.; Moretti, A.; Giamattei, M.T.; Ortolani, S.; Tarantino, U.; Brandi, M.L. Physical exercise and sarcopenia in older people: Position paper of the Italian Society of Orthopaedics and Medicine (OrtoMed). *Clin. Cases Miner. Bone Metab.* **2014**, *11*, 215–221. [CrossRef]
32. Clausen, J.S.R.; Marott, J.L.; Holtermann, A.; Gyntelberg, F.; Jensen, M.T. Midlife Cardiorespiratory Fitness and the Long-Term Risk of Mortality: 46 Years of Follow-Up. *J. Am. Coll. Cardiol.* **2018**, *72*, 987–995. [CrossRef] [PubMed]
33. Kim, T.N.; Park, M.S.; Kim, Y.J.; Lee, E.J.; Kim, M.-K.; Kim, J.M.; Ko, K.S.; Rhee, B.D.; Won, J.C. Association of Low Muscle Mass and Combined Low Muscle Mass and Visceral Obesity with Low Cardiorespiratory Fitness. *PLoS ONE* **2014**, *9*, e100118. [CrossRef] [PubMed]
34. Distefano, G.; Standley, R.A.; Zhang, X.; Carnero, E.A.; Yi, F.; Cornnell, H.H.; Coen, P.M. Physical activity unveils the relationship between mitochondrial energetics, muscle quality, and physical function in older adults. *J. Cachexia Sarcopenia Muscle* **2018**, *9*, 279–294. [CrossRef] [PubMed]
35. Laurin, J.L.; Reid, J.J.; Lawrence, M.M.; Miller, B.F. Long-term aerobic exercise preserves muscle mass and function with age. *Curr. Opin. Physiol.* **2019**, *10*, 70–74. [CrossRef]
36. Ryan, A.S. Exercise in aging: Its important role in mortality, obesity and insulin resistance. *Aging Health* **2010**, *6*, 551–563. [CrossRef] [PubMed]

37. Thinggaard, M.; Jacobsen, R.; Jeune, B.; Martinussen, T.; Christensen, K. Is the Relationship Between BMI and Mortality Increasingly U-Shaped With Advancing Age? A 10-Year Follow-up of Persons Aged 70–95 Years. *J. Gerontol. A Biol. Sci. Med. Sci.* **2010**, *65A*, 526–531. [CrossRef]
38. Yoo, S.Z.; No, M.H.; Heo, J.W.; Park, D.H.; Kang, J.H.; Kim, S.H.; Kwak, H.B. Role of exercise in age-related sarcopenia. *J. Exerc. Rehabil.* **2018**, *14*, 551–558. [CrossRef]
39. Takeshima, N.; Rogers, M.E.; Islam, M.M.; Yamauchi, T.; Watanabe, E.; Okada, A. Effect of concurrent aerobic and resistance circuit exercise training on fitness in older adults. *Eur. J. Appl. Physiol.* **2004**, *93*, 173–182. [CrossRef]
40. Guðlaugsson, J.; Aspelund, T.; Guðnason, V.; Ólafsdóttir, A.S.; Jónsson, P.V.; Arngrímsson, S.Á.; Jóhannsson, E. The effects of 6 months' multimodal training on functional performance, strength, endurance, and body mass index of older individuals. Are the benefits of training similar among women and men? *Læknablaðið* **2013**, *2013*, 331–337. [CrossRef]
41. Fien, S.; Henwood, T.; Climstein, M.; Rathbone, E.; Keogh, J.W.L. Exploring the feasibility, sustainability and the benefits of the GrACE + GAIT exercise programme in the residential aged care setting. *PeerJ* **2019**, *2019*, e6973. [CrossRef] [PubMed]
42. Yoshiko, A.; Tomita, A.; Ando, R.; Ogawa, M.; Kondo, S.; Saito, A.; Tanaka, N.I.; Koike, T.; Oshida, Y.; Akima, H. Effects of 10-week walking and walking with home-based resistance training on muscle quality, muscle size, and physical functional tests in healthy older individuals. *Eur. Rev. Aging Phys. Act.* **2018**, *15*. [CrossRef] [PubMed]
43. Cadore, E.L.; Rodríguez-Mañas, L.; Sinclair, A.; Izquierdo, M. Effects of different exercise interventions on risk of falls, gait ability, and balance in physically frail older adults: A systematic review. *Rejuvenation Res.* **2013**, *16*, 105–114. [CrossRef] [PubMed]
44. Ehsani, A.A.; Spina, R.J.; Peterson, L.R.; Rinder, M.R.; Glover, K.L.; Villareal, D.T.; Binder, E.F.; Holloszy, J.O. Attenuation of cardiovascular adaptations to exercise in frail octogenarians. *J. Appl. Physiol.* **2003**, *95*, 1781–1788. [CrossRef] [PubMed]
45. Fernández-Huerta, L.; Córdova-León, K. Reliability of two gait speed tests of different timed phases and equal non-timed phases in community-dwelling older persons. *Medwave* **2019**, *19*, e7611. [CrossRef]
46. Urbaniak, G.C.; Plous, S. Research Randomizer (Version 4.0). Available online: https://www.randomizer.org/ (accessed on 1 September 2019).
47. Shubert, T.E.; Smith, M.L.; Jiang, L.; Ory, M.G. Disseminating the Otago exercise program in the United States: Perceived and actual physical performance improvements from participants. *J. Appl. Gerontol.* **2018**, *37*, 79–98. [CrossRef]
48. Gearhart, R.F.; Lagally, K.M.; Riechman, S.E.; Andrews, R.D.; Robertson, R.J. Safety of using the adult OMNI Resistance Exercise Scale to determine 1-RM in older men and women. *Percept. Mot. Skills* **2011**, *113*, 671–676. [CrossRef]
49. Farinatti, P.T.V.; Monteiro, W.D. Walk-run transition in young and older adults: With special reference to the cardio-respiratory responses. *Eur. J. Appl. Physiol.* **2010**, *109*, 379–388. [CrossRef]
50. Guralnik, J.M.; Ferrucci, L.; Pieper, C.F.; Leveille, S.G.; Markides, K.S.; Ostir, G.V.; Studenski, S.; Berkman, L.F.; Wallace, R.B. Lower extremity function and subsequent disability: Consistency across studies, predictive models, and value of gait speed alone compared with the short physical performance battery. *J. Gerontol. A Biol. Sci. Med. Sci.* **2000**, *55*. [CrossRef]
51. Guralnik, J.M.; Ferrucci, L.; Simonsick, E.M.; Salive, M.E.; Wallace, R.B. Lower-extremity function in persons over the age of 70 years as a predictor of subsequent disability. *N. Engl. J. Med.* **1995**, *332*, 556–562. [CrossRef]
52. Ostir, G.V.; Volpato, S.; Fried, L.P.; Chaves, P.; Guralnik, J.M. Reliability and sensitivity to change assessed for a summary measure of lower body function: Results from the Women's Health and Aging Study. *J. Clin. Epidemiol.* **2002**, *55*, 916–921. [CrossRef]
53. Penninx, B.W.; Ferrucci, L.; Leveille, S.G.; Rantanen, T.; Pahor, M.; Guralnik, J.M. Lower extremity performance in nondisabled older persons as a predictor of subsequent hospitalization. *J. Gerontol. A Biol. Sci. Med. Sci.* **2000**, *55*, M691–M697. [CrossRef] [PubMed]
54. Beaudart, C.; Reginster, J.Y.; Slomian, J.; Buckinx, F.; Dardenne, N.; Quabron, A.; Slangen, C.; Gillain, S.; Petermans, J.; Bruyère, O. Estimation of sarcopenia prevalence using various assessment tools. *Exp. Gerontol.* **2015**, *61*, 31–37. [CrossRef] [PubMed]

55. Lauretani, F.; Russo, C.R.; Bandinelli, S.; Bartali, B.; Cavazzini, C.; Di Iorio, A.; Corsi, A.M.; Rantanen, T.; Guralnik, J.M.; Ferrucci, L. Age-associated changes in skeletal muscles and their effect on mobility: An operational diagnosis of sarcopenia. *J. Appl. Physiol.* **2003**, *95*, 1851–1860. [CrossRef] [PubMed]
56. Salzman, S.H. The 6-min walk test: Clinical and research role, technique, coding, and reimbursement. *Chest* **2009**, *135*, 1345–1352. [CrossRef]
57. Pinto-Plata, V.M.; Cote, C.; Cabral, H.; Taylor, J.; Celli, B.R. The 6-min walk distance: Change over time and value as a predictor of survival in severe COPD. *Eur. Respir. J.* **2004**, *23*, 28–33. [CrossRef]
58. Enright, P.L. The six-minute walk test. *Respir. Care* **2003**, *48*, 783–785.
59. Rikli, R.E.; Jones, C.J. *Senior Fitness Test Manual*; Human Kinetics: New York, NY, USA, 2013; ISBN 1450411185.
60. Sperandio, E.F.; Arantes, R.L.; Matheus, A.C.; Silva, R.P.; Lauria, V.T.; Romiti, M.; Gagliardi, A.R.T.; Dourado, V.Z. Intensity and physiological responses to the 6-minute walk test in middle-aged and older adults: A comparison with cardiopulmonary exercise testing. *Braz. J. Med. Biol. Res.* **2015**, *48*, 349–353. [CrossRef]
61. Giannitsi, S.; Bougiakli, M.; Bechlioulis, A.; Kotsia, A.; Michalis, L.K.; Naka, K.K. 6-minute walking test: A useful tool in the management of heart failure patients. *Ther. Adv. Cardiovasc. Dis.* **2019**, *13*. [CrossRef]
62. Steffen, T.M.; Hacker, T.A.; Mollinger, L. Age- and gender-related test performance in community-dwelling elderly people: Six-Minute Walk Test, Berg Balance Scale, Timed Up & Go Test, and gait speeds. *Phys. Ther.* **2002**, *82*, 128–137. [CrossRef]
63. Dantas, E.H.M.; Figueira, H.A.; Emygdio, R.F.; Vale, R.G.S. Functional Autonomy GdlAm Protocol Classification Pattern in Elderly Women. *Indian J. Appl. Res.* **2014**, *4*, 262–266. [CrossRef]
64. Vale, R.G.; Soares-Pernambuco, C.; Ferreira Emygdio, R.; Marcos-Pardo, P.J.; Martín-Dantas, E.H. Avaliação da autonomia funcional. In *Manual de Avaliaçao do Idoso*; Vale, R.G.S., Soares-Pernambuco, C., Martín-Dantas, E.H., Eds.; Icone Editora: São Paulo, Brazil, 2016; pp. 149–162. ISBN 978-85-274-1285-8.
65. Rikli, R.E.; Jones, C.J. Development and validation of a functional fitness test for community-residing older adults. *J. Aging Phys. Act.* **1999**, *7*, 129–161. [CrossRef]
66. Rikli, R.E.; Jones, C.J. Development and validation of criterion-referenced clinically relevant fitness standards for maintaining physical independence in later years. *Gerontologist* **2013**, *53*, 255–267. [CrossRef] [PubMed]
67. Hopkins, W.G.; Marshall, S.W.; Batterham, A.M.; Hanin, J. Progressive statistics for studies in sports medicine and exercise science. *Med. Sci. Sports Exerc.* **2009**, *41*, 3–12. [CrossRef]
68. Preacher, K.J.; Hayes, A.F. Asymptotic and resampling strategies for assessing and comparing indirect effects in multiple mediator models. *Behav. Res. Methods* **2008**, *40*, 879–891. [CrossRef] [PubMed]
69. Preacher, K.J.; Kelley, K. Effect size measures for mediation models: Quantitative strategies for communicating indirect effects. *Psychol. Methods* **2011**, *16*, 93–115. [CrossRef] [PubMed]
70. Sobel, M.E. Asymptotic Confidence Intervals for Indirect Effects in Structural Equation Models. *Sociol. Methodol.* **1982**, *13*, 290. [CrossRef]
71. De Resende-Neto, A.G.; do Nascimento, M.A.; da Silva, D.R.P.; Netto, R.S.M.; de Santana, J.M.; Silva, M.E.D.G. Effects of Multicomponent Training on Functional Fitness and Quality of Life in Older Women: A Randomized Controlled Trial. *Int. J. Sport. Exerc. Med.* **2019**, *5*. [CrossRef]
72. De Matos, D.G.; Mazini Filho, M.L.; Moreira, O.C.; De Oliveira, C.E.; De Oliveira Venturini, G.R.; Da Silva-Grigoletto, M.E.; Aidar, F.J. Effects of eight weeks of functional training in the functional autonomy of elderly women: A pilot study. *J. Sports Med. Phys. Fit.* **2017**, *57*, 272–277. [CrossRef]
73. Carrasco-Poyatos, M.; Rubio-Arias, J.A.; Ballesta-García, I.; Ramos-Campo, D.J. Pilates vs. muscular training in older women. Effects in functional factors and the cognitive interaction: A randomized controlled trial. *Physiol. Behav.* **2019**, *201*, 157–164. [CrossRef]
74. Cao, L.; Jiang, Y.; Li, Q.; Wang, J.; Tan, S. Exercise Training at Maximal Fat Oxidation Intensity for Overweight or Obese Older Women: A Randomized Study. *J. Sports Sci. Med.* **2019**, *18*, 413.
75. De Liao, C.; Tsauo, J.Y.; Lin, L.F.; Huang, S.W.; Ku, J.W.; Chou, L.C.; Liou, T.H. Effects of elastic resistance exercise on body composition and physical capacity in older women with sarcopenic obesity. *Medicine* **2017**, *96*. [CrossRef]
76. Huang, C.; Sun, S.; Tian, X.; Wang, T.; Wang, T.; Duan, H.; Wu, Y. Age modify the associations of obesity, physical activity, vision and grip strength with functional mobility in Irish aged 50 and older. *Arch. Gerontol. Geriatr.* **2019**, *84*. [CrossRef] [PubMed]

77. Tay, J.; Goss, A.M.; Locher, J.L.; Ard, J.D.; Gower, B.A. Physical Function and Strength in Relation to Inflammation in Older Adults with Obesity and Increased Cardiometabolic Risk. *J. Nutr. Heal. Aging* **2019**, *23*, 949–957. [CrossRef] [PubMed]
78. Peterson, M.D.; Rhea, M.R.; Sen, A.; Gordon, P.M. Resistance exercise for muscular strength in older adults: A meta-analysis. *Ageing Res. Rev.* **2010**, *9*, 226–237. [CrossRef]
79. Barbat-Artigas, S.; Pinheiro Carvalho, L.; Rolland, Y.; Vellas, B.; Aubertin-Leheudre, M. Muscle Strength and Body Weight Mediate the Relationship Between Physical Activity and Usual Gait Speed. *J. Am. Med. Dir. Assoc.* **2016**, *17*, 1031–1036. [CrossRef] [PubMed]
80. Ibeneme, S.C.; Ekanem, C.; Ezuma, A.; Iloanusi, N.; Lasebikan, N.N.; Lasebikan, O.A.; Oboh, O.E. Walking balance is mediated by muscle strength and bone mineral density in postmenopausal women: An observational study. *BMC Musculoskelet. Disord.* **2018**, *19*, 1–10. [CrossRef] [PubMed]
81. Ibeneme, S.; Ezeigwe, C.; Ibeneme, G.C.; Ezuma, A.; Okoye, I.; Nwankwo, J.M. Response of Gait Output and Handgrip Strength to Changes in Body Fat Mass in Pre- and Postmenopausal Women. *Curr. Ther. Res. Clin. Exp.* **2019**, *90*, 92–98. [CrossRef] [PubMed]
82. Binder, E.F.; Schechtman, K.B.; Ehsani, A.A.; Steger-May, K.; Brown, M.; Sinacore, D.R.; Yarasheski, K.E.; Holloszy, J.O. Effects of exercise training on frailty in community-dwelling older adults: Results of a randomized, controlled trial. *J. Am. Geriatr. Soc.* **2002**, *50*, 1921–1928. [CrossRef]

International Journal of
Environmental Research and Public Health

Article

Relationships between Exercise Modality and Activity Restriction, Quality of Life, and Hematopoietic Profile in Korean Breast Cancer Survivors

MunHee Kim [1], Wi-Young So [2] and Jiyoun Kim [3,*]

[1] Department of Health Science, Korea National Sport University, Seoul 05541, Korea; kmoonhee@hanmail.net
[2] Sports and Health Care Major, College of Humanities and Arts, Korea National University of Transportation, Chungju-si 27469, Korea; wowso@ut.ac.kr
[3] Department of Exercise Rehabilitation and Welfare, Gachon University, Incheon 21936, Korea
* Correspondence: eve14jiyoun@gachon.ac.kr; Tel.: +82-32-820-4237; Fax: +82-32-820-4449

Received: 21 August 2020; Accepted: 17 September 2020; Published: 21 September 2020

Abstract: This study aimed to examine the relationships between activity restriction, quality of life (QoL), and hematopoietic profile in breast cancer survivors according to exercise modality. The subjects in this study were 187 female breast cancer survivors among a total of 32,631 participants in the Korea National Health and Nutrition Examination Survey, which was conducted from 2016 to 2018. The selected subjects participated in a questionnaire survey and blood analysis. A cross-analysis was conducted to determine the relationship between participation in various modality of exercise (e.g., aerobic exercise, resistance exercise, walking exercise). The phi coefficients or Cramer's V value for activity restriction and QoL were calculated; an independent *t*-test was conducted to evaluate the differences between hematopoietic profiles based on the modality of exercise. Statistically significant correlations were seen between obesity and aerobic exercise and walking frequency, as well as between diabetes and aerobic exercise and activity restriction. With respect to QoL, there was a statistically significant correlation between participation in aerobic exercise and exercise ability, participation in aerobic exercise and anxiety/depression, participation in resistance exercise and subjective health status, participation in resistance exercise and exercise ability, and participation in weekly walking exercise and self-care ability. Regarding hemodynamic changes, red blood cells increased significantly in breast cancer survivors who participated in weekly resistance exercise compared to in those who did not. In conclusion, exercise participation had a positive effect on activity restriction, QoL, and hematopoietic profile in breast cancer survivors; in particular, some modalities of aerobic exercise were more effective.

Keywords: aerobic exercise; obesity; resistance exercise; subjective health status; walking

---

1. Introduction

Breast cancer is the most common cancer among Korean women. According to a report from the Korea Central Cancer Registry under the Ministry of Health and Welfare, the age-adjusted cancer incidence rate in 2016 was 62.6 out of 100,000 women, which is significantly higher than 54.7 in 2014 and 56.1 in 2015. When noted according to age group, breast cancer is most common among women in their 40s (44.3%), followed by those in their 50s (30.2%) and 60s (16.1%) [1]. The Korea Ministry of Health and Welfare reported that, while the number of patients with cancer who survived for more than 5 years after cancer diagnosis exceeded 1 million in 2017, and the cancer survival rate reached 70%, 40% of patients with breast cancer developed depression [1,2]. Breast cancer survivors experience many activity restrictions because of sexual problems, infertility, fatigue, appearance, separation or divorce from their spouses, fear of cancer recurrence and death, etc. [2–6].

Because of these psychological stresses, those with breast cancer tend not to actively participate in many activities, which leads to the deterioration of cardiovascular health, muscle strength, and bone health, thus increasing the risk of osteoporosis and cardiovascular disease [7]. These diseases eventually cause activity restrictions for breast cancer survivors, including discomfort in daily life and absenteeism due to concurrent diseases [8]. In fact, 48.4% of breast cancer survivors in Korea are obese, which increases the risk of breast cancer recurrence and mortality to 35–40%, and may cause insulin resistance, metabolic syndrome, and type 2 diabetes [9]. Since obesity and diabetes also increase the risk of other cancers, more care is required [10]. In recent studies, the expression and activity of iron-related proteins (ferritin, hepcidin, and ferroportin) in breast cancer cells affected the prognosis of breast cancer [11]. In particular, poor iron metabolism (anemia) in patients with breast cancer is a common phenomenon based on tumor stage and anticancer treatment used, and about 43–47% of patients with breast cancer develop anemia [12,13]. In addition, patients with cancer experience inflammatory reactions in their bodies due to obesity, as their level of activity decreases because of fatigue [14,15].

The prevention of cancer is of primary importance; however, women who already have cancer need proper physical and emotional care to maintain their quality of life (QoL). For cancer survivors, the ongoing management of lifestyle (nutrition, physical activity, sleep, stress) is important. Among lifestyle factors, physical activity is widely recognized as an effective non-pharmacological treatment for patients with cancer [16–18]. In order to manage or prevent breast cancer, various modalities of exercise are used. Therefore, this study, utilizing the Korea National Health and Nutrition Examination Survey (KNHANES) conducted from 2016 to 2018, aimed to examine the relationships between participation in various modalities of exercise and activity restriction, QoL, and hemodynamic changes in breast cancer survivors, and to determine which modality of exercise is more effective for breast cancer survivors.

2. Materials and Methods

2.1. Study Design and Participants

The KNHANES is a national survey conducted by trained experts every year under the supervision of the Korea Centers for Disease Control and Prevention. This study was conducted with 187 live female patients with breast cancer of 32,631 participants who participated in the KNHANES for 3 years. The mean age of the subjects and mean age at diagnosis were post-menopausal. For more information about the physical characteristics of the subjects, please see Table 1.

Table 1. The characteristics of the study subjects.

Variables	Mean ± Standard Deviation
Breast cancer survivor (*n*)	187
Age (years)	60.8 ± 11.1
Breast cancer diagnosis age (years)	54.5 ± 11.3
Menarche age (years)	14.3 ± 1.8
Menopause age (years)	48.00 ± 4.9
Height (cm)	155.8 ± 6.1
Weight (kg)	57.5 ± 8.3

2.2. Physical Activity Assessment

The KNHANES physical activity levels were measured using 3 exercise categories: (1) aerobic exercise only (medium intensity aerobic activity of 150 min per week or high-intensity aerobic activity for 75 min per week); (2) resistance exercise only (>1 time per week); (3) walking exercise only

(1–2/times per week, 3–5/times per week, 6–7/times per week). Each participant could belong to multiple exercise categories.

2.3. *Activity Restriction*

The questions asked during the health interview survey regarding activity restrictions consisted of: the presence of activity restriction, causes of activity restriction, diseases, and experience of absenteeism. To ascertain activity restriction, 5 questions (discomfort in the last 2 weeks, disease in the past month, disease in the last year, absenteeism in the last month, absenteeism in the last year) were asked, and were designed to be answered with a "yes" or "no." In addition, the causes of activity restriction, the presence of obesity, diabetes, and anemia (dizziness), which are closely associated with breast cancer, were examined.

2.4. *Subjective Health Status and QoL*

For subjective health status, the question "How do you think your health is in normal times?" was asked, and answers were scored from 1 point for "Very Bad" to 5 points for "Very Good" using a 5-point Likert scale. The higher the score, the better the perceived health status by the subject. The EuroQoL-5 dimension (EQ-5D) developed by the EuroQoL Group was used to measure QoL related to overall health. It consists of 5 multiple-choice questions concerning exercise ability, self-care, daily activities, pain/discomfort, and anxiety/depression. Each of the 5 questions can be answered with 1 of 3 responses: "No problem at all", "There are some problems", and "There are serious problems". In this study, Cronbach's alpha for the instrument was 0.78.

2.5. *Blood Analysis*

Fasting blood samples from all participants in the Korea National Health and Nutrition Examination Survey were collected; white blood cells, red blood cells (RBCs), hemoglobin, platelets, hematocrit, and hs-C-reactive protein (CRP) levels were analyzed. For diabetes, stage 3 diabetes (normal, impaired glucose tolerance (IGT) and diabetes) was based on blood glucose after fasting for more than 8 h. Anemia was determined based on a hemoglobin level <12 g/dL.

2.6. *Ethics Statement*

The KNHANES was conducted as an interview survey in which the investigators interviewed the subjects and collect responses to the questions. In this study, the raw data from the seventh survey (2016–2018) that met the criteria of the study were downloaded from the KNHANES website (http://knhanes.cdc.go.kr/). In order to use the data, protocols for using the raw data from the KNHANES website were followed. Since the KNHANES is considered a public welfare study conducted by the Korean government, this study was conducted without the prior approval of the Research Ethics Review Committee.

2.7. *Statistical Analysis*

Phi coefficients or Cramer's V value were calculated using cross-analysis to determine whether there was a relationship between exercise participation, activity restriction, subjective health status, and QoL by exercise modality. An independent *t*-test was conducted to examine the differences between exercise modality participation by hematopoietic profile, and alpha (α) was set to 0.05. The reasons for the different case numbers for each variable is due to missing data from those who did not respond to the questionnaire. All analyses were conducted using SPSS version 18.0 (IBM Corp., Armonk, NY, USA).

3. Results

This study examined the relationships between exercise modality, activity restriction, subjective health status, QoL, and hematopoietic profile in breast cancer survivors who participated in the 2016–2018 KNHANES. The results of the cross-analysis, conducted to determine the correlation between exercise participation and activity restriction-related variants by exercise modality (aerobic exercise, resistance exercise, walking exercise) in the breast cancer survivors, are presented in Table 2. There were no statistically significant correlations between participation in various modalities of exercise and activity restriction (discomfort in the past 2 weeks, disease in the last month, disease in the last year, absenteeism in the last month, absenteeism in the last year) in the breast cancer survivors.

Among activity restriction due to disease, there was a statistically significant correlation between obesity and aerobic exercise participation ($p < 0.046$) and walking exercise frequency ($p < 0.029$). However, there was an exception; one subject who participated in aerobic and resistance exercises had a higher obesity rate than those who did not participate. There was also a significant correlation between diabetes and aerobic exercise participation at the level of $p < 0.038$. The subjects who participated in aerobic exercise showed a lower prevalence of diabetes compared to those who did not participate in aerobic exercise.

The results of the cross analysis, conducted to examine the correlation between subjective health status and QoL by exercise modality (aerobic exercise, resistance exercise, walking exercise) in breast cancer survivors, are shown in Table 3. There was a statistically significant correlation between subjective health status and resistance exercise participation at the level of $p < 0.180$. There was a statistically significant correlation between mobility, aerobic exercise participation, and resistance exercise participation at the levels of $p < 0.028$ and $p < 0.026$. There were also a statistically significant correlation between self-care and walking exercise frequency, and anxiety/depression and aerobic exercise participation at the levels of $p < 0.037$ and $p < 0.017$.

The independent *t*-test conducted to examine the effect of exercise participation on the hematopoietic profile by exercise modality in breast cancer survivors is presented in Table 4. The RBC was significantly higher at the level of $p < 0.028$ for those who participated in resistance exercise compared to those who did not.

Table 2. Correlation between exercise modality and activity restriction-related variants in breast cancer survivors.

		Aerobic Exercise % (Frequency)		Resistance Exercise % (Frequency)		Walking % (Frequency)		
		Yes	No	Yes	No	1–2/Wk.	3–5/Wk.	6–7/Wk.
Discomfort in the last 2 weeks	Yes	10.8% (20)	18.3% (34)	4.81% (9)	24.5% (46)	12.6% (15)	10.9% (13)	9.2% (11)
	No	32.3% (60)	38.7% (72)	13.9% (26)	56.9% (106)	21.0% (25)	27.7% (33)	18.5% (22)
	Phi coefficients	0.077		0.039			0.084	
Presence of disease in the last month	Yes	5.9% (11)	4.3% (8)	2.7% (5)	8.0% (15)	5.0% (6)	5.9% (7)	2.5% (3)
	No	37.1% (69)	53.7% (98)	16.0% (30)	73.2% (137)	28.6% (34)	32.8% (39)	25.2% (30)
	Phi coefficients	0.101		0.056			0.079	
Presence of disease in the last year	Yes	13.1% (8)	16.4% (10)	3.2% (2)	24.2% (16)	15.8% (6)	13.2% (5)	5.3% (2)
	No	39.3% (24)	31.1% (19)	24.6% (15)	45.9% (28)	13.2% (5)	23.7% (9)	28.9% (11)
	Phi coefficients	0.104		0.242			0.130	
Presence of absenteeism in the last month	Yes	5.3% (4)	2.7% (2)	2.7% (2)	5.3% (4)	3.8% (2)	3.8% (2)	3.8% (2)
	No	36.0% (27)	56.0% (42)	13.3% (10)	78.7% (59)	32.1% (17)	29.3% (15)	28.3% (15)
	Phi coefficients	−0.152		−0.101			0.019	
Presence of absenteeism in the last year	Yes	0.0% (0)	11.1% (3)	0.0% (0)	16.7% (5)	15.0% (3)	0.0% (0)	0.0% (0)
	No	37.0% (10)	51.9% (14)	10.0% (3)	63.3% (19)	25.0% (5)	25.0% (5)	35.0% (7)
	Phi coefficients	0.356		0.210			0.437	
Obesity	Underweight	0.5% (1)	2.2% (4)	0.0% (0)	3.2% (3)	0.8% (1)	2.3% (4)	0.0% (0)
	Normal	25.8% (48)	29.5% (55)	8.4% (8)	44.2% (42)	16.8% (20)	22.7% (27)	16.8% (20)
	Pre-obesity	11.8% (22)	9.1% (17)	5.3% (5)	15.8% (15)	5.9% (7)	5.9% (7)	10.0% (12)
	Obesity	4.8% (9)	16.1% (30)	2.1% (7)	21.1% (20)	10.1% (12)	6.7% (8)	0.8% (1)
	Cramer's V	0.246 *		0.158			0.290 *	
Presence of diabetes	Normal	25.5% (40)	36.3% (57)	14.5% (23)	47.7% (75)	18.4% (19)	30.1% (31)	22.3% (23)
	IGT	15.9% (25)	8.9% (14)	5.0% (8)	19.6% (31)	8.7% (9)	7.7% (8)	2.9% (3)
	Diabetes	5.1% (8)	8.3% (13)	0.6% (1)	12.7% (20)	2.9% (3)	3.9% (4)	2.9% (3)
	Cramer's V.	0.204 *		0.177			0.487	
Presence of anemia	Yes	39.1% (63)	50.3% (81)	7.5% (28)	72.5% (116)	26.7% (28)	35.2% (37)	30.5% (32)
	No	6.2% (10)	4.3% (7)	1.8% (3)	8.7% (14)	2.9% (3)	4.7% (5)	0.0% (0)
	Phi coefficients	0.093		0.159			0.141	

IGT: impaired glucose tolerance; * $p < 0.05$; cross-analysis.

Table 3. Correlation between exercise modality, subjective health status and QoL in breast cancer survivors.

			Aerobic Exercise % (Frequency)		Resistance Exercise % (Frequency)		Walking % (Frequency)		
			Yes	No	Yes	No	1–2/Wk.	3–5/Wk.	6–7/Wk.
Subjective health status		Very Good	2.7% (5)	1.1% (2)	1.1% (2)	2.7% (5)	0.0% (0)	0.0% (0)	0.0% (0)
		Good	7.0% (13)	7.0% (15)	1.6% (3)	13.4% (25)	4.2% (5)	8.4% (10)	3.4% (4)
		Normal	25.3% (47)	28.5% (53)	14.4% (27)	39.0% (73)	16.8% (20)	18.5% (22)	15.7% (19)
		Bad	5.4% (10)	12.4% (23)	1.1% (2)	16.6% (31)	6.7% (6)	8.4% (10)	5.9% (7)
		Very bad	2.7% (5)	7.0% (13)	0.5% (1)	9.6% (18)	7.6% (9)	3.4% (4)	2.5% (3)
		Cramer's V	0.194		0.252 *			0.401	
QoL	Mobility	No difficulty walking	37.0% (69)	40.3% (75)	17.6% (33)	59.4% (111)	22.7% (27)	31.0% (37)	21.8% (26)
		Walking is a bit difficult	5.9% (11)	15.1% (28)	1.1% (2)	19.8% (37)	7.6% (9)	7.6% (9)	5.8% (7)
		Lying all day	0.0% (0)	1.6% (3)	0.0% (0)	2.1% (4)	3.4% (4)	0.0% (0)	0.0% (0)
		Cramer's V	0.196 *		0.198 *			0.189	
	Self-care	No problems taking a bath/dressing	53.2% (99)	42.5% (79)	18.7% (35)	76.5% (143)	28.6% (34)	37.8% (45)	26.9% (32)
		Taking a bath/dressing somewhat hindered	3.2% (8)	0.5% (1)	0.0% (0)	4.3% (8)	5.0% (6)	0.8% (1)	0.8% (1)
		Not able to take a bath/dress alone	0.5% (1)	0.0% (0)	0.0% (0)	0.5% (1)	0.0% (0)	0.0% (0)	0.0% (0)
		Cramer's V	0.197		0.108			0.037 *	
	Usual activities	No ADL disruption	38.1% (71)	48.9% (91)	17.6% (33)	68.9% (129)	25.2% (30)	35.3% (42)	24.4% (29)
		ADL is somewhat impeded	4.8% (9)	8.0% (15)	1.0% (2)	12.3% (23)	8.4% (10)	3.3% (4)	3.3% (4)
		Phi coefficients	−0.043		−0.108			0.093	
	Pain/ discomfort	No pain/ discomfort	27.9% (52)	37.6% (70)	13.9% (26)	51.3% (96)	18.5% (22)	27.7% (33)	17.6% (21)
		Some pain/ discomfort	13.4% (25)	16.3% (30)	4.3% (8)	25.7% (48)	10.9% (13)	10.1% (12)	7.5% (9)
		Severe pain/ discomfort	1.6% (3)	3.2% (6)	0.5% (1)	4.3% (8)	4.2% (5)	0.8% (1)	2.5% (3)
		Cramer's V	0.785		0.093			0.136	
	Anxiety/ depression	No anxiety/depression	39.2% (73)	45.1% (84)	66.3% (33)	66.3% (124)	26.0% (31)	34.4% (41)	23.5% (28)
		Some anxiety/ depression	2.1% (4)	10.7% (20)	1.1% (2)	12.3% (23)	5.9% (7)	4.2% (5)	2.5% (3)
		Severe anxiety/ depression	1.6% (3)	1.0% (2)	0.0% (0)	2.7% (5)	1.7% (2)	0.0% (0)	1.7% (2)
		Cramer's V	0.209 *		0.138			0.132	

* $p < 0.05$, tested by cross-analysis.

Table 4. Differences between exercise modality participation in relation to hemodynamic variables in breast cancer survivors.

Variables	Aerobic Exercise % (Frequency)		Resistance Exercise % (Frequency)		Walking % (Frequency)		
	Yes	No	Yes	No	1–2/Wk.	3–5/Wk.	6–7/Wk.
White blood cell (Thous/uL)	5.51 ± 1.30	5.57 ± 1.65	5.42 ± 1.71	5.54 ± 1.49	5.86 ± 1.92	5.12 ± 1.35	5.60 ± 1.22
Red blood cell (Mil/uL)	4.27 ± 0.40	4.29 ± 0.33	4.41 ± 0.38 *	4.25 ± 0.35	4.30 ± 0.37	4.33 ± 0.36	4.40 ± 0.38
Hemoglobin (g/dL)	12.97 ± 0.97	13.19 ± 0.87	12.81 ± 2.50	13.05 ± 0.91	13.1 ± 1.00	13.0 ± 0.93	13.35 ± 0.90
Platelets (Thous/uL)	246.34 ± 64.85	233.15 ± 62.36	242.65 ± 71.42	237.75 ± 64.73	220.45 ± 53.27	240.45 ± 65.52	252.00 ± 73.49
Hematocrit (%)	39.70 ± 3.00	40.36 ± 2.77	39.94 ± 5.48	39.86 ± 2.84	40.08 ± 3.06	40.09 ± 2.95	40.84 ± 3.03
C-reactive protein (mg/L)	1.55 ± 2.75	1.14 ± 3.01	1.18 ± 2.07	0.73 ± 0.30	0.93 ± 0.92	0.78 ± 0.86	1.42 ± 3.71

* $p < 0.05$; tested by an independent t-test.

4. Discussion

Breast cancer is affected by genetic and environmental factors, such as menarche, menopause, childbirth, and lactation experience; it is reported that a Western diet and inactive lifestyle increase the incidence [19]. Women who have undergone surgery because of the development of breast cancer, ovarian cancer, and uterine cancer may develop depression because they feel deprived of femininity, which may lead to family or social problems. In addition, it has been reported that the risk of myopathy, osteoporosis, and cardiovascular disease increases in breast cancer survivors [7].

Cancer survivors are recommended to participate in various modalities of exercise to prevent daily fatigue and cancer recurrence. In this study, there was no significant correlation between exercise participation and activity restriction-related discomfort or disease, and absenteeism for the last 2 weeks; however, there was a correlation between obesity and diabetes and activity restriction. Specifically, obesity and diabetes were significantly correlated with aerobic exercise participation and walking exercise frequency in breast cancer survivors. In this study, aerobic exercise and walking exercise showed a significantly positive correlation. It is suggested that aerobic exercise and walking (6–8 repetitions per week) are good solutions for obesity in breast cancer survivors. In addition, diabetes showed a correlation with aerobic exercise, showing that participation in aerobic exercise has a lower prevalence of diabetes compared with no participation in aerobic exercise. A meta-analysis conducted by Protani et al. [20], reported that the risk of cancer recurrence or death was 30% higher in breast cancer survivors who were obese than in breast cancer survivors of normal weight. It has also been reported that excess fat tissue caused by obesity increases the recurrence rate of breast cancer [21], and aerobic exercise (walking exercise) reduces the size of fat cells [22] and improves immune function [23]. However, increased fatigue due to a rapid increase in the level of activity may lower the immunity in patients with cancer, so care must be taken during exercise [24]. In addition, patients with breast cancer tend to lose muscle strength because of changes in body composition during anticancer treatment; resistance exercise has a positive effect on maintaining body composition and strength [25], indicating that breast cancer survivors need to participate in various modalities of exercise to further reduce cancer-related risk factors and prevent concurrent diseases.

In this study, there was a significant correlation between resistance exercise participation and subjective health status in breast cancer survivors. With respect to QoL, mobility and anxiety/depression were significantly correlated with aerobic exercise participation self-care, and walking exercise frequency. These results are similar to those of a study that reported a significant increase in QoL, fatigue, and depression symptoms after cancer survivors participated in exercise [26].

Regarding the examination of the relationship between exercise modality and the hematopoietic profile of the breast cancer survivors in this study, RBC significantly increased depending upon weekly resistance exercise participation; thus, those who participated in resistance exercise had higher RBC counts than those who did not. An increase in RBC count is closely associated with the prevalence of anemia. The blood cells of patients with cancer do not pass through blood vessels because of the deformation of red blood cells, which forms congestion and causes anemia. This phenomenon has been reported in more than 40–64% of patients with cancer [27,28]. Anemia can cause dizziness, weakness, and fatigue in everyday life, which may, in turn, deteriorate the QoL and restrict the activities of breast cancer survivors [29]. Although not statistically significant, it was found that the prevalence of anemia was higher in those who participated in all modality of exercise than in those who did not, as shown in Table 1, which is considered to be closely associated with increased RBCs, even though they were within the normal range. Mohamady et al. [30] and Drouin et al. [31] reported that participation in a 7-week exercise program prevented the increase in RBC and hemoglobin in patients with breast cancer who were undergoing radiation therapy. It has also been reported that exercise improves systemic inflammation in cancer survivors [32–35].

However, in this study, the inflammatory index hs-CRP was within the normal range for all exercise modality, and there was no difference. The results of this study found that participation in physical activities (aerobic exercise, resistance exercise, walking exercise) lowered the prevalence of obesity and diabetes affecting patients with breast cancer. Physical activity participation improved subjective health status and exercise ability, and reduced depression and anxiety, thus improving the quality of life of breast cancer survivors in Korea. Among the modalities of exercise assessed, aerobic exercise had a greater positive correlation, indicating that it may be more effective.

There are several limitations to this study. First, the amount of exercise participation was not directly measured by objective observance, but surveyed indirectly by using a questionnaire. Second, there was a lack of a methodological approach for measuring the proper amount of exercise according to the grade of breast cancer and cancer therapy method, suggesting the necessity of a follow-up study. Third, this study was conducted only with patients with breast cancer; thus, the findings cannot be generalized to other cancer patients. Fourth, additional physical activity evaluations, such as activities of daily living or instrumental activities of daily living, were not conducted, suggesting the necessity for a further study with additional variables. However, combining resistance exercise and aerobic exercise to lessen muscle weakening is recommended. The habit of performing exercise on a regular basis is considered most important for breast cancer survivors for the prevention of cancer recurrence and for cancer recovery.

5. Conclusions

This study found that, for breast cancer survivors, participation in physical activity, such as aerobic exercise, resistance exercise, and walking exercise may lower the prevalence of diseases such as obesity and diabetes. Furthermore, physical activity can reduce depression and anxiety and improve subjective health status, exercise ability, and quality of life. In particular, aerobic exercise was shown to be effective in positively affecting a number of variables, but resistance training is also recommended to prevent muscle loss. The effort to establish regular exercise habits, regardless of modality, seems to be important for the mental and physiological health of breast cancer survivors.

Author Contributions: Study design: M.K., J.K., and W.-Y.S., Study conduct: M.K., J.K., and W.-Y.S., Data collection: M.K., J.K., and W.-Y.S., Data analysis: M.K., J.K., and W.-Y.S., Data interpretation: M.K., J.K., and W.-Y.S., Drafting manuscript: M.K., J.K., and W.-Y.S., Revising the manuscript content: M.K., J.K., and W.-Y.S. All authors have read and agreed to the published version of the manuscript.

Funding: This work was supported by the Ministry of Education of the Republic of Korea and the National Research Foundation of Korea (NRF-2019S1A5B6102784).

Conflicts of Interest: The authors declare no conflict of interest.

References

1. Korea Ministry of Health and Welfare. Korea Central Cancer Registry. Korea Ministry of Health and Welfare. 2019. Available online: http://www.cancer.go.kr (accessed on 20 September 2020).
2. Vahdaninia, M.; Omidvari, S.; Montazeri, A. What do predict anxiety and depression in breast cancer patients? A follow-up study. *Soc. Psychiatr. Psychiatr. Epidemiol.* **2010**, *45*, 355–361. [CrossRef] [PubMed]
3. Van Wyk, J.; Carbonatto, C. The social functioning of women with breast cancer in the context of the life world: A social work perspective. *Soc. Work* **2016**, *52*, 439–458. [CrossRef]
4. Otto, A.K.; Szczesny, E.C.; Soriano, E.C.; Laurenceau, J.P.; Siegel, S.D. Effects of a randomized gratitude intervention on death-related fear of recurrence in breast cancer survivors. *Health Psychol.* **2016**, *35*, 1320–1328. [CrossRef] [PubMed]
5. Howard-Anderson, J.; Ganz, P.A.; Bower, J.E.; Stanton, A.L. Quality of life fertility concerns and behavioral health outcomes in younger breast cancer survivors: A systematic review. *J. Natl. Cancer Inst.* **2012**, *104*, 386–405. [CrossRef]
6. Damodar, G.; Smitha, T.; Gopinath, S.; Vijayakumar, S.; Rao, Y.A. Assessment of quality of life in breast cancer patients at a tertiary care hospital. *Arch. Pharma Pract.* **2013**, *4*, 1. [CrossRef]
7. Ording, A.G.; Garne, J.P.; Nystrom, P.M.; Froslev, T.; Sorensen, H.T.; Lash, T.L. Comorbid diseases interact with breast cancer to affect mortality in the first year after diagnosis. A Danish nationwide matched cohort study. *PLoS ONE* **2013**, *9*, e76013. [CrossRef]

8. Su, M.; Hua, X.; Wang, J.; Yao, N.; Zhao, D.; Liu, W.; Zou, Y.; Anderson, R.; Sun, X. Health-related quality of life among cancer survivors in rural China. *Qual. Life Res.* **2019**, *28*, 695–702. [CrossRef]
9. Seo, J.S.; Hyun-Ah Park, H.A.; Kang, J.H.; Kim, K.W.; Young-Gyu Cho, Y.G.; Hur, Y.I.; Park, Y.R. Obesity and obesity-related lifestyles of korean breast cancer survivors. *Korean J. Health Promot.* **2014**, *14*, 93–102. [CrossRef]
10. Kang, C.; LeRoith, D.; Gallagher, E.J. Diabetes obesity and breast cancer. *Endocrinology* **2018**, *1159*, 3801–3812. [CrossRef]
11. Lamy, P.J.; Durigova, A.; Jacot, W. Iron homeostasis and anemia markers in early breast cancer. *Clin. Chim. Acta* **2014**, *1434*, 34–40. [CrossRef]
12. Tas, F.; Eralp, Y.; Basaran, M.; Alici, S.; Argon, A.; Bulutlar, G.; Camlica, H.; Aydiner, A.; Topuz, E. Anemia in oncology practice: Relation to diseases and their therapies. *Am. J. Clin. Oncol.* **2002**, *25*, 371–379. [CrossRef] [PubMed]
13. Escoffery, C.; Rodgers, K.; Kegler, M.C.; Haardörfer, R.; Howard, D.; Roland, K.B.; Wilson, K.M.; Castro, G.; Rodriguez, J. Key informant interviews with coordinators of special events conducted to increase cancer screening in the United States. *Health Educ. Res.* **2014**, *29*, 730–739. [CrossRef] [PubMed]
14. Eguchi, Y.; Hyogo, H.; Ono, M.; Mizuta, T.; Ono, N.; Fujimoto, K.; Chayama, K.; Saibara, T. JSG-NAFLD. Prevalence and associated metabolic factors of nonalcoholic fatty liver disease in the general population from 2009 to 2010 in Japan: A multicenter large retrospective study. *J. Gastroenterol.* **2012**, *47*, 586–595. [CrossRef] [PubMed]
15. Cho, K.K.; Kim, Y.H.; Kim, Y.H. Association of fitness, body circumference, muscle mass, and exercise habits with metabolic syndrome. *J. Mens Health* **2019**, *15*, 46–55.
16. Brown, J.C.; Huedo-Medina, T.B.; Pescatello, L.S.; Pescatello, S.M.; Ferrer, R.A.; Johnson, B.T. Efficacy of exercise interventions in modulating cancer-related fatigue among adult cancer survivors: A meta-analysis. *Cancer Epidemiol. Biomark. Prev.* **2011**, *20*, 123–133. [CrossRef]
17. Mishra, S.I.; Scherer, R.W.; Geigle, P.M.; Berlanstein, D.R.; Topaloglu, O.; Gotay, C.C.; Snyder, C. Exercise interventions on health-related quality of life for cancer survivors. *Cochrane Database Syst Rev.* **2012**, *8*, CD007566. [CrossRef]
18. Szymlek-Gay, E.A.; Richards, R.; Egan, R. Physical activity among cancer survivors: A literature review. *N. Z. Med. J.* **2011**, *124*, 77–89.
19. Nkondjock, A.; Ghadirian, P. Risk factors and risk reduction of breast cancer. *Med. Sci.* **2005**, *21*, 175–180.
20. Protani, M.; Coory, M.; Martin, J.H. Effect of obesity on survival of women with breast cancer: Systematic review and meta-analysis. *Breast Cancer Res. Treat.* **2010**, *123*, 627–635. [CrossRef]
21. Vona-Davis, L.C.; Rose, D.P. Angiogenesis, adipokines and breast cancer. *Cytokine Growth Factor Rev.* **2009**, *20*, 193–201. [CrossRef]
22. Bruno, E.; Roveda, E.; Vitale, J.; Montaruli, A.; Berrino, F.; Villarni, A.; Venturelli, E.; Gargano, G.; Galasso, L.; Caumo, A.; et al. Effect of aerobic exercise intervention on markers of insulin resistance in breast cancer women. *Eur. J. Cancer Care* **2018**, *27*, e12617. [CrossRef] [PubMed]
23. Kreutz, C.; Schmidt, M.E.; Steindorf, K. Effects of physical and mind-body exercise on sleep problems during and after breast cancer treatment: A systematic review and meta- analysis. *Breast Cancer Res. Treat.* **2019**, *176*, 1–15. [CrossRef] [PubMed]
24. Meneses-Echávez, J.F.; González-Jiménez, E.; Ramírez-Vélez, R. Effects of supervised exercise on cancer-related fatigue in breast cancer survivors: A systematic review and meta-analysis. *BMC Cancer* **2015**, *15*, 77. [CrossRef] [PubMed]
25. Hasenoehrl, T.; Keilani, M.; Palma, S.; Crevenna, R. Resistance exercise and breast cancer related lymphedema—A systematic review update. *Support Care Cancer* **2019**, *42*, 26–35. [CrossRef] [PubMed]
26. Cramp, F.; Byron-Daniel, J. Exercise for the management of cancer-related fatigue in adults. *Cochrane Database Syst. Rev.* **2012**, *14*, CD006145. [CrossRef]
27. Steinberg, D. Anemia and cancer. *CA Cancer J. Clin.* **1989**, *39*, 296–304. [CrossRef]
28. Gaspar, B.L.; Sharma, P.; Das, R. Anemia in malignancies: Pathogenetic and diagnostic considerations. *Hematology* **2015**, *20*, 18–25. [CrossRef] [PubMed]
29. Hess, L.M.; Insel, K.C. Chemotherapy-related change in cognitive function: A conceptual model. *Oncol. Nurs. Forum* **2007**, *34*, 981–994. [CrossRef]

30. Mohamady, H.M.; Elsisi, H.F.; Aneis, Y.M. Impact of moderate intensity aerobic exercise on chemotherapy-induced anemia in elderly women with breast cancer a randomized controlled clinical trial. *J. Adv. Res.* **2017**, *8*, 7–12. [CrossRef]
31. Drouin, J.S.; Young, T.J.; Beeler, J.; Byrne, K.; Birk, T.J.; Hryniuk, W.M.; Hryniuk, L.E. Random control clinical trial on the effects of aerobic exercise training on erythrocyte levels during radiation treatment for breast cancer. *Cancer* **2006**, *107*, 2490–2495. [CrossRef]
32. Allgayer, H.; Nicolaus, S.; Schreiber, S. Decreased interleukin-1 receptor antagonist response following moderate exercise in patients with colorectal carcinoma after primary treatment. *Cancer Detect Prev.* **2004**, *28*, 208–213. [CrossRef]
33. Hutnick, N.A.; Williams, N.I.; Kraemer, W.J.; Orsega-Smith, E.; Dixon, R.H.; Bleznak, A.D.; Mastro, A.M. Exercise and lymphocyte activation following chemotherapy for breast cancer. *Med. Sci. Sports Exerc.* **2005**, *37*, 1827–1835. [CrossRef] [PubMed]
34. Singh, M.P.; Singh, G.; Singh, S.M. Role of host's antitumor immunity in exercise-dependent regression of murine T-cell lymphoma. *Comp. Immunol. Microbiol. Infect Dis.* **2005**, *28*, 231–248. [CrossRef] [PubMed]
35. Meneses-Echávez, J.F.; Ramírez Vélez, R.; González Jiménez, E.; Schmidt Río-Valle, J.; Sánchez Pérez, M.J.; Molina Montes, E. Exercise training, inflammatory cytokines, and other markers of low-grade inflammation in breast cancer survivors: A systematic review and meta-analysis. *J. Clin. Oncol.* **2017**, *32*, 121. [CrossRef]

International Journal of Environmental Research and Public Health

Int. J. Environ. Res. Public Health **2020**, *17*, 6604

Article

Body Water Content and Morphological Characteristics Modify Bioimpedance Vector Patterns in Volleyball, Soccer, and Rugby Players

Francesco Campa [1,*], Analiza M. Silva [2], Catarina N. Matias [2], Cristina P. Monteiro [2], Antonio Paoli [3], João Pedro Nunes [4], Jacopo Talluri [5], Henry Lukaski [6] and Stefania Toselli [7]

1 Department for Life Quality Studies, University of Bologna, 47921 Rimini, Italy
2 Exercise and Health Laboratory, CIPER, Faculdade Motricidade Humana, Universidade de Lisboa, 1499-002 Cruz Quebrada, Portugal; analiza@fmh.ulisboa.pt (A.M.S.); cmatias@fmh.ulisboa.pt (C.N.M.); cmonteiro@fmh.ulisboa.pt (C.P.M.)
3 Department of Biomedical Science, University of Padova, 35100 Padova, Italy; antonio.paoli@unipd.it
4 Metabolism, Nutrition, and Exercise Laboratory, Physical Education and Sports Center, Londrina State University, 86057 Londrina, Brazil; joaonunes.jpn@hotmail.com
5 Department of clinical research and development, Akern Ltd., 56121 Pisa, Italy; jacopo.talluri@akern.com
6 Department of Kinesiology and Public Health Education, Hyslop Sports Center, University of North Dakota, Grand Forks, ND 58202, USA; henry.lukaski@und.edu
7 Department of Biomedical and Neuromotor Sciences, University of Bologna, 40126 Bologna, Italy; stefania.toselli@unibo.it
* Correspondence: francesco.campa3@unibo.it

Received: 3 August 2020; Accepted: 9 September 2020; Published: 10 September 2020

Abstract: Background: Bioimpedance vector analysis (BIVA) is a widely used method based on the interpretation of raw bioimpedance parameters to evaluate body composition and cellular health in athletes. However, several variables contribute to influencing BIVA patterns by militating against an optimal interpretation of the data. This study aims to explore the association of morphological characteristics with bioelectrical properties in volleyball, soccer, and rugby players. **Methods:** 164 athletes belonging to professional teams (age 26.2 ± 4.4 yrs; body mass index (BMI) 25.4 ± 2.4 kg/m^2) underwent bioimpedance and anthropometric measurements. Bioelectric resistance (R) and reactance (Xc) were standardized for the athlete's height and used to plot the vector in the R-Xc graph according to the BIVA approach. Total body water (TBW), phase angle (PhA), and somatotype were determined from bioelectrical and anthropometric data. **Results:** No significant difference ($p > 0.05$) for age and for age at the start of competition among the athletes was found. Athletes divided into groups of TBW limited by quartiles showed significant differences in the mean vector position in the R-Xc graph ($p < 0.001$), where a higher content of body fluids resulted in a shorter vector and lower positioning in the graph. Furthermore, six categories of somatotypes were identified, and the results of bivariate and partial correlation analysis highlighted a direct association between PhA and mesomorphy ($r = 0.401$, $p < 0.001$) while showing an inverse correlation with ectomorphy ($r = -0.416$, $p < 0.001$), even adjusted for age. On the contrary, no association was observed between PhA and endomorphy ($r = 0.100$, $p = 0.471$). **Conclusions:** Body fluid content affects the vector length in the R-Xc graph. In addition, the lateral displacement of the vector, which determines the PhA, can be modified by the morphological characteristics of the athlete. In particular, higher PhA values are observed in subjects with a high mesomorphic component, whereas lower values are found when ectomorphy is dominant.

Keywords: body composition; BIVA; phase angle; R-Xc graph; somatotype; total body water; vector length

1. Introduction

In recent years, the bioimpedance vector analysis (BIVA) has been widely used in the sports field for the assessment of body composition and cellular health in athletes [1]. This is because BIVA is not subject to errors related to prediction equations since it interprets the raw bioimpedance values [resistance (R) and reactance (Xc)], and it is an easy to use and non-invasive method. Its application is based on the bivariate interpretation of R and Xc standardized for height on a graph. Vector displacements identify increases or losses in total body water (TBW) or in the ratio between intra (ICW) and extracellular (ECW) fluids for which increases or decreases correspond to shifts to the left or to the right of the R-Xc graph, respectively [2]. In addition, it is possible to obtain an immediate analysis of the subject's body composition by comparing the vector position with tolerance ellipses built from the data of the reference population [3–6].

Recent studies have focused on evaluating factors influencing the vector position of athletes, including maturity status, [7–10] dehydration [11–14], and fitness level [15]. Since the bioelectric properties of body tissues depend on body fluids and cells' membrane integrity [16], the main determinant of the vector position in the R-Xc graph is the TBW and the distribution of the fluids among the two compartments (ICW and ECW). In fact, the vector length is inversely proportional to TBW, while the lateral displacements of the vector are directly correlated with the ICW/ECW ratio [17–20]. In addition to these body composition variables, Campa et al. [21] have recently suggested that the somatotype also influences vector position in the R-Xc graph, where athletes with higher mesomorphic and endomorphic components are positioned more to the lower-left than athletes with a dominant ectomorphy. However, as the content of total body fluids is greatly associated with the vector, correct discrimination in the R-Xc graph based on somatotype categories may be compromised. Indeed, athletes with high body weight can still be located at the bottom of the graph regardless of their morphology.

The bioelectric values and the vector position reflect body composition; in particular, the phase angle (PhA), obtained as the arctangent of Xc/R, correctly mirrors the ICW/ECW ratio [17–20]. In fact, athletes with a high PhA are positioned to the left portion in the R-Xc graph [21,22], and increases in muscle mass, body cell mass, and, therefore, ICW lead to vector shifts further to the left over time [23,24]. A high muscularity is observed in subjects with a high PhA or in those whose somatotype shows a dominant mesomorphic component [21,22], as mesomorphy characterizes skeletal muscle features [25]; moreover, both PhA and somatotype can be modified with nutrition and exercise [23,26]. However, to the best of our knowledge, no study has explored the associations between somatotype and PhA, while analyzing the influence of morphology on the vector position for similar TBW values in athletes.

Therefore, this study aimed to analyze the associations of morphological characteristics with bioelectrical properties using the BIVA approach, according to different levels of body water content in male volleyball, soccer, and rugby players. Our hypothesis was that PhA was associated with morphological characteristics in the athletes.

2. Methods

2.1. Subjects

This was a cross-sectional observational study conducted in 164 athletes engaged in 7 professional Italian teams participating in Series A2, Series B, and Series A divisions of volleyball, soccer, and rugby, respectively (age 26.2 ± 4.4 yrs; body mass index (BMI) 25.4 ± 2.4 kg/m^2; age at start competition

14.2 ± 1.3 yrs). The following inclusion criteria were used: (i) A minimum of 10 h of training per week; (ii) tested negative for performance-enhancing drugs; and (iii) not taking any medication. The athletes were tested in the morning (9:00 AM) in the facilities of the teams. All measurements were performed under resting conditions in the second off-season period. All participants gave informed consent after receiving a detailed description of the study procedures. The project was conducted in accordance with the guidelines of the Declaration of Helsinki and was approved by the local Bioethics Committee of the University of Bologna. (Ethical Approval Code: 25027).

2.2. Procedures

All athletes were tested to ensure a well-hydrated state using the urine specific gravity test (refractometer Urisys 1100; Roche Diagnostics), according to Armstrong et al. [27]. A urine specific gravity value < 1.022 for the first urine was used to identify an euhydration state.

The anthropometric traits were body mass, height, humerus and femur breadths, contracted arm and calf girths, and 4-skinfold thicknesses (triceps, subscapular, supraspinal, and medial calf). All anthropometric measurements were taken by a certified anthropometrist according to standard methods in the literature [28], whose technical error was 5% and 1.5% for skinfolds and all other measurements, respectively. Height was recorded to the nearest 0.1 cm using a stadiometer (Raven Equipment Ltd., Great Donmow, UK) and body mass was measured to the nearest 0.1 kg using a high-precision mechanical scale (Seca, Basel, Switzerland). BMI was calculated as the ratio of body weight to height squared (kg/m^2). Girths were taken to the nearest 0.1 cm using a tape measure (GMP, Zürich, Switzerland). Breadths were measured to the nearest 0.1 cm using a sliding caliper (GMP, Zürich, Switzerland). Skinfold thicknesses were measured to the nearest 0.1 mm using a Lange skinfold caliper (Beta technology Inc., Cambridge, MD, USA).

Bioimpedance analysis (BIA) was performed by a phase-sensitive single-frequency bioimpedance analyzer (101 Anniversary, Akern, Florence, Italy), which applied an alternating current of 400 microamperes at 50 kHz. Vector length (VL) was calculated as (adjusted R^2 + adjusted $Xc^2)^{0.5}$ and PhA as the arctangent of Xc/R × 180/π. BIVA was applied to normalize V, R, and Xc for height (H) in meters [29]. TBW was calculated from bioimpedance values, according to specific equations developed for athletes using a 4-compartment model as a criterion method [30] then the athletes were divided into quartiles.

Somatotype components were calculated according to the Heath and Carter method [25] as follow:

Endomorphy = − 0.7182 + 0.1451 (X) − 0.00068 (X 2) + 0.0000014 (X 3), where X = (sum of triceps, subscapular and supraspinal skinfolds) multiplied by (170.18/H in cm);

Mesomorphy = 0.858 × humerus breadth + 0.601 × femur breadth + 0.188 × corrected arm girth + 0.161 × corrected calf girth − H 0.131 + 4.5;

Ectomorphy = 0.732 HWR − 28.58, where HWR = (height divided by the cube root of weight).

From the 13 initial proposed categories by Heath and Carter [25], the athletes were grouped in 6 somatotype categories:

- Endomorphic mesomorph (EnM): Mesomorphy is dominant and endomorphy is greater than ectomorphy.
- Balanced mesomorph (BM): Mesomorphy is dominant and endomorphy and ectomorphy are equal.
- Ectomorphic mesomorph (EcM): Mesomorphy is dominant and ectomorphy is greater than endomorphy.
- Mesomorph-ectomorph (M-Ec): Mesomorphy and ectomorphy are equal, and endomorphy is smaller.
- Mesomorphic ectomorph (MEc): Ectomorphy is dominant and mesomorphy is greater than endomorphy.
- Balanced ectomorph (Bec): Ectomorphy is dominant and endomorphy and mesomorphy are equal.

2.3. Statistical Analysis

To verify the normality of the data, the Shapiro-Wilk test was applied. The athletes were divided into groups limited by quartiles of TBW and the one-way ANOVA was performed to evaluate the difference in BIVA patterns (PhA and VL/H). When a significant F ratio was obtained, the Bonferroni post hoc test was used to assess the differences between the 4 groups, setting the significance at $p < 0.008$. The two-sample Hotelling's T^2 test was used to compare the mean impedance vectors among the athletes grouped according to quartiles of TBW. Bivariate and partial (controlling for age) correlations were performed to evaluate the associations between PhA and the somatotype components. The mean standard deviation was calculated for each variable. Data were analyzed with IBM SPSS Statistics, version 24.0 (IBM Corp., Armonk, NY, USA).

3. Results

No significant difference ($p > 0.05$) for age and for age at start of competition among the athletes was found. The soccer, volleyball, and rubgy players showed an average EcM (endomorphy: 1.6 ± 0.3; mesomorphy: 4.7 ± 0.9; ectomorphy: 2.9 ± 0.8), EcM (endomorphy: 2.0 ± 0.7; mesomorphy: 4.0 ± 1.3; ectomorphy: 3.2 ± 1.1), and EnM (endomorphy: 2.1 ± 0.7; mesomorphy: 6.0 ± 1.1; ectomorphy: 0.9 ± 0.3) somatotype, respectively (Figure 1).

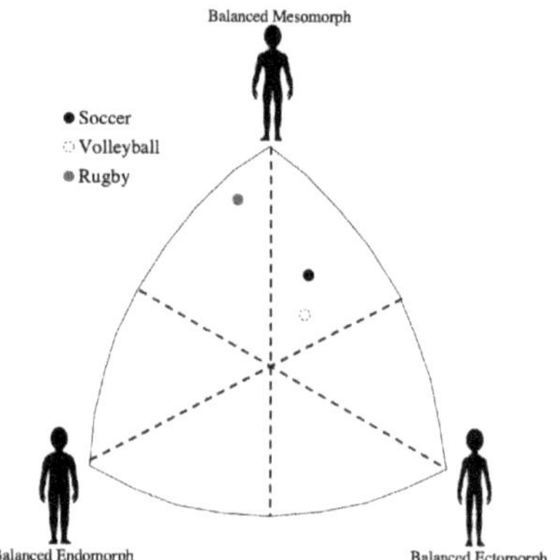

Figure 1. Representation of the athletes' somatotype.

Descriptive body fluids and bioelectrical characteristics are presented in Table 1, while the mean impedance vectors of the athletes divided according to quartiles of TBW are shown in Figure 1. Forty-two athletes were included in the first group (Q1) (endomorphy: 1.8 ± 0.6, mesomorphy: 4.4 ± 1.1, ectomorphy: 2.9 ± 1.0), 40 in the second group (Q2) (endomorphy: 2.2 ± 0.7, mesomorphy: 4.6 ± 1.5, ectomorphy 2.8 ± 1.3), 41 in the third group (Q3) (endomorphy: 2.2 ± 0.7, mesomorphy: 4.3 ± 1.4, ectomorphy 2.8 ± 1.3) and 41 in the fourth group (Q4) (endomorphy: 2.4 ± 0.7, mesomorphy: 5.2 ± 1.6, ectomorphy 2.0 ± 1.5). Six somatotype categories were identified, and their absolute frequencies for each group are presented in Figure 2. The results of the two-sample Hotelling t^2 test showed significant differences between all the groups (Q1 vs. Q2, $t = 21.1$, $p < 0.001$; Q1 vs. Q3, $t = 105.8$, $p < 0.001$; Q1 vs. Q4, $p < 0.001$; $t = 201.6$, $p < 0.001$; Q2 vs. Q3, $t = 39.4$, $p < 0.001$; Q2 vs. Q4, $t = 98.1$, $p < 0.001$; Q3 vs. Q4, $t = 22.7$, $p < 0.001$) indicating that the athletes with higher TBW were positioned to the lower left in the R-Xc graph than those with a lower TBW, as displayed in Figure 2. In addition, significant differences ($p < 0.008$) were found between the 4 groups for VL/H but not for PhA, as reported in Table 2.

Figure 3 illustrates the mean vectors of the athletes subdivided by somatotype in each TBW group. For each TBW group, somatotype categories with a dominant mesomorphy (EnM, BM, and EcM) showed a vector tending to be positioned more to the left than those with a greater ectomorphy

(MEc and BEc). Moreover, as displayed in Figure 4, PhA was directly correlated with the mesomorphic component ($r = 0.401$, $p < 0.001$; Panel A) and inversely with the ectomorphic component ($r = -0.416$, $p < 0.001$; Panel B), even when corrected for age ($p < 0.001$). On the contrary, no association was observed between PhA and endomorphy ($r = 0.100$, $p = 0.471$; Panel C).

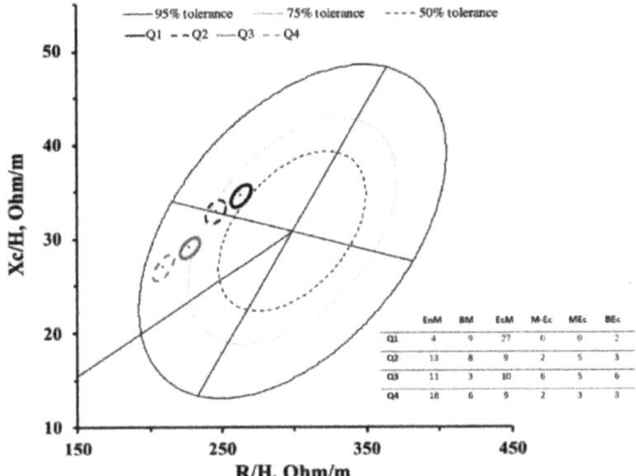

	EnM	BM	EcM	M-Ec	MEc	BEc
Q1	4	9	27	0	0	2
Q2	13	8	9	2	5	3
Q3	11	3	10	6	5	6
Q4	18	6	9	2	3	3

Figure 2. Scattergram of the mean impedance vectors of the athletes divided by the total body water groups plotted on the 50%, 75%, and 95% tolerance ellipses of the healthy male Italian population [31]; on the right side, the absolute frequencies of the somatotype categories for each quartile is shown. EnM = Endomorphic Mesomorph, BM = Balanced Mesomorph, EcM = Ectomorphic Mesomorph, M-Ec = Mesomorph Ectomorph, MEc = Mesomorphic Ectomorph, BEc = Balanced Ectomorph.

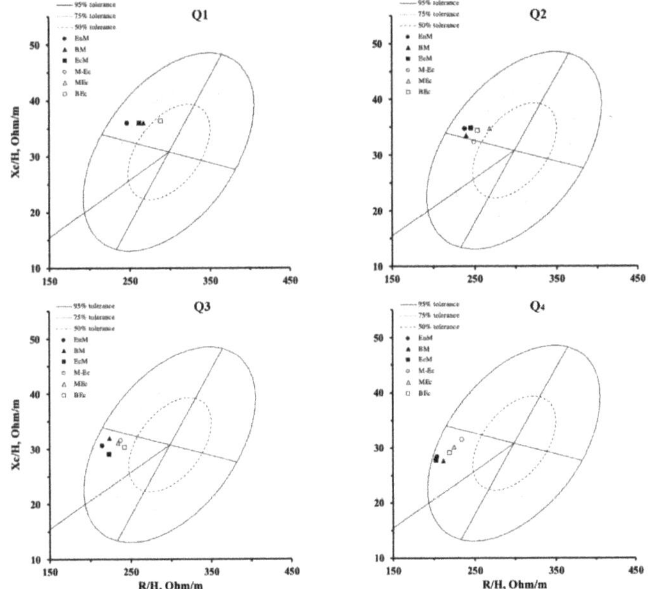

Figure 3. Scattergram of the mean impedance vectors of the athletes are categorized by somatotype and divided according to groups of TBW. EnM = Endomorphic Mesomorph, BM = Balanced Mesomorph, EcM = Ectomorphic Mesomorph, M-Ec = Mesomorph Ectomorph, MEc = Mesomorphic Ectomorph, BEc = Balanced Ectomorph.

Table 1. Descriptive statistics for the athletes divided by total body water quartile [first (Q1), second (Q2), third (Q3), and forth (Q4) quartile] and somatotype.

Variable	Somatotype	Q1 (n = 42)	Q2 (n = 40)	Q3 (n = 41)	Q4 (n = 41)
TBW (L)	EnM	49.2 ± 0.3	52.4 ± 1.5	56.6 ± 1.4	67.5 ± 3.8
	BM	45.9 ± 1.6	51.6 ± 1.6	58.1 ± 1.1	65.1 ± 5.2
	EcM	45.9 ± 3.1	52.5 ± 0.9	56.7 ± 1.3	63.9 ± 4.7
	M-Ec	-	53.7 ± 0.8	56.7 ± 0.9	62.4 ± 1.2
	MEc	-	52.4 ± 0.7	57.7 ± 1.5	61.2 ± 1.2
	BEc	44.2 ± 3.6	51.7 ± 0.9	56.9 ± 1.3	61.6 ± 0.7
	Whole sample	46.1 ± 2.9	52.3 ± 1.3	56.9 ± 1.3	65.2 ± 4.3
R/H (Ohm/m)	EnM	247.0 ± 8.8	238.5 ± 14.1	214.9 ± 18.8	203.9 ± 18.1
	BM	266.8 ± 17.3	238.2 ± 16.9	223.5 ± 6.0	211.6 ± 17.6
	EcM	262.2 ± 15.9	245.5 ± 18.4	223.5 ± 10.8	203.1 ± 12.5
	M-Ec	-	249.2 ± 16.1	237.7 ± 12.8	234.6 ± 29.0
	MEc	-	268.0 ± 9.5	234.5 ± 12.5	224.2 ± 4.9
	BEc	287.8 ± 17.6	253.9 ± 9.5	243.1 ± 17.2	219.6 ± 14.0
	Whole sample	262.9 ± 17.1	245.5 ± 17.3	227.5 ± 17.2	209.0 ± 17.9
Xc/H (Ohm/m)	EnM	35.9 ± 1.4	34.7 ± 3.3	30.7 ± 2.8	28.3 ± 4.0
	BM	36.0 ± 2.7	33.4 ± 2.6	31.9 ± 3.6	27.6 ± 3.9
	EcM	35.9 ± 3.4	34.7 ± 4.1	29.0 ± 2.2	27.7 ± 2.9
	M-Ec	-	32.2 ± 2.3	31.5 ± 2.6	31.4 ± 3.1
	MEc	-	34.7 ± 2.3	31.1 ± 2.6	30.2 ± 2.2
	BEc	36.1 ± 3.3	34.2 ± 1.5	30.4 ± 3.1	29.0 ± 3.8
	Whole sample	35.9 ± 3.0	34.3 ± 3.1	30.5 ± 2.7	28.4 ± 3.5

Note: Data are presented as mean ± SD. TBW = total body water, R/H = resistance standardized for height, Xc/H = reactance standardized for height, EnM = Endomorphic Mesomorph, BM = Balanced Mesomorph, EcM = Ectomorphic Mesomorph, M-Ec = Mesomorph Ectomorph, MEc = Mesomorphic Ectomorph, BEc = Balanced Ectomorph.

Table 2. Descriptive statistics for the athletes divided by total body water quartile [first (Q1), second (Q2), third (Q3), and forth (Q4) quartile] and somatotype.

Variable	Somatotype	Q1 (n = 42)	Q2 (n = 40)	Q3 (n = 41)	Q4 (n = 41)	df	F *	p
PhA (°)	EnM	8.3 ± 0.1	8.3 ± 0.7	8.1 ± 0.5	8.0 ± 0.7			
	BM	7.7 ± 0.5	7.9 ± 0.6	8.2 ± 1.1	7.4 ± 0.6			
	EcM	7.8 ± 0.6	8.1 ± 0.8	7.4 ± 0.4	7.8 ± 0.5			
	M-Ec	-	7.4 ± 0.5	7.5 ± 0.6	7.7 ± 0.2			
	MEc	-	7.4 ± 0.5	7.6 ± 0.4	7.7 ± 0.4			
	BEc	7.1 ± 0.2	7.7 ± 0.2	7.1 ± 0.3	7.5 ± 0.6			
	Whole sample	7.8 ± 0.5	7.9 ± 0.6	7.7 ± 0.6	7.8 ± 0.6	3	1.8	0.144
VL/H (Ohm/m)	EnM	249.6 ± 8.8	241.1 ± 14.2	217.1 ± 18.9	205.9 ± 18.2			
	BM	269.3 ± 17.3	241.2 ± 16.8	225.8 ± 5.5	213.4 ± 17.9			
	EcM	264.6 ± 16.1	247.9 ± 18.5	225.4 ± 10.9	204.9 ± 12.7			
	M-Ec	-	251.3 ± 16.2	239 ± 12.8	236.6 ± 29.2			
	MEc	-	270.3 ± 9.5	236.6 ± 12.7	226.3 ± 4.2			
	BEc	290.4 ± 17.8	256.2 ± 9.6	244.9 ± 17.4	221.5 ± 18.0			
	Whole sample	265.4 ± 17.1 [2,3,4]	247.9 ± 17.3 [1,3,4]	229.5 ± 17.3 [1,2,4]	210.9 ± 18.1 [1,2,3]	3	74.7	<0.001

Note: Data are presented as mean ± SD. PhA = phase angle, VL/H = vector length standardized for height, EnM = Endomorphic Mesomorph, BM = Balanced Mesomorph, EcM = Ectomorphic Mesomorph, M-Ec = Mesomorph Ectomorph, MEc = Mesomorphic Ectomorph, BEc = Balanced Ectomorph. * Results of the one-way ANOVA considering the athletes as a whole sample; [1] Different ($p < 0.008$) from Q1; [2] Different ($p < 0.008$) from Q2; [3] Different ($p < 0.008$) from Q3; [4] Different ($p < 0.008$) from Q4.

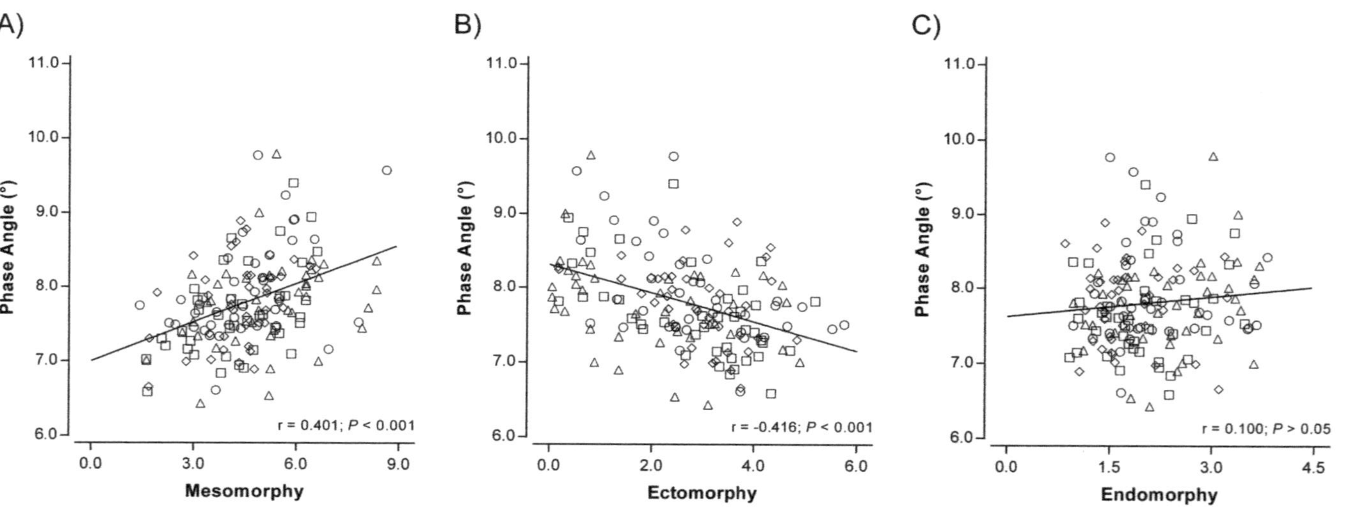

Figure 4. Correlation between phase angle and mesomorphy (**A**), ectomorphy (**B**), and endomorphy (**C**). ◇ = subjects from Q1. ○ = subjects from Q2. □ = subjects from Q3. △ = subjects from Q4.

4. Discussion

The aim of this study was to analyze the associations of morphological characteristics with bioelectrical properties in volleyball, soccer, and rugby players. An important finding has emerged from our results regarding the role of body fluids on vector length. In addition, this study has shown, for the first time, the associations between the somatotype components and PhA. As hypothesized, when considering subjects with a similar TBW, the differences in vector position may reflect morphological peculiarities; this was possible to observe due to the data analysis carried out in this study, in which the athletes were divided into separate groups limited by quartiles of TBW.

We observed that body fluids content was a determining factor for vector length, extending the findings of previous research studies [17,20]. In particular, athletes with a higher TBW (Q4) showed a mean vector positioned lower than the other athletes (Figure 2). This is in line with previous studies that observed vector length and its changes to accurately reflect changes in TBW using dilution techniques to assess water and its compartments [17,20,21]. In addition, when the mean vectors for each somatotype category were plotted on the R-Xc graph, it was possible to observe how athletes with a higher mesomorphic component showed a vector tending to be more left in the graph. Conversely, athletes with a dominant ectomorphy presented a vector displacement to the right. In this regard, our results have shown that PhA correlates directly with the mesomorphic component, while an inverse association was observed with the ectomorphic component of the somatotype. These findings are in line with previous investigations that observed in athletes with a higher muscle mass, including bodybuilders, a vector position at the limits of the 95th percentile to the left of the reference ellipses of the normal population [3,22].

The athletes belonging to the six identified somatotype categories were distributed among the four groups of TBW except for the first group where mesomorph ectomorph, and mesomorphic ectomorph athletes, were not present. As a result, it was possible to explore the association between PhA and morphological features. The endomorphic mesomorph, balanced mesomorph, and ectomorphic mesomorph somatotypes are characterized by a dominant mesomorphy, due to a muscular related body shape. On the contrary, mesomorphic ectomorph and balanced ectomorph categories imply a dominant ectomorphy, and therefore, athletes tend to be taller with a lower muscle mass than the other somatotype categories [25]. Rakovi'c et al. [32] showed how mesomorphic features are linked to individual sports that require higher muscle strength, while ectomorphy is predominant in runners [33], especially those involved in long distance. In previous research [21], it was highlighted how R/H and Xc/H were able to discriminate somatotypes. However, if we consider this new and more individual approach, considering the TBW values, probably some of those athletes needed to be revised, since body fluids have a great influence on the vector position. For this reason, the athletes' somatotypes were analyzed according to groups of body fluids to reduce the differences attributed to TBW, and consequently, to better understand how somatotype is associated with the vector position. Due to this approach, it was possible to observe how the vector position changes based on the morphological features. Indeed, when the athletes were divided into TBW groups, significant differences were found in vector length, but not for PhA, which instead represents the lateral displacement of the vector. This suggests that athletes with a similar PhA could have a greater TBW and, therefore, a different vector position and body composition characteristics. In fact, when comparing the four groups, athletes with a shorter mean vector were those with a higher TBW. A recent literature review on PhA in sports [34] concluded that it was not clear whether PhA differs among athletes engaged in different sports. On the contrary, studies on athletes practicing the same sports, but at different competitive levels, have shown that elite athletes show a higher PhA than those engaged in lower-level categories [4–6]. In particular, Micheli et al. [6] suggested that this is due to a lower R/H in relation to Xc/H and reflects a condition of greater muscularity and body cell mass content in the athletes that compete in the higher levels. In this study, it was shown how the interpretation of the vector position in the R-Xc graph overcomes the limits linked to the interpretation of the PhA alone. In this regard, Reis et al. [35] recently showed that the bioimpedance vector position varied in response to changes in the macrocycle training load in swimmers of both sexes. The authors identified an accumulation of fluids and, therefore, a shortening of the vector following a first phase characterized by a high training load, and then a subsequent shift of the vector position to the left side as a result of muscle adaptations that occurred after a recovery period. Similarly, Mascherini et al., in 2014 [24], for the first time studied vector changes over the competitive season in soccer players and showed that PhA decreased during the preparatory phase and then increased near the beginning of the competition. In line with this, Nabuco et al. [36] highlighted a moderate and inverse association between PhA and the values obtained from a fatigue assessment test.

The results of the present study add important and useful information for a correct interpretation of BIVA. The careful evaluation and monitoring of body composition allow the athlete to be predisposed to achieving high peak performance [37]. Through BIVA, it is possible to evaluate the progress in the athlete's physical condition during the season [38] in response to a training program or a nutritional intervention [37], avoiding incurring decreases in physical performance. This method, in addition to monitoring body fluids, also allows information about other body composition variables at a whole-body level [39]. In fact, it is not always possible to collect the various anthropometric measures that allow the evaluation of the somatotype; in this regard BIVA, if correctly interpreted, can provide important information, minimizing the need for skinfolds and girths collection by a certified anthropometrist. While this method requires further study, especially concerning monitoring the bioimpedance parameters in the short term, the innovation of this study lies in the fact that it provides useful information for the correct interpretation of the vector position in the R-Xc graph, specifically the role of body water as a mediator of the morphological associations with bioelectrical parameters.

Some limitations of this study need to be addressed. First of all, our results are only generalized for male athletes. Secondly, the findings of this study are only applicable to single-frequency BIA equipment. In fact, different results in measuring raw BIA parameters are obtained using devices that work on single- or multi-frequency [40]. On the other hand, the fact that it was ensured that the athletes were in a state of euhydration is a strength of the present work. Future research should study the potential of BIVA patterns as a biomarker of physical condition during the training process, as in specific microcycles. Indeed, as bioimpedance analysis (BIA) equipment provides an easy, simple, and low-cost application, it allows for a frequent assessment to optimize monitoring of the athlete's physical condition.

5. Conclusions

BIVA provides meaningful information on body composition assessment in athletes. This study showed that body fluid content affects the vector length in the R-Xc graph, while the lateral displacement of the vector can be modified by the morphological characteristics of the athlete. In particular, the mesomorphic component of the somatotype is related to higher PhA, an important marker of cellular integrity and overall physical function.

Author Contributions: Conceptualization, A.M.S., C.N.M., J.T., H.L. and S.T.; data curation, F.C. and C.P.M.; formal analysis, F.C., C.P.M.and J.P.N.; investigation, A.P., J.T., and S.T.; methodology, A.M.S., C.N.M., J.P.N., and H.L.; supervision, H.L. and S.T.; visualization, S.T.; writing—original draft, F.C., A.M.S., C.N.M., and J.P.N.; writing—review and editing, F.C., A.M.S, C.N.M., C.P.M., A.P., and S.T. All authors have read and agreed to the published version of the manuscript.

Funding: This research received no external funding.

Acknowledgments: The authors are grateful to all the athletes who took part in this study.

Conflicts of Interest: The authors declare no conflict of interest.

References

1. Castizo-Olier, J.; Irurtia, A.; Jemni, M.; Carrasco-Marginet, M.; Fernández-García, R.; Rodríguez, F.A. Bioelectrical impedance vector analysis (BIVA) in sport and exercise: Systematic review and future perspectives. *PLoS ONE* **2018**, *13*, e0197957. [CrossRef] [PubMed]
2. Piccoli, A.; Rossi, B.; Pillon, L.; Bucciante, G. A new method for monitoring body fluid variation by bioimpedance analysis: The RXc graph. *Kidney Int.* **1994**, *46*, 534–539. [CrossRef]
3. Campa, F.; Matias, C.; Gatterer, H.; Toselli, S.; Koury, J.C.; Andreoli, A.; Melchiorri, G.; Sardinha, L.B.; Silva, A.M. Classic bioelectrical impedance vector reference values for assessing body composition in male and female athletes. *Int. J. Environ. Res. Public Health* **2019**, *16*, 5066. [CrossRef] [PubMed]
4. Campa, F.; Toselli, S. Bioimpedance vector analysis of elite, subelite, and low-level male volleyball players. *Int. J. Sports Physiol. Perform.* **2018**, *13*, 1250–1253. [CrossRef] [PubMed]
5. Giorgi, A.; Vicini, M.; Pollastri, L.; Lombardi, E.; Magni, E.; Andreazzoli, A.; Orsini, M.; Bonifazi, M.; Lukaski, H.; Gatterer, H. Bioimpedance patterns and bioelectrical impedance vector analysis (BIVA) of road cyclists. *J. Sports Sci.* **2018**, *36*, 2608–2613. [CrossRef]
6. Micheli, M.L.; Pagani, L.; Marella, M.; Gulisano, M.; Piccoli, A.; Angelini, F.; Burtscher, M.; Gatterer, H. Bioimpedance and impedance vector patterns as predictors of league level in male soccer players. *Int. J. Sports Physiol. Perform.* **2014**, *9*, 532–539. [CrossRef] [PubMed]
7. Campa, F.; Silva, A.M.; Iannuzzi, V.; Mascherini, G.; Benedetti, L.; Toselli, S. The role of somatic maturation on bioimpedance patterns and body composition in male elite youth soccer players. *Int. J. Environ. Res. Public Health* **2019**, *16*, 4711. [CrossRef]
8. De Araújo Jerônimo, A.F.; Batalha, N.; Collado-Mateo, D.; Parraca, J.A. Phase Angle from Bioelectric Impedance and Maturity-Related Factors in Adolescent Athletes: A Systematic Review. *Sustainability* **2020**, *12*, 4806.
9. Koury, J.C.; Trugo, N.M.F.; Torres, A.G. Phase angle and bioelectrical impedance vectors in adolescent and adult male athletes. *Int. J. Sports Physiol. Perform.* **2014**, *9*, 798–804. [CrossRef]
10. Toselli, S.; Marini, E.; Maietta Latessa, P.; Benedetti, L.; Campa, F. Maturity Related Differences in Body Composition Assessed by Classic and Specific Bioimpedance Vector Analysis among Male Elite Youth Soccer Players. *Int. J. Environ. Res. Public Health* **2020**, *17*, 729. [CrossRef]
11. Campa, F.; Piras, A.; Raffi, M.; Trofè, A.; Perazzolo, M.; Mascherini, G.; Toselli, S. The effects of dehydration on metabolic and neuromuscular functionality during cycling. *Int. J. Environ. Res. Public Health* **2020**, *17*, 1161. [CrossRef] [PubMed]
12. Campa, F.; Gatterer, H.; Lukaski, H.; Toselli, S. Stabilizing bioimpedance-vector-analysis measures with a 10-minute cold shower after running exercise to enable assessment of body hydration. *Int. J. Sports Physiol. Perform.* **2019**, *14*, 1006–1009. [CrossRef] [PubMed]
13. Carrasco-Marginet, M.; Castizo-Olier, J.; Rodríguez-Zamora, L.; Iglesias, X.; Rodríguez, F.A.; Chaverri, D.; Brotons, D.; Irurtia, A. Bioelectrical impedance vector analysis (BIVA) for measuring the hydration status in young elite synchronized swimmers. *PLoS ONE* **2017**, *12*, e0178819. [CrossRef]
14. Gatterer, H.; Schenk, K.; Laninschegg, L.; Lukaski, H.; Burtscher, M. Bioimpedance identifies body FluidLoss after exercise in the heat: A pilot study with body cooling. *PLoS ONE* **2014**, *9*, e109729. [CrossRef] [PubMed]
15. Mundstock, E.; Amaral, M.A.; Baptista, R.R.; Sarria, E.E.; Dos Santos, R.R.; Detoni Filho, A.; Rodrigues, C.A.; Forte, G.C.; Castro, L.; Padoin, A.V.; et al. Association between phase angle from bioelectrical impedance analysis and level of physical activity: Systematic review and meta-analysis. *Clin. Nutr.* **2019**, *38*, 1504–1510. [CrossRef] [PubMed]
16. Kyle, U.G.; Bosaeus, I.; De Lorenzo, A.D.; Deurenberg, P.; Elia, M.; Gómez, J.M.; Heitmann, B.L.; Kent-Smith, L.; Melchior, J.C.; Pirlich, M.; et al. Bioelectrical impedance analysis—part I: Review of principles and methods. *Clin. Nutr.* **2004**, *23*, 1226–1243. [CrossRef] [PubMed]
17. Campa, F.; Matias, C.N.; Marini, E.; Heymsfield, S.B.; Toselli, S.; Sardinha, L.B.; Silva, A.M. Identifying athlete body-fluid changes during a competitive season with bioelectrical impedance vector analysis. *Int. J. Sports Physiol. Perform.* **2020**, *15*, 361–367. [CrossRef]
18. Francisco, R.; Matias, C.N.; Santos, D.A.; Campa, F.; Minderico, C.S.; Rocha, P.; Heymsfield, S.B.; Lukaski, H.; Sardinha, L.B.; Silva, A.M. The predictive role of raw bioelectrical impedance parameters in water compartments and fluid distribution assessed by dilution techniques in athletes. *Int. J. Environ. Res. Public Health* **2020**, *17*, 759. [CrossRef]

19. González-Correa, C.H. Body Composition by Bioelectrical Impedance Analysis. In *Bioimpedance in Biomedical Applications and Research*; Simini, F., Bertemes-Filho, P., Eds.; Springer: Cham, Switzerland, 2018; ISBN 978-3-319-74387-5.
20. Marini, E.; Campa, F.; Bua, R.; Stagi, S.; Matias, C.N.; Toselli, S.; Sardinha, L.B.; Silva, A.M. Phase angle and bioelectrical impedance vector analysis in the evaluation of body composition in athletes. *Clin. Nutr.* **2020**, *39*, 447–454. [CrossRef]
21. Campa, F.; Silva, A.M.; Talluri, J.; Matias, C.N.; Badicu, G.; Toselli, S. Somatotype and Bioimpedance Vector Analysis: A New Target Zone for Male Athletes. *Sustainability* **2020**, *12*, 4365. [CrossRef]
22. Piccoli, A.; Pastori, G.; Codognotto, M.; Paoli, A. Equivalence of information from single frequency v. bioimpedance spectroscopy in bodybuilders. *Br. J. Nutr.* **2007**, *97*, 182–192. [CrossRef] [PubMed]
23. Castizo-Olier, J.; Carrasco-Marginet, M.; Roy, A.; Chaverri, D.; Iglesias, X.; Pérez-Chirinos, C.; Rodríguez, F.; Irurtia, A. Bioelectrical Impedance Vector Analysis (BIVA) and Body Mass Changes in an Ultra-Endurance Triathlon Event. *J. Sports Sci. Med.* **2018**, *17*, 571–579.
24. Mascherini, G.; Gatterer, H.; Lukaski, H.; Burtscher, M.; Galanti, G. Changes in hydration, body-cell massand endurance performance of professional soccer players through a competitive season. *J. Sports Med. Phys. Fit.* **2015**, *55*, 749–755.
25. Carter, J.E.L. *The Heath-Carter Somatotype Method*, 3rd ed.; San Diego State University Syllabus Service: San Diego, CA, USA, 1980.
26. Carter, J.E.; Phillips, W.H. Structural changes in exercising middle-aged males during a 2-year period. *J. Appl. Physiol.* **1969**, *27*, 787–794. [CrossRef]
27. Armstrong, L.E.; Pumerantz, A.C.; Fiala, K.A.; Roti, M.W.; Kavouras, S.A.; Casa, D.J.; Maresh, C.M. Human hydration indices: Acute and longitudinal reference values. *Int. J. Sport Nutr. Exerc. Metab.* **2010**, *20*, 145–153. [CrossRef] [PubMed]
28. Lohman, T.G.; Roche, A.F.; Martorell, R. *Anthropometric Standardization Reference Manual*; Human Kinetics Books: Champain, IL, USA, 1988.
29. Lukaski, H.C.; Piccoli, A. Bioelectrical impedance vector analysis for assessment of hydration in physiological states and clinical conditions. In *Handbook of Anthropometry*; Preedy, V., Ed.; Springer: Berlin, Germany, 2012; pp. 287–305.
30. Matias, C.N.; Santos, D.A.; Júdice, P.B.; Magalhães, J.P.; Minderico, C.S.; Fields, D.A.; Lukaski, H.C.; Sardinha, L.B.; Silva, A.M. Estimation of total body water and extracellular water with bioimpedance in athletes: A need for athlete-specific prediction models. *Clin. Nutr.* **2016**, *35*, 468–474. [CrossRef] [PubMed]
31. Piccoli, A.; Nigrelli, S.; Caberlotto, A.; Bottazzo, S.; Rossi, B.; Pillon, L.; Maggiore, Q. Bivariate normal values of the bioelectrical impedance vector in adult and elderly populations. *Am. J. Clin. Nutr.* **1995**, *61*, 269–270. [CrossRef]
32. Rakovic, A.; Savanovic, V.; Stankovic, D.; Pavlovic, R.; Simeonov, A.; Petkovic, E. Analysis of the elite athletes'somatotypes. *Acta Kinesiol.* **2015**, *9*, 47–53.
33. Sánchez, M.C.; Muros, J.J.; Lópezm, B.Ó.; Zabala, M. Anthropometric characteristics, body composition and somatotype of elite male young runners. *Int. J. Environ. Res. Public Health* **2020**, *17*, 674. [CrossRef]
34. Di Vincenzo, O.; Marra, M.; Scalfi, L. Bioelectrical impedance phase angle in sport: A systematic review. *J. Int. Soc. Sports Nutr.* **2019**, *16*, 49. [CrossRef]
35. Reis, J.F.; Matias, C.N.; Campa, F.; Morgado, J.P.; Franco, P.; Quaresma, P.; Almeida, N.; Curto, D.; Toselli, S.; Monteiro, C.P. Bioimpedance Vector Patterns Changes in Response to Swimming Training: An Ecological Approach. *Int. J. Environ. Res. Public Health* **2020**, *17*, 4851. [CrossRef] [PubMed]
36. Nabuco, H.C.G.; Silva, A.M.; Sardinha, L.B.; Rodrigues, F.B.; Tomeleri, C.M.; Ravagnani, F.C.; Cyrino, E.S.; Ravagnani, C.F. Phase Angle is Moderately Associated with Short-term Maximal Intensity Efforts in Soccer Players. *Int. J. Sports Med.* **2019**, *40*, 739–743. [CrossRef] [PubMed]
37. Silva, A.M.; Nunes, C.L.; Matias, C.N.; Rocha, P.M.; Minderico, C.S.; Heymsfield, S.B.; Lukaski, H.; Sardinha, L.B. Usefulness of raw bioelectrical impedance parameters in tracking fluid shifts in judo athletes. *Eur. J. Sport Sci.* **2019**, *28*, 1–10. [CrossRef]
38. Pollastri, L.; Lanfranconi, F.; Tredici, G.; Schenk, K.; Burtscher, M.; Gatterer, H. Body fluid status and physical demand during the Giro d'Italia. *Res. Sports Med.* **2016**, *24*, 30–38. [CrossRef] [PubMed]

39. Silva, A.M. Structural and functional body components in athletic health and performance phenotypes. *Eur. J. Clin. Nutr.* **2019**, *73*, 215–224. [CrossRef] [PubMed]
40. Silva, A.M.; Matias, C.N.; Nunes, C.L.; Santos, D.A.; Marini, E.; Lukaski, H.C.; Sardinha, L.B. Lack of agreement of in vivo raw bioimpedance measurements obtained from two single and multi-frequency bioelectrical impedance devices. *Eur. J. Clin. Nutr.* **2019**, *73*, 1077–1083. [CrossRef]

International Journal of
Environmental Research and Public Health

Article

Effects of Aerobic and Anaerobic Fatigue Exercises on Postural Control and Recovery Time in Female Soccer Players

Özkan Güler [1], Dicle Aras [1], Fırat Akça [1], Antonino Bianco [2,*], Gioacchino Lavanco [2], Antonio Paoli [3] and Fatma Neşe Şahin [1]

1. Faculty of Sports Sciences, Ankara University, Gölbaşı, Ankara 06830, Turkey; oguler@ankara.edu.tr (Ö.G.); daras@ankara.edu.tr (D.A.); fakca@ankara.edu.tr (F.A.); nesesahin@ankara.edu.tr (F.N.Ş.)
2. Department of Psychological, Pedagogical, Educational Science and Human Movement, University of Palermo, 90144 Palermo, Italy; gioacchino.lavanco@unipa.it
3. Department of Biomedical Science, University of Padova, 35122 Padova, Italy; antonio.paoli@unipd.it

* Correspondence: antonino.bianco@unipa.it; Tel.: +39-091-23896910

Received: 4 August 2020; Accepted: 24 August 2020; Published: 28 August 2020

Abstract: Sixteen female soccer players (age = 20.19 ± 1.52 years; body mass = 56.52 ± 4.95 kg; body height = 164.81 ± 4.21 cm) with no history of lower extremity injury participated in the study. The Biodex SD Balance system was used to determine the non-dominant single-leg stability. In anaerobic exercise, each subject performed four maximal cycling efforts against a resistance equivalent to 0.075 kg/body mass for 30 s with three-minute rest intervals. In aerobic exercise, subjects performed the Bruce protocol on a motorized treadmill. After each exercise, subjects subsequently performed a single-leg stability test and then repeated the same test for four times with five-minute passive rest periods. In accordance with the results, it was found that the impairment observed right after the aerobic loading was higher ($p < 0.001$) compared to the anaerobic one. However, the time-related deterioration in both aerobic and anaerobic loadings was similar. The B-pre value was lower than B_{post} and B_5 ($p < 0.01$) and B_{10} ($p < 0.05$) in both conditions. Subjects could reach the initial balance level at B_{15} after aerobic and anaerobic loadings. The lactate level did not reach resting value even after 20 min of both fatigue protocols. Although the fatigue after aerobic and aerobic exercise negatively affects a single-leg dynamic balance level, single leg balance ability returns to the baseline status after 10 min of passive recovery duration.

Keywords: balance; fatigue; female; support leg; recovery

1. Introduction

The popularity of soccer among females is increasing each passing day. It is estimated that around 30 million females are actively playing licensed soccer in more than 100 countries around the world. Studies indicate that as the participation of females in soccer increases, the incidence of injury increases at a high rate [1–3]. In a soccer match, soccer players engage in many moves such as high-intensity acceleration, deceleration, sudden change of direction, bounce, and other soccer-oriented movements. Along with these moves, soccer players often experience various injuries when using one leg for stopping and cutting during pressure, while using the other leg to tackle the ball [4]. In addition to these, injuries in soccer are caused by sudden acceleration and deceleration without impact, rapid disorientation, and exposure to high loads while maintaining the stability of the knee joint in unpredictable movements [5–8]. When the injuries experienced in soccer were evaluated according to sex differences, It was reported that female athletes had a higher incidence of anterior cruciate ligament (ACL) experience in lower extremity injuries due to biomechanical and neuromuscular differences [5,6] than males [9].

Furthermore, previous studies indicate that female soccer player has the risk of ACL injury nine times greater than males [10]. Several risk factors cause these injuries in female soccer players. These risk factors in female soccer players include a previous history of injury [11], as well as a decline in hip strength [12] due to accumulated fatigue [11] and deterioration of lower extremity dynamic balance [13]. Epidemiological studies have pointed out that 50% of the injuries occur at the end of competitions or sports activities, and 58% of these injuries are due to non-impact conditions. That fatigue is an essential element of sensory-motor changes associated with injury [14,15]. Ekstrand et al. [9] report in their study that traumatic injuries occur more often in the last minutes of both halves of a soccer match [9]. In addition to these, it was reported in another study that lower extremity injuries were commonly seen at the last minutes of competition in sports such as soccer, which includes high-intensity moves and multi-directional sprints [9,16]. Therefore, it can be stated that non-contact injuries caused by fatigue occurred in the last fifteen minutes of play in both the first and second half of games. In non-contact injuries, neuromuscular fatigue is seen as a risk factor [17–20]. Neuromuscular fatigue is divided into two, according to the intensity and duration of exercise, like peripheral and central nervous system fatigue. Long-term activities affect the central nervous system, while short-term high-intensity activities cause peripheral fatigue [21,22]. Peripheral fatigue arises when there is not adequate energy provided to the muscles, despite the increasing energy need [22].

Moreover, it has been reported that muscle fatigue affects both peripheral and central proprioceptive processes [23–25]. Balance is defined as being able to hold the body center of gravity within the center of support [26]. In order to maintain balance visual, vestibular, and proprioceptive systems play a crucial role, and these systems are affected by many factors [27–32]. The proprioceptive system consists of the Golgi tendon organ, the muscle spindle, the Pacini corpuscle, free nerve endings, and the receptors in the joint capsules and skin [32–34]. It ensures maintenance of the balance with the information collected from these structures [32,34]. The proprioceptive system is affected by fatigue, aging, sarcopenia, neurological disease fibromyalgia, cancer, and rheumatological diseases and may result in impaired balance [35–38]. Many researchers have shown that fatigue negatively affects dynamic postural control [39–42]. Fatigue is an essential factor that acutely affects balance ability. In a study in which the center of pressure (COP) was measured before, in halftime and immediately after a soccer match, it was determined that the balance skill of the support leg was impaired in the post-match measurement [41]. The return of balance ability to average values after fatigue depends on many factors. The return of post-fatigue balance ability to initial level depends on the duration, intensity, and type of intensity of the fatigue protocol performed [43]. Deficits in dynamic postural control is a risk factor in experiencing falls and lower extremity injuries [13,42,44–47]. The deterioration of dynamic balance is associated with reactive and compensating movements, and it is stated that this impairment has been linked to falling risk and lower extremity injuries [18,48,49]. Since fatigue increases the rate of injury in athletes [9,16], and the lack of postural control is a lower extremity injury risk factor [13,50,51], it can be expected that the rate of injury as a result of fatigue-induced postural control (fatigue-induced balance deficits) may increase.

Soccer is classified as both an aerobic and an anaerobic sport. In the game, players may experience fatigue from time to time as a result of aerobic and anaerobic activities. There are studies in the literatüre on the effects of anaerobic fatigue on balance performance in soccer players. However, up to date, no previous studies have examined the effects of both types of fatigue in soccer players. Besides, many activities such as passing, kicking, and jumping in soccer are carried out on the support leg. This research will be the first to examine the effects of fatigue on support leg balance performance. Therefore, this study aims to determine the effects of different fatigue protocols on balance performance of the support-leg in female soccer players and to understand the time required for the balance to recover after loading.

2. Materials and Methods

Sixteen sub-elite female soccer players (with mean age of 20.19 ± 1.52 years, body mass 56.52 ± 4.95 kg, body height 164.81 ± 4.21 cm, percent body fat 22.63 ± 2.42%, and maxVO_2 52.33 ± 5.74 mL.kg.$^{-1}$min^{-1}) participated in the study voluntarily. Players who suffered lower extremity injuries for the last six months were not included in the study. Participants were instructed not to perform exercises that may cause exhaustion 48 h before the tests and not to use stimulants such as alcohol, caffeine, or drugs in the last 24 h before the study. The study was conducted according to the Declaration of Helsinki and was approved by the Ethics Committee of Ankara University, Approval code 21-1300-17, released in December 2017.

2.1. Balance Test

The participants were invited to participate in the Biodex SD Balance System (Biodex, Shirley, NY, USA) athletic single-leg testing protocol [52]. As noted, the Biodex Balance System (BBS) uses a circular platform that is free to move in the anterior-posterior and medial-lateral axes simultaneously. The BBS measures, in degrees, the tilt about each axis during dynamic conditions and calculates an overall stability index (OSI). A high score in the OSI indicates poor balance. The platform stability ranges from 1–12, with 1 representing the most significant instability.

The familiarization protocol was implemented before the experiments. All participants were instructed to perform the balance test on five different days of the targeted week.

The athletic single leg test protocol consisted of 3 trials of 20 s of upright stance on support-leg with 10 s of rest intervals between trials. Participants were asked to place their feet with the malleolar axis aligned with the midpoint of the platform over the center dot of the platform in a comfortable position. An athletic single-leg test was conducted on BBS with the platform set at level 4. Balance tests were carried out on the non-dominant leg. The same test protocol was performed before (B_{pre}) and right after (B_{post}) both aerobic and anaerobic fatigue protocols, and repeated at the 5th (B_5), 10th (B_{10}), 15th (B_{15}), and 20th (B_{20}) minutes. The participants were allowed to rest passively during the 5 min of recovery periods. There was a 2-day period between aerobic and anaerobic fatigue protocols (Figure 1).

2.2. Aerobic Fatigue Protocol

The Bruce protocol was performed on a motorized treadmill (Cosmed, Rome, Italy) in order to create aerobic fatigue in soccer players [53]. The participants continued the test until they were exhausted. At the end of the test, the maxVO_2 consumption values of the participants were calculated and recorded. Rating of perceived exertion (RPE) was obtained using the 6–12 point Borg scale at the end of every load [54]. MaxVO_2 was defined as the highest 30 s average in oxygen uptake and maximal heart rate (HRmax) as the highest every 10 s average during the Bruce protocol. A test was considered maximal when four of the following criteria were completed: VO2 plateau at peak exercise, respiratory exchange ratio ≥ 1.10 greater age-predicted maximal heart rate (220-age), and an indication of 18–20 rating on the Borg RPE scale [55].

2.3. Anaerobic Loading Protocol

Anaerobic fatigue protocol was performed using a bicycle ergometer (Monark Erogomedic 894 E Peak Bike Vansbro Sweden). The Wingate test protocol was used for anaerobic fatigue. In the Wingate protocol, participants were asked to pedal at maximal speed for 30 s. As the intensity of the training, a weight equivalent to 7.5% of the participants' body mass was placed on the load scale. Once the participants started pedaling, the scale dropped when the bike's wheel revolution went up to 150 rpm and the maximal pedaling for 30 s. Soccer players were verbally motivated during training. Participants repeated the Wingate test protocol a total of 6 times with intervals of 4 min.

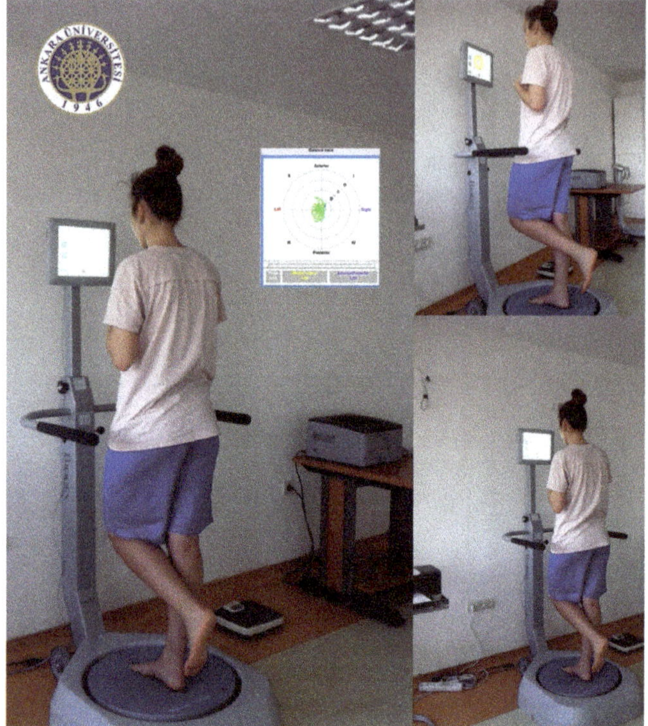

Figure 1. One subject during the data collection phase. Female soccer player.

2.4. Lactate Testing

Participants' blood lactate values were determined immediately after the fatigue protocol and during the recovery period (L_{pre}) just before the balance tests at 0th (L_{post}), 5th (L_5), 10th (L_{10}), 15th (L_{15}), and 20th (L_{20}) minutes. During the lactate test, participants' fingertips were wiped with alcohol-based tissue paper, and their capillary blood samples were taken with a lancet pen. Blood lactate levels of the participants were determined by an Accutrend Plus lactate device (Roche Diagnostics, Basel, Switzerland).

2.5. Statistical Analysis

In all statistical analyses, SPSS version 20 was used (SPSS Inc., Chicago, IL, USA). First, because the number of participants was below 50, the normality of the data was analyzed with the Shapiro-Wilk test. Depending on the distribution, lactate and balance values obtained at different times following aerobic and anaerobic fatigue protocols were compared by the Paired Sample t-Test or Wilcoxon Test. For the intergroup analyzes, either the Repeated Measurements Analysis of Variance (Aerobic Lpre, L5, L10, L15, L20; Bpre, B5, B15, B20; Anaerobic Lpost, L5, L10, L15, L20; Bpre, B5, B15, B20) or the Friedman test (Aerobic Lpost, Bpost, B10; Anaerobic Lpre, Bpost, B10) was used again depending on the distribution. In the case of the dataset exhibited both non-normally distributed and norma distributed data, we proceeded with the non-parametric analysis. In all statistical analyses, the alpha value was considered to be 0.05.

3. Results

The following variables were shown to be not normally distributed (Aerobic Lpost, Bpost, B10; Anaerobic Lpre, Bpost, B10).

The lactate values obtained from the participants before, during, and after aerobic and anaerobic fatigue protocols are shown in Table 1.

Table 1. Mean comparisons of participants with lactate and balance, which vary depending on aerobic and anaerobic loading. In horizontal, the repeated measures *p* values, in vertical the *p* values for aerobic vs. anaerobic comparisons.

Fatigue Protocol	L_{pre}	L_{post}	L_5	L_{10}	L_{15}	L_{20}	p_-
Aerobic	1.20 ± 0.36	11.70 ± 2.53	11.81 ± 2.51	9.86 ± 2.58	7.84 ± 2.15	6.93 ± 1.87	0.000 **
Anaerobic	1.18 ± 0.33	15.43 ± 2.40	15.09 ± 2.27	13.76 ± 2.50	11.25 ± 2.14	9.49 ± 2.51	0.000 **
p_-	0.823	0.001 **	0.000 **	0.000 **	0.000 **	0.001 **	-
Fatigue Protocol	B_{pre}	B_{post}	B_5	B_{10}	B_{15}	B_{20}	p_-
Aerobic	0.89 ± 1.39	2.29 ± 1.04	1.35 ± 0.29	1.08 ± 0.35	1.01 ± 0.24	1.04 ± 0.26	0.000 **
Anaerobic	0.90 ± 1.40	1.58 ± 0.57	1.16 ± 0.21	1.02 ± 0.28	0.95 ± 0.15	0.98 ± 0.18	0.000 **
p_-	0.745	0.001 **	0.014 *	0.308	0.410	0.509	-

L: Lactate; B: Balance; * $p < 0.05$; ** $p < 0.01$.

According to the results, there was no significant difference between participants' resting lactate concentration values obtained before aerobic and anaerobic loading ($p > 0.05$). However, the lactate values obtained immediately after, 5th, 10th, 15th, and 20th min were statistically significant ($p < 0.01$) according to the fatigue conditions. After anaerobic loading, lactate values were found to be higher. Besides, it was understood that 20 min was not sufficient for the recovery, regardless of the type of loading.

Lactate values obtained at six different phases of the aerobic fatigue protocol were significantly different ($p < 0.01$). The resting lactate value was determined to be lower than all others reached after loading. Also, a significant difference was found between L_{post} and L_{10}, L_{15} and L_{20}; L_5 and L_{10}, L_{post} L_{15}, and L_{20}; L_{10} and L_{15} and L_{20} ($p < 0.01$); and with L_5 and L_{20} ($p < 0.05$).

In lactate values measured after anaerobic loading, the resting value was determined to be significantly lower than in all other measurements. The difference between L_{post} and L_5 was not significant, similar to the aerobic one. However, a significant difference at the level of $p < 0.01$ was found between L_{post} and L_{10}, L_{15}, and L_{20}, between L_5 and L_{10}, L_{15}, and L_{20} and between L_{15} and L_{20}.

When the results related to the balance were examined, a significant difference was found at the level of $p < 0.01$ between the balance values obtained immediately after aerobic and anaerobic loading and at the 5th minute. It was understood that there is more deterioration in the balance after aerobic loading. No significant difference was observed in the values obtained after 10th, 15th, and 20th min, depending on the type of loading (Figure 2).

When the values obtained due to aerobic loading are taken into account, it was understood that the differences between B_{pre} and B_{post}, B_5 ($p < 0.01$), and between B_{pre} and B_{10} were ($p < 0.05$) statistically significant. However, there was no significant difference with the values reached between B_{15}, B_{20}, and B_{pre}. Accordingly, it can be stated that recovery takes place after aerobic load in the state of balance between 10th and 15th minutes. In the other results, on the other hand, a significant difference was found between B_{post} and B_5, B_{10}, B_{15}, and B_{20} ($p < 0.01$). The measurement of B_5 was found to be significantly higher than the measurements of B_{10}, B_{15} and B_{20} ($p < 0.01$).

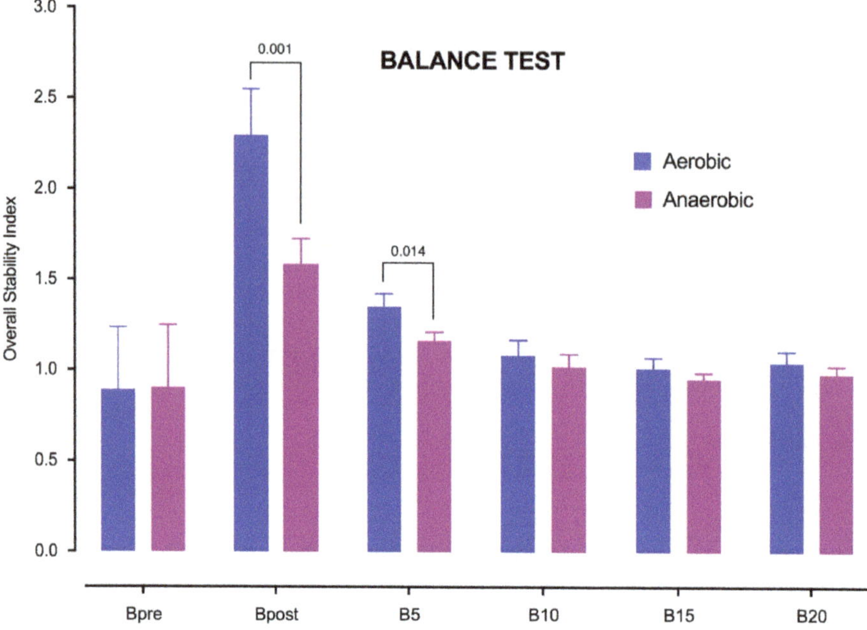

Figure 2. Overall stability index across the experimentation. Aerobic condition vs. anaerobic condition.

4. Discussion

This study aimed to investigate the acute effects of aerobic and anaerobic exercises on dynamic balance skill and recovery time in female soccer players. In order to control the fatigue level, lactate concentrations of the subjects were also collected. According to the lactate test results, both fatigue protocols were found to be successful in creating fatigue. Although it was higher after the anaerobic exercise, the lactate level did not return to the initial level within 20 min after both fatigue conditions. Besides, as a result of strenuous aerobic or anaerobic exercise, female soccer players' ability to balance the support leg was affected negatively. After both aerobic and anaerobic loading, the recovery time of balance skill lasted about 10 min. The deterioration in the athletic single-leg stability test was observed to be higher after the aerobic fatigue protocols in all of the measurements. Furthermore, the difference was statically significant in B_{post} ($p < 0.01$) and B_5 ($p < 0.05$) values. In the literature, many studies suggest that aerobic and anaerobic fatigue negatively affect balance ability [41–43,56–61]. There are similar studies on this subject in the literature. In a study in which both aerobic and anaerobic fatigue protocol was implemented, and the balance level was determined with the Balance Error Scoring System (BESS), no difference was found between the balance performance and its time of recovery after both fatigue protocols. However, athletes returned to their initial balance performance values within 8–13 min after both fatigue protocols [58]. Steinberg et al. investigated the balance level after a Yo-Yo test with the Interactive Balance System (Tetrax) device and reported that balance skill returned to the initial level within 10 min after fatigue [62]. In the current study, balance skills returned to the initial level after approximately 10–15 min after both aerobic and anaerobic fatigue. Moreover, the present study found no difference between balance levels during the recovery time after aerobic or anaerobic fatigue. In a study conducted on a bicycle ergometer, participants performed two maximal Wingate tests lasting 30 s with a rest interval of 2 min. At the end of high-intensity activation, it was determined that balance skill was affected negatively and returned to the baseline level within 10 min [63]. Ishizuka et al. applied the functional fatigue protocol to 14 male and 9 female college-level soccer players and determined the balance with the Biodex Limit of Stability

Test. As a result, the subjects were found to have returned their initial level within 10 min after the fatigue [59]. In a study by Matsuda et al., a functional fatigue protocol was applied to 100 recreationally active college students. After the functional fatigue protocol, it was reported in the measurements made with the balance error score system that the balance performance returned to its initial level within approximately 20 min [64]. Although a similar fatigue protocol was implemented, different balance performance recovery times were observed between the study mentioned above and our study. This difference is thought to be since soccer players have a better balance skill than other athletes and sedentary people. Contrary to the result of the current study, it is reported in some studies that the recovery time of balance ability lasts longer than 10 min, while in some studies, it is reported that balance ability is not affected after fatigue. In a study where the fatigue protocol involving sports activities was applied, it was reported that the participants' balance values returned to their initial values after 20 min as a result of the balance test conducted with the balance error score system [56]. In another study, in which balance measurements were made after a 25-min treadmill run, it took approximately 15 min for the athletes participating in the study to return their balance performance to its initial level [65]. In another study, where balance measurement was performed with the biodex balance system before and during a soccer match, it was determined that the dominant leg balance performance of the players was impaired while no change was observed in the support leg balance performance [66]. In contrast to this, soccer players' support leg balance skills were found to be negatively affected following the fatigue protocol in the present study. The reason for the difference is thought to be due to the degree of difficulty differences between the protocols of balance tests. In the literature, some studies did not observe any changes experienced in balance performance after fatigue. After soccer-specific fatigue [67] and after soccer training [68], balance measurements made with the Biodex Balance System have reported that the balance performance of soccer players is not affected by fatigue. Paillard reported that the return of post-fatigue balance ability to initial values depends on the duration, density, and intensity of the fatigue protocol performed [43]. In these studies, it is thought that the reason why there is no difference after fatigue is due to insufficient density and intensity of the fatigue protocol for impaired balance ability. In addition, longer times to return to initial values of balance performance were reported in these studies compared to the current study. The reason for these varied results may have been the difference between the branches of the athletes participating in the studies or the difference in balance measurement methods.

5. Conclusions

As a result of this study, it was clearly observed that balance performance is impaired in soccer players after both aerobic and anaerobic fatigue. The impairment of fatigue and balance performance are seen as significant risk factors. Although there is not enough data on the effects of fatigue on balance ability in soccer players, it is stated in several studies that fatigue increases the incidence of injury experienced [69] and that deterioration in balance performance may increase ankle injuries [70]. Many researchers also suggest that balance training should be performed to prevent injuries [1,71,72]. Therefore, trainers should give importance to balance training in order to prevent non-contact injuries caused by loss of balance. In future studies, it is suggested to investigate the effects of fatigue on the balance ability in athletes performing balance training.

The study was performed exclusively on healthy young adult female soccer players (who suffer from a higher prevalence of non-contact ACL injuries). A limitation of the study could be seen as a lack of control of the menstrual cycle. In addition, in this study, anaerobic fatigue protocol was performed with a bicycle ergometer and aerobic fatigue with a treadmill. The measurement of balance performance after a real soccer match is thought to provide a clearer picture of the effects of soccer-specific fatigue mechanisms on balance performance.

Author Contributions: Conceptualization, Ö.G.; Data curation, D.A. and F.A.; Formal analysis, Ö.G. and F.N.Ş.; Funding acquisition, G.L.; Investigation, D.A.; Methodology, Ö.G., F.A., and F.N.Ş.; Project administration, A.B. and F.N.Ş.; Supervision, G.L. and A.P.; Visualization, A.P.; Writing—original draft, D.A., F.A., and F.N.Ş.;

Writing—review & editing, A.B. and A.P. All authors have read and agreed to the published version of the manuscript.

Funding: This research received no external funding.

Conflicts of Interest: The authors declare no conflict of interest.

References

1. Emery, C.; Meeuwisse, W.H.; Hartmann, S.E. Evaluation of risk factors for injury in adolescent soccer: Implementation and validation of an injury surveillance system. *Am. J. Sports Med.* **2005**, *33*, 1882–1891. [CrossRef] [PubMed]
2. Faude, O.; Junge, A.; Kindermann, W.; Dvorak, J. Injuries in female soccer players: A prospective study in the German national league. *Am. J. Sports Med.* **2005**, *33*, 1694–1700. [CrossRef] [PubMed]
3. Mtshali, P.; Mbambo-Kekana, N.; Stewart, A.; Musenge, E. Common lower extremity injuries in female high school soccer players in Johannesburg east district. *S. Afr. J. Sports Med.* **2009**, *21*, 163–166. [CrossRef]
4. Waldén, M.; Krosshaug, T.; Bjørneboe, J.; Andersen, T.E.; Faul, O.; Hägglund, M. Three distinct mechanisms predominate in non-contact anterior cruciate ligament injuries in male professional football players: A systematic video analysis of 39 cases. *Br. J. Sports Med.* **2015**, *49*, 1452–1460. [CrossRef]
5. Hewett, T.E.; Myer, G.D.; Ford, K.R.; Heidt, R.S.; Colosimo, A.J.; McLean, S.G.; Bogert, A.V.D.; Paterno, M.V.; Succop, P. Biomechanical measures of neuromuscular control and valgus loading of the knee predict anterior cruciate ligament injury risk in female athletes: A prospective study. *Am. J. Sports Med.* **2005**, *33*, 492–501. [CrossRef]
6. McLean, S.G.; Samorezov, J.E. Fatigue-induced ACL injury risk stems from a degradation in central control. *Med. Sci. Sports Exerc.* **2009**, *41*, 1662–1673. [CrossRef]
7. Smith, H.C.; Vacek, P.; Johnson, R.J.; Slauterbeck, J.R.; Hashemi, J.; Shultz, S.; Beynnon, B.D. Risk factors for anterior cruciate ligament injury: A review of the literature—Part 1: Neuromuscular and anatomic risk. *Sports Health* **2012**, *4*, 69–78. [CrossRef]
8. Olchowik, G.; Czwalik, A. Effects of soccer training on body balance in young female athletes assessed using computerized dynamic posturography. *Appl. Sci.* **2020**, *10*, 1003. [CrossRef]
9. Ekstrand, J.; Hägglund, M.; Walden, M. Epidemiology of muscle injuries in professional football (soccer). *Am. J. Sports Med.* **2011**, *39*, 1226–1232. [CrossRef]
10. Gwinn, D.E.; Wilckens, J.H.; McDevitt, E.R.; Ross, G.; Kao, T.C. The relative incidence of anterior cruciate ligament injury in men and women at the united states naval academy. *Am. J. Sports Med.* **2000**, *28*, 98–102. [CrossRef]
11. Dvorák, J.; Junge, A.; Chomiak, J.; Graf-Baumann, T.; Peterson, L.; Rösch, D.; Hodgson, R. Risk factor analysis for injuries in football players. *Am. J. Sports Med.* **2000**, *28*, 69–74. [CrossRef] [PubMed]
12. Khayambashi, K.; Ghoddosi, N.; Straub, R.K.; Powers, C.M. Hip muscle strength predicts non-contact anterior cruciate ligament injury in male and female athletes: A prospective study. *Am. J. Sports Med.* **2016**, *44*, 355–361. [CrossRef] [PubMed]
13. Plisky, P.J.; Rauh, M.J.; Kaminski, T.W.; Underwood, F.B. Star excursion balance test as a predictor of lower extremity injury in high school basketball players. *J. Orthop. Sports Phys. Ther.* **2006**, *36*, 911–919. [CrossRef]
14. Hawkins, R.D.; Hulse, M.A.; Wilkinson, C.; Hodson, A.; Gibson, M. The association football medical research programme: An audit of injuries in professional football. *Br. J. Sports Med.* **2001**, *35*, 43–47. [CrossRef] [PubMed]
15. Woods, C.; Hawkins, R.; Hulse, M.; Hodson, A. The Football Association Medical Research Programme: An audit of injuries in professional football: An analysis of ankle sprains. *Br. J. Sports Med.* **2003**, *37*, 233–238. [CrossRef] [PubMed]
16. Lundblad, M.; Walden, M.; Magnusson, H.; Karlsson, J.; Ekstrand, J. The UEFA injury study: 11-year data concerning 346 MCL injuries and time to return to play. *Br. J. Sports Med.* **2013**, *47*, 759–762. [CrossRef]
17. Yu, B.; Kirkendall, D.; Taft, T.; Garrett, W., Jr. Lower extremity motor control-related and other risk factors for non-contact anterior cruciate ligament injuries. *Instr. Course Lect.* **2002**, *51*, 315.
18. Alentorn-Geli, E.; Myer, G.D.; Silvers, H.J.; Samitier, G.; Romero, D.; Lázaro-Haro, C.; Cugat, R. Prevention of non-contact anterior cruciate ligament injuries in soccer players. Part 1: Mechanisms of injury and underlying risk factors. *Knee Surg. Sports Traumatol. Arthrosc.* **2009**, *17*, 705–729. [CrossRef]

19. Verschueren, J.; Tassignon, B.; De Pauw, K.; Proost, M.; Teugels, A.; Van Cutsem, J.; Roelands, B.; Verhagen, E.; Meeusen, R. Does acute fatigue negatively affect intrinsic risk factors of the lower extremity injury risk profile? A systematic and critical review. *Sports Med.* **2019**, *50*, 767–784. [CrossRef]
20. Han, J.; Anson, J.; Waddington, G.S.; Adams, R.; Liu, Y. The role of ankle proprioception for balance control in relation to sports performance and injury. *BioMed Res. Int.* **2015**, *2015*, 1–8. [CrossRef]
21. Millet, G.Y.; Lepers, R. Alterations of neuromuscular function after prolonged running, cycling and skiing exercises. *Sports Med.* **2004**, *34*, 105–116. [CrossRef] [PubMed]
22. Glaister, M. Multiple sprint work. *Sports Med.* **2005**, *35*, 757–777. [CrossRef] [PubMed]
23. Forestier, N.; Teasdale, N.; Nougier, V. Alteration of the position sense at the ankle induced by muscular fatigue in humans. *Med. Sci. Sports Exerc.* **2002**, *34*, 117–122. [CrossRef]
24. Vromans, M.; Faghri, P.D. Electrical stimulation frequency and skeletal muscle characteristics: Effects on force and fatigue. *Eur. J. Transl. Myol.* **2017**, *27*, 239–245. [CrossRef] [PubMed]
25. Gentil, P.; Campos, M.H.; Soares, S.; Costa, G.D.C.T.; Paoli, A.; Bianco, A.; Bottaro, M. Comparison of elbow flexor isokinetic peak torque and fatigue index between men and women of different training level. *Eur. J. Transl. Myol.* **2017**, *27*, 246–250. [CrossRef]
26. Yaggie, J.A.; Campbell, B.M. Effects of balance training on selected skills. *J. Strength Cond. Res.* **2006**, *20*, 422–428.
27. Patti, A.; Bianco, A.; Şahin, N.; Sekulic, D.; Paoli, A.; Iovane, A.; Messina, G.; Gagey, P.M.; Palma, A. Postural control and balance in a cohort of healthy people living in Europe. *Medicine* **2018**, *97*, e13835. [CrossRef]
28. Hrysomallis, C. Balance ability and athletic performance. *Sports Med.* **2011**, *41*, 221–232. [CrossRef]
29. Eisen, T.C.; Danoff, J.V.; Leone, J.E.; Miller, T.A. The effects of multiaxial and uniaxial unstable surface balance training in college athletes. *J. Strength Cond. Res.* **2010**, *24*, 1740–1745. [CrossRef]
30. Coscia, F.; Gigliotti, P.V.; Piratinskij, A.; Pietrangelo, T.; Verratti, V.; Foued, S.; Diemberger, I.; Illic, G.F. Effects of a vibrational proprioceptive stimulation on recovery phase after maximal incremental cycle test. *Eur. J. Transl. Myol.* **2019**, *29*, 8373. [CrossRef]
31. Lin, S.; Sun, Q.; Wang, H.; Xie, G. Influence of transcutaneous electrical nerve stimulation on spasticity, balance, and walking speed in stroke patients: A systematic review and meta-analysis. *J. Rehabilit. Med.* **2018**, *50*, 3–7. [CrossRef]
32. Proske, U. Exercise, fatigue and proprioception: A retrospective. *Exp. Brain Res.* **2019**, *237*, 2447–2459. [CrossRef] [PubMed]
33. Puh, U.; Dečman, M.; Palma, P. Vsebina in učinki programov proprioceptivne vadbe za spodnje ude-pregled literature. *Fizioterapija* **2016**, *24*, 50–58.
34. Proske, U.; Allen, T. The neural basis of the senses of effort, force and heaviness. *Exp. Brain Res.* **2019**, *237*, 589–599. [CrossRef]
35. Vantieghem, S.; Bautmans, I.; Tresignie, J.; Provyn, S. Self-perceived fatigue in adolescents in relation to body composition and physical outcomes. *Pediatr. Res.* **2017**, *83*, 420–424. [CrossRef] [PubMed]
36. Larsson, L.; Degens, H.; Li, M.; Salviati, L.; Lee, Y.I.; Thompson, W.; Kirkland, J.L.; Sandri, M. Sarcopenia: Aging-related loss of muscle mass and function. *Physiol. Rev.* **2019**, *99*, 427–511. [CrossRef] [PubMed]
37. Coletti, D. Chemotherapy-induced muscle wasting: An update. *Eur. J. Transl. Myol.* **2018**, *28*, 153–157. [CrossRef]
38. Edmunds, K.; Gíslason, M.; Sigurðsson, S.; Guðnason, V.; Harris, T.; Carraro, U.; Gargiulo, P. Advanced quantitative methods in correlating sarcopenic muscle degeneration with lower extremity function biometrics and comorbidities. *PLoS ONE* **2018**, *13*, e0193241. [CrossRef]
39. Steib, S.; Hentschke, C.; Welsch, G.H.; Pfeifer, K.; Zech, A. Effects of fatiguing treadmill running on sensorimotor control in athletes with and without functional ankle instability. *Clin. Biomech.* **2013**, *28*, 790–795. [CrossRef]
40. Wright, K.E.; Lyons, S.; Navalta, J.W. Effects of exercise-induced fatigue on postural balance: A comparison of treadmill versus cycle fatiguing protocols. *Graefes Arch. Clin. Exp. Ophthalmol.* **2012**, *113*, 1303–1309. [CrossRef]
41. Pau, M.; Mereu, F.; Melis, M.; Leban, B.; Corona, F.; Ibba, G. Dynamic balance is impaired after a match in young elite soccer players. *Phys. Ther. Sport* **2016**, *22*, 11–15. [CrossRef] [PubMed]
42. Lacey, M.; Donne, B. Does fatigue impact static and dynamic balance variables in athletes with a previous ankle injury? *Int. J. Exerc. Sci.* **2019**, *12*, 1121–1137. [PubMed]

43. Paillard, T. Effects of general and local fatigue on postural control: A review. *Neurosci. Biobehav. Rev.* **2012**, *36*, 162–176. [CrossRef] [PubMed]
44. Šarabon, N.; Löfler, S.; Hosszu, G.; Hofer, C. Mobility test protocols for the elderly: A methodological note. *Eur. J. Transl. Myol.* **2015**, *25*, 253–256. [CrossRef]
45. De La Motte, S.J.; Lisman, P.; Gribbin, T.C.; Murphy, K.; Deuster, P.A. Systematic review of the association between physical fitness and musculoskeletal injury risk. *J. Strength Cond. Res.* **2019**, *33*, 1723–1735. [CrossRef]
46. Šarabon, N.; Smajla, D.; Kozinc, Ž.; Kern, H. Speed-power based training in the elderly and its potential for daily movement function enhancement. *Eur. J. Transl. Myol.* **2020**, *30*, 8898. [CrossRef]
47. Ponce-Bravo, H.; Ponce, C.; Feriche, B.; Padial, P. Influence of two different exercise programs on physical fitness and cognitive performance in active older adults: Functional resistance-band exercises vs. recreational oriented exercises. *J. Sports Sci. Med.* **2015**, *14*, 716–722.
48. Nejc, S.; Loefler, S.; Cvecka, J.; Sedliak, M.; Kern, H. Strength training in elderly people improves static balance: A randomized controlled trial. *Eur. J. Transl. Myol.* **2013**, *23*, 85–89. [CrossRef]
49. De Venuto, D.; Mezzina, G. High-specificity digital architecture for real-time recognition of loss of balance inducing fall. *Sensors* **2020**, *20*, 769. [CrossRef]
50. McGuine, T.A.; Greene, J.J.; Best, T.; Leverson, G. Balance as a predictor of ankle injuries in high school basketball players. *Clin. J. Sport Med.* **2000**, *10*, 239–244. [CrossRef]
51. Wang, H.K.; Chen, C.H.; Shiang, T.Y.; Jan, M.H.; Lin, K.H. Risk-factor analysis of high school basketball–player ankle injuries: A prospective controlled cohort study evaluating postural sway, ankle strength, and flexibility. *Arch. Phys. Med. Rehabilit.* **2006**, *87*, 821–825. [CrossRef] [PubMed]
52. Riemann, B.L.; Davies, G.J. Limb, sex, and anthropometric factors influencing normative data for the biodex balance system SD athlete single leg stability test. *Athl. Train. Sports Health Care* **2013**, *5*, 224–232. [CrossRef]
53. Hamlin, M.J.; Draper, N.; Blackwell, G.; Shearman, J.P.; Kimber, N.E. Determination of maximal oxygen uptake using the bruce or a novel athlete-led protocol in a mixed population. *J. Hum. Kinet.* **2012**, *31*, 97–104. [CrossRef] [PubMed]
54. Borg, G.A. Psychophysical bases of perceived exertion. *Med. Sci. Sports Exerc.* **1982**, *14*, 377–381. [CrossRef]
55. Cunha, F.; Midgley, A.; Monteiro, W.D.; Farinatti, P. Influence of cardiopulmonary exercise testing protocol and resting VO2 assessment on %HRmax, %HRR, %VO2 max and %VO2R relationships. *Int. J. Sports Med.* **2010**, *31*, 319–326. [CrossRef]
56. Susco, T.M.; McLeod, T.C.V.; Gansneder, B.M.; Shultz, S.J. Balance recovers within 20 min after exertion as measured by the balance error scoring system. *J. Athl. Train.* **2004**, *39*, 241–246.
57. Cetin, N.; Bayramoglu, M.; Aytar, A.; Sürenkök, Ö.; Yemisci, O.U. Effects of lower-extremity and trunk muscle fatigue on balance. *Open Sports Med. J.* **2008**, *2*, 16–22. [CrossRef]
58. Fox, Z.G.; Mihalik, J.P.; Blackburn, J.T.; Battaglini, C.L.; Guskiewicz, K.M. Return of postural control to baseline after anaerobic and aerobic exercise protocols. *J. Athl. Train.* **2008**, *43*, 456–463. [CrossRef]
59. Ishizuka, T.; Hess, R.A.; Reuter, B.; Federico, M.S.; Yamada, Y. Recovery of time on limits of stability from functional fatigue in division ii collegiate athletes. *J. Strength Cond. Res.* **2011**, *25*, 1905–1910. [CrossRef]
60. Pau, M.; Laconi, I.; Leban, B. Effect of fatigue on postural sway in sport-specific positions of young rhythmic gymnasts. *Sport Sci. Health* **2020**, *17*, 1–8. [CrossRef]
61. Farzami, A.; Anbarian, M. The effects of fatigue on plantar pressure and balance in adolescent volleyball players with and without history of unilateral ankle injury. *Sci. Sports* **2020**, *35*, 29–36. [CrossRef]
62. Steinberg, N.; Eliakim, A.; Zaav, A.; Pantanowitz, M.; Halumi, M.; Eisenstein, T.; Meckel, Y.; Nemet, D. Postural balance following aerobic fatigue tests: A longitudinal study among young athletes. *J. Mot. Behav.* **2016**, *48*, 332–340. [CrossRef] [PubMed]
63. Yaggie, J.; Armstrong, W.J. Effects of lower extremity fatigue on indices of balance. *J. Sport Rehabilit.* **2004**, *13*, 312–322. [CrossRef]
64. Matsuda, S.; Demura, S.; Uchiyama, M. Centre of pressure sway characteristics during static one-legged stance of athletes from different sports. *J. Sports Sci.* **2008**, *26*, 775–779. [CrossRef]
65. Nardone, A.; Tarantola, J.; Giordano, A.; Schieppati, M. Fatigue effects on body balance. *Electroencephalogr. Clin. Neurophysiol. Mot. Control.* **1997**, *105*, 309–320. [CrossRef]

66. Yamada, R.K.F.; Arliani, G.G.; Almeida, G.P.L.; Venturine, A.M.; Dos Santos, C.V.; Astur, D.C.; Cohen, M. The effects of one-half of a soccer match on the postural stability and functional capacity of the lower limbs in young soccer players. *Clinics* **2012**, *67*, 1361–1364. [CrossRef]
67. Greig, M.; Walker-Johnson, C. The influence of soccer-specific fatigue on functional stability. *Phys. Ther. Sport* **2007**, *8*, 185–190. [CrossRef]
68. Gioftsidou, A.F.; Malliou, P.; Pafis, G.; Beneka, A.; Godolias, G. Effects of a soccer training session fatigue on balance ability. *J. Hum. Sport Exerc.* **2011**, *6*, 521–527. [CrossRef]
69. Frisch, A.; Urhausen, A.; Seil, R.; Croisier, J.L.; Windal, T.; Theisen, D. Association between preseason functional tests and injuries in youth football: A prospective follow-up. *Scand. J. Med. Sci. Sports* **2011**, *21*, e468–e476. [CrossRef]
70. Willems, T.M.; Witvrouw, E.; Delbaere, K.; Mahieu, N.; De Bourdeaudhuij, L.; De Clercq, D. Intrinsic risk factors for inversion ankle sprains in male subjects: A prospective study. *Am. J. Sports Med.* **2005**, *33*, 415–423. [CrossRef] [PubMed]
71. Kraemer, R.; Knobloch, K. A soccer-specific balance training program for hamstring muscle and patellar and achilles tendon injuries. *Am. J. Sports Med.* **2009**, *37*, 1384–1393. [CrossRef] [PubMed]
72. Hübscher, M.; Zech, A.; Pfeifer, K.; Hänsel, F.; Vogt, L.; Banzer, W. Neuromuscular training for sports injury prevention. *Med. Sci. Sports Exerc.* **2010**, *42*, 413–421. [CrossRef] [PubMed]

MDPI
St. Alban-Anlage 66
4052 Basel
Switzerland
Tel. +41 61 683 77 34
Fax +41 61 302 89 18
www.mdpi.com

International Journal of Environmental Research and Public Health Editorial Office
E-mail: ijerph@mdpi.com
www.mdpi.com/journal/ijerph

www.ingramcontent.com/pod-product-compliance
Lightning Source LLC
LaVergne TN
LVHW070136170726
843515LV00007B/1809

* 9 7 8 3 0 3 6 5 4 2 5 2 2 *